RESONANCE IONIZATION SPECTROSCOPY 1996

RESONANCE IONIZATION SPECTROSCOPY 1996

SPECTROSCOPY 1996

Eighth International Symposium

State College, PA   June 30–July 5 1996

EDITORS
Nicholas Winograd
The Pennsylvania State University
University Park, Pennsylvania

James E. Parks
The University of Tennessee
Knoxville, Tennessee

American Institute of Physics

**AIP CONFERENCE
PROCEEDINGS 388**

Woodbury, New York

L.C. Catalog Card No. 96-80324
ISBN 1-56396-611-5
ISSN 0094-243X
DOE CONF- 960686

Printed in the United States of America

CONTENTS

INTRODUCTION

SESSION I: MOLECULAR RIS

SESSION II: FEMTOSECOND RIS

SESSION III: ATOMIC RIS I

SESSION IV: ENVIRONMENTAL APPLICATIONS

SESSION V: SURFACE ANALYTICAL APPLICATIONS

SESSION VIII: RIS AND ULTRATRACE ANALYSIS

SESSION IX: RIS AND MOLECULAR CLUSTERS

SESSION X: RIS APPLIED TO NUCLEAR AND PARTICLE PHYSICS

SESSION XI: ATOMIC RIS II

SESSION XII: BIOLOGICAL AND MEDICAL APPLICATIONS OF RIS

POSTER SESSIONS

POSTER SESSION I: ATOMIC RIS

POSTER SESSION II: MOLECULAR RIS

APPENDIX

PREFACE

The international symposia on Resonance Ionization Spectroscopy and Its Applications have established a tradition of excellence by hosting programs that highlight the latest developments in the technology itself and serve as a guide for its future. This tradition, begun at the inaugural symposium in 1980, has continued for seven subsequent meetings. Originally, RIS applications were few and at best were considered only as suggestions for further RIS research. This approach has encouraged resonance ionization spectroscopy scientists to delve into other research fields; including high energy physics, nuclear physics, medical and biological science, geological and astronomical science, and other engineering and technological efforts where RIS might prove relevant for making ultrasensitive measurements. The interaction of RIS scientists with those in other fields has resulted in an information exchange that benefits both groups of researchers, with each adding relevance to the others' scientific pursuits.

The commitment to excellent research efforts within a diverse atmosphere of disciplines and applications permeated the Eighth International Symposium on RIS and Its Applications, held June 30 to July 5, 1996. These proceedings of the symposium attest to a superior program carefully outlined by the meeting planners with guidance from the international advisory committee. The symposium addressed 11 scientific subjects: molecular RIS, femtosecond RIS, atomic RIS, environmental applications, surface analytical applications, new laser sources and applications, RIS and laser desorption, RIS and ultratrace analysis, RIS and molecular clusters, RIS applied to nuclear and particle physics, and biological and medical applications of RIS. The program's highlights were invited presentations of exemplary RIS work across disciplines, as well as discussions of how the technology may be influenced by potential applications in other fields.

The 1996 symposium was preceded by a one-day short course on RIS to introduce the fundamentals to those who were less familiar with the technology. This course was designed with a practical orientation and included an introduction to RIS and its applications, a "seat-of-the-pants" approach to RIS theory, high resolution RIS, molecular RIS and clusters, femtosecond and other laser systems, and a concluding presentation on brainstorming RIS. Course presenters were J. E. Parks, W. M. Fairbank, Jr., B. A. Bushaw, C. S. Feigerle, P. Bado, and G. S. Hurst, and their efforts were well-received by the 62 participants. Interest in the short course exceeded all expectations and generated the momentum for the remainder of the conference. A condensation of the presentation by Fairbank, "A Seat-of-the-Pants Approach to RIS Theory," was selected as an appendix to these proceedings as an example of the pragmatic instructional approach provided by the workshop.

The symposium's international advisory committee has long been committed to the notion that the future of RIS depends on the education and training of graduate students. As with previous meetings, conference organizers set aside support to finance graduate student attendance at the meeting. In addition, the IAC also honored the best work presented at the symposium by a graduate student. Criteria for the award include originality, innovation, analysis and interpretation of results, thoroughness of work, and the significance of the contribution to RIS. The University of Tennessee Institute of Resonance Ionization Spectroscopy sponsored the cash award, claimed by Erno Vandeweert of the Laboratorium voor Vaste-Stoffysika en Magnetisme in Leuven, Belgium, for his work entitled, "Metastable State Population and Kinetic Energy Distributions of Sputtered Ni and Co Atoms Studied by Resonance Ionization Mass Spectrometry."

Poster presentations are an integral part of the symposia and provide a forum to spark discussions and exchange ideas. To recognize outstanding work presented in a poster session, an award was given to J. A. Whitby, J. D. Gilmour, and G. Turner of the Department of Earth Sciences, Manchester University in Manchester, UK, for their work entitled, "A Study of Xenon Isotopes in a Martian Meteorite Using the RELAX Ultrasensitive Mass Spectrometer." Their work, as well as Vandeweert's, is included in these proceedings. Both entries are highly recommended as examples of the exceptional efforts currently being pursued in RIS research.

The International Symposia on RIS and its Applications is coordinated through The University of Tennessee Institute of Resonance Ionization Spectroscopy. The Eberly College of Science at The Pennsylvania State University hosted the 1996 symposium. The conference venue was The Penn State Scanticon Conference Center Hotel, which is located in the Research Park adjacent to the Materials Research Center. These logistics provided easy access to observe state-of-the-art RIS research laboratories for studying surface interactions and to view some of the latest innovations in ultrafast laser system design and surface analysis equipment. The hotel's facilities, coupled with the proximity of the labs, provided a relaxed atmosphere for learning, discussion, and collegiality. For one of us (JEP), this aspect of the meeting was inspiring and much appreciated as an example of excellence in research and development in an educational environment.

The success of the symposia is always due to the hard work and advice of the RIS international advisory committee and the local organizing committee. Much of the attention to detail, and ultimately the meeting's success, can be attributed to Ms. Sabrina Glasgow, Conference Secretary. Her pleasant personality and efficiency were greatly appreciated and a special thanks is offered to her for her work. The organizational skills of Ms. Catherine Longmire, Senior Editorial Assistant for The University of Tennessee Institute of RIS, have provided continuity among the symposia for the efficient coordination of all aspects of the conference planning and operations. Her work from the first announcement to the completion of the editing of these proceedings is truly appreciated and a special thanks is given to her. The support of our respective universities, the U.S. Department of Energy, the National Science Foundation, and our other corporate sponsors is gratefully acknowledged. With their ongoing support and the help of the RIS community, research and development in RIS will continue to flourish.

Nicholas Winograd
James E. Parks

SPONSORS

Financial support from the following organizations is gratefully acknowledged:

United States Department of Energy
Office of Health and Environmental Research

National Science Foundation

The Pennsylvania State University
University Park, Pennsylvania, USA

The University of Tennessee
Knoxville, Tennessee, USA

Clark-MXR, Inc.
Dexter, Michigan

Physical Electronics, Inc.
Eden Prairie, Minnesota

Positive Light
Marlboro, Massachusetts

HOSTS

The Eberly College of Science of The Pennsylvania State University is the host of RIS-96.

The International Symposium on Resonance Ionization Spectroscopy and Its Applications Conference Series is coordinated through the Institute of Resonance Ionization Spectroscopy of The University of Tennessee.

CONFERENCE CHAIRMAN

J. E. Parks
N. Winograd

INTERNATIONAL ADVISORY COMMITTEE

E. Arimondo	Italy	V. S. Letokhov	Russia
O. Axner	Sweden	T. B. Lucatorto	USA
P. Benetti	Italy	N. Mikami	Japan
P. Camus	France	C. M. Miller	USA
G. Goldstein	USA	J. C. Miller	USA
W. Hogervorst	The Netherlands	V. I. Mishin	Russia
H. Hotop	Germany	N. Omenetto	Italy
G. S. Hurst	USA	J. E. Parks	USA
H. J. Kluge	Germany	T. J. Whitaker	USA
P. Lambropoulos	Germany	N. Winograd	USA
K.W.D. Ledingham	Scotland	X. Xu	P.R. China

LOCAL ORGANIZING COMMITTEE

A.W. Castleman, Jr.
S. Glasgow
N. Winograd

CONFERENCE SECRETARY

S. Glasgow

INTRODUCTION

Laser-Experiments with Single Atoms and the Test of Basic Quantum Phenomena

H. Walther

Sektion Physik der Universität München and
Max-Planck-Institut für Quantenoptik
85748 Garching, Fed. Rep. of Germany

INTRODUCTION

Today, ion traps have become an important tool for precision spectroscopy. After the pioneering work in single ion spectroscopy performed end of the seventies (1) it got clear that this system allows to observe the particle in an unperturbed environment with practically no influence of the trapping fields. Another advantage is that the interaction time with the laser beam is not limited by the motion of the particle, therefore no transit time broadening is present. The fact that the motion of the ions can be cooled by laser light allows to reach the so-called Dicke limit (2,3) which means that the residual motional amplitude of the ion is much smaller than a fraction of the light wavelength used to probe the resonance line. In this limit the Doppler broadening disappears and since the temperature is in the range of mK or below also the second-order Doppler effect is excluded. Thus the ultimate resolution is determined by the natural lifetime of the investigated transition.

High resolution experiments have also been performed in ion traps in the microwave region. In this case the Doppler effect does not show up due to the low frequency of the investigated transitions and the number of ions stored in the trap can be large (for details see a recent review (4)). Besides for high resolution spectroscopy single trapped ions have recently also been used to investigate basic quantum phenomena. A prominent example is the investigation of the quantized motion of an ion in a trap observed through the Jaynes-Cummings dynamics and the realization of a Schrödinger cat (5,6). In this connection see also the review by Blatt (7).

In this paper two groups of experiments will be briefly discussed being performed in our laboratory with the aim to realize a new frequency standard and to investigate basic phenomena in radiation - atom interaction.

THE IN⁺ ION EXPERIMENT

For the experiments with single laser-cooled In^+ ions a Paul-Straubel-trap with a 1 mm diameter ring electrode was used (8), driven by an AC voltage of 1000 V at 10.7 MHz. Together with a small indium oven, an electron source for ionization and the detection optics for monitoring the fluorescence of the ions, the trap is mounted inside a stainless steel ultrahigh vacuum chamber.

The used Paul-Straubel trap produces a steep potential for a good localization of the stored ion and simultaneously the laser straylight is very small owing to the open electrode structure (8). Without damping through a buffergas the trap still can store several ions but at a typical well depth of 4 eV these can be evaporated from the trap by heating them with the laser detuned to the blue side of the resonance. After this procedure usually a single ion remains which is then trapped with high stability; it can be kept for hours even when it is not permanently laser cooled. We also observed Coulomb crystals (10) of two laser cooled indium ions separated by 3 μm.

In order to laser cool the ion on the narrow $^1S_0 \rightarrow {}^3P_1$ transition its hyperfine component F = 9/2 → 11/2 was excited with circular polarized light in zero magnetic field. This effectively prepares a two level system (F = m_F = 9/2) → (F = m_F = 11/2) by means of optical pumping.

If we look at the temporal behavior of the resonance fluorescence of a single indium ion on the transition $^1S_0 \rightarrow {}^3P_1$ we observe dark periods whenever the metastable $^3P_0$ level is populated. This can happen either by direct laser excitation from the ground state, as it will be done for the frequency standard, or through a magnetic dipole decay from the $^3P_1$ level. The possibility of observing quantum jumps in such a three level system with only one driven transition has been discussed theoretically (see e.g.(11)) and has been realized e.g. for the Hg^+ ion (12).

The evaluation of the duration of the dark periods (for details see Ref. (13.14)) leads to the lifetime of the $^3P_0$ level being τ ($^3P_0$) = 0.14(2)s. The uncertainty results from the number of observed decays.

The experiments show that the indium ion is a possible candidate for an optical frequency standard. Although the cooling transition in indium is a relatively slow intercombination line with only 360 kHz natural linewidth we were able to cool a single ion using this transition and to obtain a rate of fluorescence photons that is high enough to detect the dark periods due to electron shelving in the $^3P_0$ level. Besides these technical difficulties the small linewidth of the cooling transition enabled us to reach the strong binding regime using moderate trapping field strength. This opens the possibility of cooling the ion into low quantum levels or even the ground state of the single ion oscillator.

A frequency standard will benefit from the insensitivity of the indium ion's $^1S_0 \rightarrow {}^3P_0$ transition frequency against external perturbations as well as from the availability of a frequency stable oscillator like a diode laser pumped Nd:YAG laser. A suit-

4

able source has been developed in our laboratory for this purpose. In addition the accurate wavelength of the transition has been determined.

The Resonance Fluorescence of a Single Trapped Ion

In the following a recent experiment on the frequency distribution of a single trapped Mg^+ Ion will be discussed which gives some basic insights in the radiation-atom interaction (15).

Resonance fluorescence of atoms as the basic process in radiation-atom interactions has always generated considerable interest. The methods of experimental investigation have changed continuously due to the availability of new experimental tools (16,17). The spectrum of the fluorescence radiation of an atom is given by the Fourier transform of the first order correlation function of the field operators. Present theory on the spectra of fluorescent radiation following monochromatic laser excitation can be summarized as follows: fluorescence radiation obtained with low incident intensity is also monochromatic owing to energy conservation. In this case, elastic scattering dominates the spectrum and thus one should measure a monochromatic line at the same frequency as the driving laser field. The atom stays in the ground state most of the time and absorption and emission must be considered as a single process. This case was treated on the basis of a quantized field many years ago by Heitler (18). With increasing intensity, upper and lower states become more strongly coupled leading to inelastic components, which increase with the square of the intensity. At low intensities, the elastic part dominates since it depends linearly on the intensity. As the intensity of the exciting light increases, the atom spends more time in the upper state and the effect of the vacuum fluctuations comes into play through spontaneous emission. The inelastic component appears in the spectrum, and the elastic component goes through a maximum where the Rabi flopping frequency $\Omega = \Gamma/\sqrt{2}$ (Γ is the natural linewidth) and then disappears with growing Ω. The inelastic part of the spectrum gradually broadens as Ω increases and for $\Omega > \Gamma/\sqrt{2}$ sidebands begin to appear. For a saturated atom, the form of the spectrum shows three well-separated Lorentzian peaks. This spectrum was first calculated by Mollow (19). For other relevant papers see the review of Cresser et al. (17).

An experimental study of this problem requires a Doppler-free observation. In order to measure the frequency distribution, the fluorescent light has to be investigated by means of a high resolution spectrometer. The first experiments of this type were performed by Schuda et al. (20) and later by Walther et al. (21), Hartig et al. (21,16) and Ezekiel et al. (22). In all these experiments, the excitation was performed by single-mode dye laser radiation, with the scattered radiation from a well collimated atomic beam observed and analyzed by Fabry-Perot interferometers. Experiments to investigate the elastic part of the resonance fluorescence giving a resolution better than the natural linewidth have been performed by Gibbs et al. (23) and Cresser et al. (17).

5

In the present experiment the laser light was scattered from a laser-cooled trapped Mg^+ ion. The trap was a modified Paul-trap, called an endcap-trap (8) (see Fig. 1a) which produces strong confinement of the trapped ion. Therefore, the number of sidebands, caused by the oscillatory motion of the laser cooled ion in the pseudopotential of the trap, is reduced.

The measurements were performed using the $3^2S_{1/2}$ - $3^2P_{3/2}$ transition of the $^{24}Mg^+$-ion at a wavelength of 280 nm. The natural width of this transition is 43 MHz. The exciting laser light was produced by frequency doubling the light from a rhodamine 110 dye laser. The laser was tuned slightly below resonance in order to Doppler-cool the secular motion of the ion. All the measurements of the fluorescent radiation described in this paper were performed with this slight detuning δ.

For the experiment described here, it is important to have the trapped ion at rest as far as possible to minimize the light lost into motional sidebands. There are two reasons which may cause motion of the ion: the first one is the periodic oscillation of the ion within the harmonic pseudopotential of the trap and the second one is micromotion which is present when the ion is not positioned exactly at the saddle point of the trap potential. Such a displacement may be caused by a contact potential resulting, for example, by a coating of the electrodes with Mg produced when the atoms are evaporated during the loading procedure of the trap. Another reason may be asymmetries due to slight misalignments of the trap electrodes. Reduction of the residual micromotion can be achieved by adjusting the position of the ion with DC-electric fields generated by additional electrodes. For the present experiment they were arranged at an angle of 120° in a plane perpendicular to the symmetry axis of the trap electrodes. By applying auxiliary voltages (U_1 and U_2) to these electrodes and U_U and U_D to the outer trap electrodes (Fig. 1a), the position of the ion can be adjusted to settle at the saddle point of the trap potential.

The micromotion of the ion can be monitored using the periodic Doppler shift at the driving frequency of the trap which results in a periodic intensity modulation in the fluorescence intensity. This modulation can be measured by means of a transient recorder, triggered by the AC-voltage applied to the trap. There are three laser beams (lasers 1-3 in Fig. 1b) passing through the trap in three independent spatial directions which allow measurement of the three components of the micromotion separately. By adjusting the compensation voltages (Fig. 1a) U_1, U_2, U_U and U_D the amplitude of the micromotion could be reduced to a value smaller than $\lambda/8$ in all spatial directions.

The amount of secular motion of the ion resulting from its finite kinetic energy cannot be tested by this method since the secular motion is not phase coupled to the trap voltage. However, the intensity modulation owing to this motion can be seen in a periodic modulation of the photon correlation signal. For all measurements presented here, this amplitude was on the order of $\lambda/5$. This corresponds to a temperature of the ion of about 1 mK and a mean vibrational quantum number $\langle$ n $\rangle \approx 3$ of the trapped ion, resulting in less than 50 % of the fluorescence energy being lost into the vibrational sidebands.

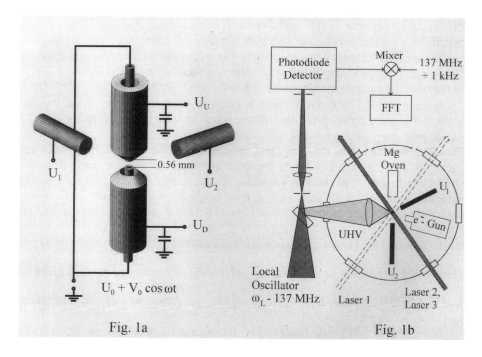

Fig. 1a   Fig. 1b

FIGURE 1a. Electrode configuration of the endcap trap. The open structure offers a large detection solid angle and good access for laser beams testing the micromotion of the ion. Micromotion is minimized by applying dc voltages: U_1, U_2, U_U, U_D.

The trap consists of two solid copper-beryllium cylinders (diameter 0.5 mm) arranged co-linearly with a separation of 0.56 mm. These correspond to the cap electrodes of a traditional Paul trap, whereas the ring electrode is replaced by two hollow cylinders, one of which is concentric with each of the cylindrical endcaps. Their inner and outer diameters are 1 and 2 mm, respectively, and they are electrically isolated from the cap electrodes. The fractional anharmonicity of this trap configuration, determined by the deviation of the real potential from the ideal quadrupole field, is below 0.1 % (8). The trap is driven at a frequency of 24 MHz with typical secular frequencies in the xy-plane of approximately 4 MHz. This requires a radio-frequency voltage with an amplitude on the order of 300 V to be applied between the cylinders and the endcaps, and with AC-grounding of the outer electrodes provided through a capacitor. **FIGURE 1b.** Scheme of heterodyne detection. The trap is omitted in the figure with only two of the compensation electrodes shown. Laser 3 is directed at an angle of 22° with respect to the drawing plane and Laser 2.

The heterodyne measurement is performed as follows. The dye laser radiation excites the trapped ion at a frequency ω_L while the fluorescence is observed in a direction of about 54° to the exciting laser beam (see Fig. 1b). Both the observation direction and the laser beam are in a plane perpendicular to the symmetry axis of the trap. Before reaching the ion, a fraction of this laser radiation is removed with a beamsplitter and then frequency shifted (by 137 MHz with an acousto-optic modulator (AOM)) to serve as the local oscillator. The local oscillator and fluorescence radiations are then overlapped and simultaneously focused onto the photodiode where the initial frequency mixing occurs. The frequency difference signal is amplified by a narrow band amplifier at 137 MHz and then further mixed down to 1 kHz so that it could be analyzed by means of a fast Fourier analyzer (FFT).

An example of a heterodyne signal is displayed in Fig. 2, where $\Delta\omega$ is the frequency difference between the heterodyne signal and the driving frequency of the AOM. Frequency fluctuations of the laser beam cancel out and do not influence the linewidth because at low intensity the fluorescence radiation always follows the frequency of the exciting laser while the local oscillator is derived directly from the same laser beam. The residual linewidth results mainly from fluctuations in the optical path length of the local oscillator or of the fluorescent beam. Both beams pass through regular air and it was observed that a forced motion of the air increased the frequency width of the heterodyne signal. The frequency resolution of the FTT was 3.75 Hz for this particular measurement. The heterodyne measurements were performed at a saturation parameter $s = \dfrac{\Omega^2 / 2}{\Delta^2 + (\Gamma^2 / 4)}$ of about 1, where Δ is the laser detuning. In this region, the elastic part of the fluorescent spectrum has its maximum (24).

The signal to noise ratio observed in the experiment is shot noise limited. The signal in Fig. 2 corresponds to a rate of the scattered photons of about 10^4 s^{-1} which is an upper limit since photons were lost from detection due to scattering into sidebands caused by the secular motion of the ion. In order to reduce this loss as much as possible, a small angle between the directions of observation and excitation was used.

Besides the fluorescence spectrum we have also studied the photon correlation under practically identical excitation conditions in order to determine the Rabi-flopping frequency at the applied laser intensity. The comparison of the results also provides a nice demonstration of complementarity since the heterodyne measurement corresponds to a „wave" detection of the radiation whereas the measurement of the photon correlation is a „particle" detection scheme. Under the same excitation conditions the wave detection provides the properties of a classical atom, i.e. a driven oscillator, whereas the particle or photon detection displays the quantum properties of the atom. Whether the atom displays classical or quantum properties thus depends solely on the method of observation.

The investigation of photon correlations employed the ordinary Hanbury-Brown and Twiss setup with two photomultipliers and a beam splitter. The setup was essentially the same as described in ref. (25). Some results are given in Fig. 3 which clearly demonstrate photon antibunching.

In conclusion, we have presented the first high-resolution heterodyne measurement of the elastic peak in resonance fluorescence. At identical experimental parameters we have also measured antibunching in the photon correlation of the scattered field. Together, both measurements show that, in the limit of weak excita-

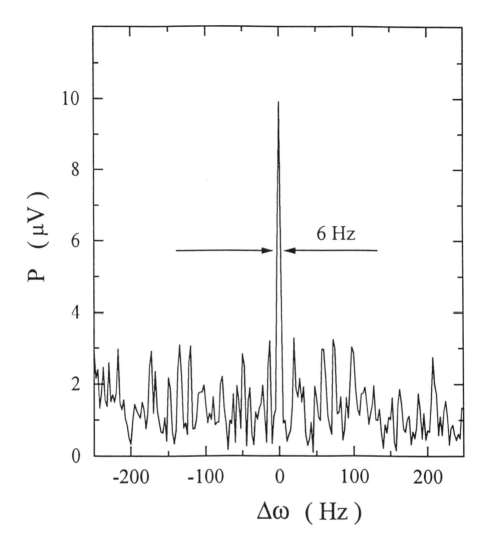

FIGURE 2. Heterodyne spectrum of a single trapped $^{24}Mg^+$-ion for $s = 0.9$, $\delta = -2.3 \, \Gamma$, $\Omega = 3.2 \, \Gamma$. Integration time: 267 ms.

tion, the fluorescence light differs from the excitation radiation in the second-order correlation but not in the first order correlation. However, the elastic component of resonance fluorescence combines an extremely narrow frequency spectrum with antibunched photon statistics, which means that the fluorescence radiation is not second-order coherent as expected from a classical point of view. This apparent

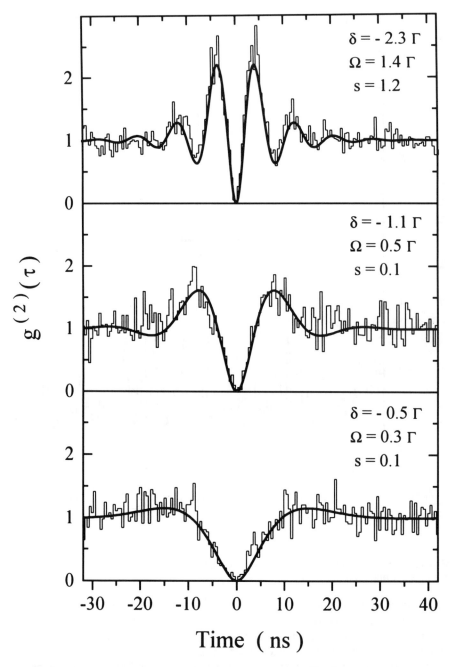

FIGURE 3. Photon correlation measurements for a single $^{24}Mg^{+}$-ion.

contradiction can be explained easily by taking into account the quantum nature of light, since first-order coherence does not imply second-order coherence for quantized fields (27). The fact that the heterodyne spectrum of a single particle can be measured opens the possibility to measure the phase-fluctuations of the radiation. It is thus possible to investigate squeezing in resonance fluorescence (28); furthermore the quantum behavior of a trapped particle can be studied by measuring the sidebands of the fluorescence (29) possibly under noise reduction as a result of squeezing. The latter would be an advantage for work aiming to realize quantum logic gates (30).

In a recent theoretical treatment of a quantized trapped particle (31) it was shown that a trapped ion in the vibrational ground state of the trap will also show the influence of the micromotion since the wavefunction distribution of the ion is pulsating at the trap frequency. This means that a trapped particle completely at rest will also scatter light into the micromotion sidebands. Investigation of the heterodyne spectrum at the sidebands may give the chance to confirm these findings. It is clear that such an experiment will not be easy since other methods are needed to verify that the ion is actually at rest at the saddle point of the potential .

REFERENCES

1. Neuhauser, W., Hohenstatt, M. Toschek, P. and Dehmelt, H., *Phys. Rev. Lett.* **41**, 233-236 (1978) and *Phys. Rev. A* **22**, 1137-1140 (1980).
2. Dicke, R.H., *Phys. Rev.* **89**, 472-473 (1953).
3. Wineland, D.J., Itano, W.M. and Van Dyck, R.S., *Advances in Atomic and Molecular Physics,* edited by D. Bates and B. Bederson, Boston: Academic Press, 1983, vol. 19, pp. 136-186.
4. Thompson, R.C., *Advances in Atomic, Molecular and Optical Physics,* edited by D. Bates and B. Bederson, Boston: Academic Press inc., 1993, vol. 30, in print.
5. Meekhof, D.N., Monroe, Ch., King, B.E., Itano, W.M., Wineland, D., *Phys.Rev.Lett.* **76**,1796-1799.
6. Monroe, Ch., Meekhof, D.N., King, B.E., Wineland, D., *Science* **272**, 1131-1136.
7. Blatt, R., *Atomic Physics* **14**, edited by D.J. Wineland, C.E. Wieman, S.J. Smith, American Institute of Physics, 219-239.
8. Schrama, C.A., Peik, E., Smith, W.W. and Walther, H., *Opt. Comm.* **101**, 32-36 (1993).
9. Lehmann, J.C. and Cohen-Tannoudji, C., *C. R. Acad. Sc. Paris* **258**, 4463 (1964).
10. Diedrich, F., Peik, E., Chen, J.M. Quint, W. and Walther, H., *Phys.Rev. Lett.* **59**, 2931-2934 (1987).
11. Javanainen, J., *Phys. Rev. A* **33**, 2121-2123 (1986), Schenzle, A., DeVoe, R.G. and Brewer, R.G., *ibid.* **33**, 2127-2130 (1986), Merz, M., Schenzle, A., *Appl. Phys. B* **50**, 115-124 (1990).
12. Bergquist, J.C., Hulet, R.G., Itano, W.M. and Wineland, D.J., *Phys.Rev. Lett.* **57**, 1699-1702 (1987).
13. Peik, E., Hollemann, G. and Walther, H., *Phys. Rev. A* **49**, 402-408 (1994).

14. Peik, E., Hollemann, G., Walther, H., Physica Scripta T59, 403-405 (1995).
15. Höffges, J.T., Baldauf, H.W., Eichler, T., Helmfrid, S.R., and Walther H., to be published.
16. Hartig, W. Rasmussen, W., Schieder, R., Walther, H. *Z. Physik A* **278**, 205-210 (1976).
17. Cresser, J.D., Häger, J., Leuchs, G., Rateike, F.M., Walther, H. *Topics in Current Physics* **27**, 21-59 (1982).
18. Heitler, W., *The Quantum Theory of Radiation*, Third Edition, Oxford University Press, 1954, pp. 196-204 .
19. Mollow, B.R., *Phys. Rev.* **188**, 1969-1975 (1969).
20. Schuda, F., Stroud, Jr., C., Hercher, M., *J. Phys.* **B1**, L198-L202 (1974).
21. Walther, H., *Lecture Notes in Physics,* **43,** 358-369 (1975).
22. Wu, F. Y., Grove, R. E., Ezekiel, S., *Phys. Rev. Lett.* **35**, 1426-1429 (1975); Grove, R.E., Wu, F. Y., Ezekiel, *Phys. Rev. A* **15**, 227-233 (1977).
23. Gibbs, H.M. and Venkatesan, T.N.C., *Opt. Comm.* **17**, 87-90 (1976).
24. Cohen-Tannoudji, C., Dupont-Roc, J., Grynberg, G., *Atom-Photon Interactions,* J. Wiley & Sons, Inc. (1992).
25. Diedrich, F., Walther, H., *Phys. Rev. Lett.,* **58**, 203-206 (1987).
26. Dagenais, M., Mandel, L., *Phys. Rev. A* **18**, 2217-2228 (1978).
27. Loudon, R., *Rep. Progr. Phys.* **43,** 913-949 (1980).
28. Loudon, R., Knight, P.L., *Journ. of Mod. Opt.* **34**, 709-759 (1987).
29. Cirac, J.I., Blatt, R., Parkins, A.S., Zoller, P., *Phys. Rev. A* **48**, 2169-2181 (1993).
30. Cirac, J.I., Zoller, P., *Phys. Rev. Lett.* **74**, 4091-4094 (1995).
31. Glauber, R., *Proceedings of the International School of Physics „Enrico Fermi", Course CXVIII Laser Manipulation of Atoms and Ions,* edited by Arimondo, E., Phillips, W.D., Strumia, F., North Holland, 1992, p. 643.

SESSION I: MOLECULAR RIS

Resonance Ionization through Unbound Intermediate States

R.J. Donovan, Z. Min, K.P. Lawley and T. Ridley

Department of Chemistry
The University of Edinburgh
West Mains Road
Edinburgh, EH9 3JJ, Scotland, U.K.

Abstract: A new two-colour optical-optical double resonance technique, involving unbound intermediate states, is described. This has been used in conjunction with mass-resolved resonance ionisation to study the molecules I_2, Cl_2, CH_3I and the CD_3 radical. Large bond extensions (≈ 1Å) in the unbound intermediate state are observed, thus significantly extending the Franck-Condon region and allowing access to a wide range of molecular ion-pair and Rydberg states.

Introduction

Optical-optical double resonance (OODR) is now a well established technique for studying the spectroscopy and dynamics of electronically excited states. By choosing a suitable vibronic level of a bound intermediate state spectra can be greatly simplified and the limitations imposed by the Franck-Condon principle, on transitions from the ground vibronic state, can be significantly relaxed. Essentially the same approach is used in the more recently developed zero kinetic energy (ZEKE) photoelectron technique which allows access to selected rovibronic states of molecular ions. However, the use of OODR requires a thorough understanding of the spectrosopy of the intermediate state and this is frequently not available without extensive further work.

In the work presented here we show that unbound or continuum intermediate states can be used for OODR, that the resulting bond stretching can be used to extend the Franck-Condon window and that little previous knowledge of the continuous intermediate state is required. We briefly describe a few examples to illustrate our recent work in this area (1-6).

Unbound Intermediate States

The excitation of unbound and repulsive states is generally of more interest to photochemists than spectroscopists, as continuous spectra provide little structural information. Furthermore, the use of continuum intermediate states for OODR appears at first sight to be rather unpromising as further excitation of the dissociating molecule (10^{-14}s) will have a low probability if nanosecond pulsed lasers are used (see figure 1). Added to this it is not immediately obvious which wavelengths will be effective for the secondary excitation step. Recording the absorption spectra of molecules undergoing dissociation, on the femtosecond timescale, is not a trivial task. Fortunately, we can appeal to microscopic reversibility to provide the necessary information. The emission from bound upper states to repulsive lower states (i.e. bound-free fluorescence) has been quite widely studied and these emission spectra carry the information that we require to predict the optimum wavelength for the reverse absorption process. Bound-free fluorescence is characterised by a long wavelength maximum (independent of v'), often referred to as the red extremum. This maximum is associated with an extremum in the Mulliken difference potential which greatly enhances the absorption cross-section for the secondary excitation of the dissociating molecule. These principles will be illustrated below, by work on I_2.

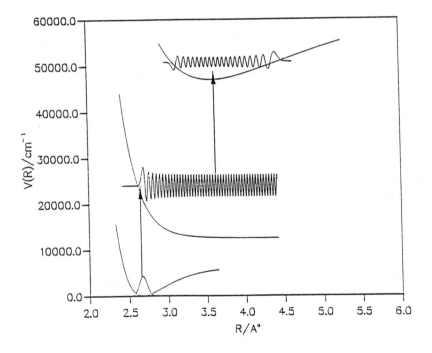

Figure 1.   Optical-optical double resonance through unbound intermediate states (i.e. bound-free-bound excitation).

Bound-Free-Bound Transitions in I$_2$

The ion-pair states of the halogens have been studied extensively using conventional OODR via bound intermediate states. We have now shown that by using the continuum of the B(0$_u^+$) state of I$_2$ these same ion-pair states can be studied, without the need to identify the precise (v,J) quantum states of the intermediate, thus greatly reducing the effort required to obtain and analyse spectra (1).

Our first observations were made using one colour excitation and relied on the fortuitous overlap of the continuum associated with transitions from the ground state to the B(0$_u^+$) state of I$_2$, with the red extremum of the E(0$_g^+$)-B(0$_u^+$) oscillatory continuum system. The overlap occurs in a relatively narrow region, (418-445 nm), and the first step, the B-X transition, is far from optimum. Nevertheless, a spectrum of the E(0$_g^+$) ion-pair state was readily recorded over the vibrational range v' = 40-70. In this case bond extensions of $\approx$ 1Å in the intermediate state are involved. (1)

Clearly, this approach is limited as one cannot in general rely on the two excitation steps overlapping in wavelength. However, this limitation is readily overcome by employing two independently tunable lasers, as for conventional OODR. Using this approach the most intense bound-free-bound spectra will be observed when the following three criteria are satisfied. Firstly, the pump wavelength should lie close to the maximum for transitions from the ground state to the continuum state. Secondly, the probe-wavelength should lie close to the red extremum in the fluorescence between the final and intermediate states. Thirdly, both transitions should be fully allowed (i.e. they should have large cross-sections). For I$_2$ the first two criteria can be satisfied using the two-colour approach but the B-X transition is semi-forbidden and the third criterion is not fully satisfied. Nevertheless we have repeated the work described above on I$_2$, using two-colour excitation to optimise both excitation steps. Spectra of the E(0$_g^+$) and f(0$_g^+$) ion-pair states were readily obtained, for a wide range of vibrational levels, using the following excitation scheme:

$$E,f(0_g^+ ; v'J') \quad \longleftarrow \quad B(0_u^+; \text{continuum}) \quad \longleftarrow \quad X(0_g^+ ; v''J'')$$

The intensity was sufficient for etalons to be placed in the pump and probe beams and fully resolved rotational spectra were obtained. An analysis of the relative strengths of the O, Q and S branches enabled the first transition to be identified as essentially parallel, showing that the B(0$_u^+$) state is the dominant intermediate, rather than the B"(1u) state, whose continuous absorption overlaps the B-X transition.

Rydberg and Ion-pair States of Cl_2

A semi-classical analysis of dissociation from unbound or repulsive intermediate states shows that the efficiency of secondary excitation is proportional to $\sqrt{\mu}$, at constant excitation energy, reflecting the transit time through the Franck-Condon region. Thus, in order to investigate systems less massive and thus less favourable than I_2, a series of studies were carried out on molecular chlorine.

The $C(^1\Pi_{1u})$ state of Cl_2 is a purely repulsive state and this was used to study the $\beta(1_g)$ ion-pair state and [½]4s,1g Rydberg state of this molecule in a one-colour, two-photon excitation scheme. The spectra observed were consistent with bond stretching of approximately 0.5Å in the C state. The E, f, and G ion-pair states were also observed over an extensive range of vibrational levels, using other repulsive intermediate states and two-colour excitation. (3)

By extending this work to higher energies (76,000-90,000 cm^{-1}) the origins of some twenty new Rydberg states were observed. (4) To our surprise it proved possible to achieve two-photon excitation of Cl_2 from the dissociating state in a one-colour experiment, yielding new information on the $\alpha(^3\Sigma)$, $\gamma(^3\Pi)$ and $^1\Sigma_u^+$ ion-pair states. (5)

Rydberg States of CH_3I (CD_3I) and the CD_3 Radical

Using the same general procedures as those described above for I_2 and Cl_2, we have studied the two-colour REMPI spectra of CH_3I and CD_3I. (6) There have been numerous one-colour (2 + 1) REMPI studies of the [½]6s Rydberg states of CH_3I and these show a short progression in v_2, the CH_3 umbrella mode (see figure 2a). However, when two-colour excitation is used, with the pump-frequency lying in the Ã-X̃ absorption continuum, extended vibrational progressions are observed in v_3, the C-I stretching mode and the ($v_2 + v_3$) combination (see figure 2b). The progression in v_3 is curtailed by predissociation rather than Franck-Condon effects. The contrast between the one-colour coherent (non-resonant) and two-colour (resonant) excitation of the [½]6s,0 state is very striking and clearly demonstrates the *multi-dimensional* nature of the potential energy surfaces involved for polyatomic molecules. (6)

By working at several different pump-wavelengths and scanning the probe frequency the relative contributions from the three known repulsive intermediate states have been explored, through changes in the strengths of the $\Omega = 0$ and 1 components of the final Rydberg states that are accessed (6).

18

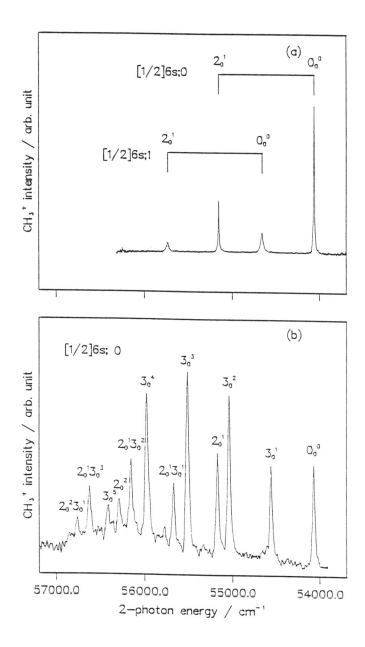

Figure 2. Comparison of the one-colour (a) and two-colour (b) REMPI spectra of CH_3I. For (b) the pump wavelength was resonant with the repulsive $^3Q_0$ intermediate state.

Finally, we have observed the two-colour excitation spectrum of the CD_3 radical, by pumping through a strongly predissociated intermediate state. Photolysis of CD_3Br in the region of 215 nm produces copious amounts of the CD_3 radical. When the pump-laser is tuned to 215.05 nm, further excitation of CD_3 to the strongly predissociated $3s(^2A_1')$ Rydberg state is achieved. The probe-laser is then used to excite the 0_0^0 band of the $4p_z(^2A_2'')$ Rydberg state.

By tuning the pump laser to 216.45 nm, which excites a hot band of CD_3, a very strong enhancement of the 2_3^1 band of the $4p_z(^2A_2'')$ Rydberg state is seen. Further work on CD_3/CH_3 and other free radicals is in progress.

References

1. Al-Kahali, M.S.N., Donovan, R.J., Lawley, K.P. and Ridley, T., Chem. Phys. Letters **220**, 225-228 (1994).

2. Donovan, R.J., Lawley, K.P., Min, Z., Ridley, T. and Yarwood, A.J., Chem. Phys. Letters **226**, 525-531 (1994).

3. Al-Kahali, M.S.N., Donovan, R.J., Lawley, K.P., Min, Z. and Ridley, T., J. Chem. Phys., **104**, 1825-1832 (1996).

4. Al-Kahali, M.S.N., Donovan, R.J., Lawley, K.P. and Ridley, T., J. Chem. Phys., **104**, 1833-1838 (1996).

5. Al-Kahali, M.S.N., Donovan, R.J., Lawley, K.P., Ridley, T. and Yarwood, A.J., J. Phys. Chem. **99**, 3978-3983 (1995).

6. Min, Z., Ridley, T., Lawley, K.P. and Donovan, R.J., to be published.

Photophysics of Fullerenes: Thermionic Emission

R. N. Compton[1,2], A. A. Tuinman[1], and J. Huang[3]

1. The University of Tennessee, Knoxville, Tennessee 37996
2. Oak Ridge National Laboratory, Oak Ridge, Tennessee 37831-6125
3. Ames Laboratory, Iowa State University, Ames, Iowa 50011

INTRODUCTION

The high degree of symmetry and large number of endohedral and exohedral π-type electrons of the hollow cage fullerene molecule give rise to many interesting photophysical properties of this new allotrope of carbon. Of the many possible isomers of a given fullerene, those satisfying the rule that the twelve pentagons be isolated (i.e. do not share a bond) are the lowest in energy. The isolated pentagon rule (IPR) isomers are also the structures found to be most prevalent in the synthesis of fullerenes. Of the 1812 possible isomers of C_{60} only the I_h form is produced in the arc discharge or laser ablation method. Reminiscent of a Georges Seurat *pointillism* painting, C_{60} at a distance appears as a spherical "atom" of K_h symmetry. Upon closer inspection the carbon atom "imperfections" give rise to the lower icosohedral symmetry of the cluster. The photophysical properties of C_{60} can be discussed by analogy to the "particle on a sphere" (see Savina et al.[1]). The energy levels for the 60 π electrons of C_{60} occupying the degenerate levels representing "electrons on a sphere," along with the filling of empty orbitals in I_h symmetry are shown below in Figure 1.

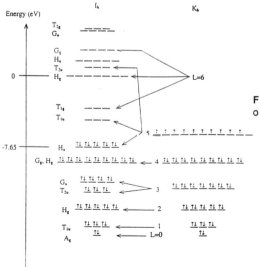

Figure 1. Energy levels occupied by 60 electrons on the surface of a sphere in K_h or I_h symmetry.

The energy level of the highest filled molecular orbital, H_u, is located at the known ionization potential of C_{60}. Although the atomic (K_h) particle on a sphere model provides a reasonable HOMO-LUMO, gap the known degeneracy of the HOMO (5-fold) and LUMO (3-fold) as well as the diamagnetism of C_{60} is accounted for by assuming I_h symmetry, as inferred from single-line NMR and four-line IR spectra. The diamagnetism of C_{60} is driven by ring currents in the twelve isolated pentagons.[2] Fullerene double bonds are more localized than that of a purely aromatic system, a property that governs its addition chemistry as well. Photoabsorption of most fullerenes occurs as a result of transitions in the UV since the high degree of symmetry renders the low-lying transitions to be forbidden. The beautiful purple color of most C_{60} solutions results from the lack of photoabsorption in the blue (424-492 nm) and red (650 nm) region of the spectrum.[3] However, perturbations with the solvent and between fullerene solutes can make transitions more allowed and change the color of the solution greatly. Solvent effects on the optical limiting action of C_{60} solutions have been considered by Koudoumas et al.[4]

Photoabsorption in C_{60} is dominated by collective electronic motion, i.e. plasmons. Bertsch et al.[5] predicted these Mie-type plasmons which have been observed in many experiments at ~ 6 eV (π plasmon) and ~ 22 eV (π plus σ plasmon). Yoo et al.[6] and Hertel et al.[7] show a weak onset at the known ionization potential (7.65 eV) of C_{60} followed by a large peak in the cross section at ~ 25 eV which contains almost all of the expected oscillator strength (60). Although less well studied, plasmons may also strongly influence the electron impact ionization of C_{60}. As shown in Figure 2, we find a weak intensity threshold in the cross section for electron impact ionization for C_{60} as well.

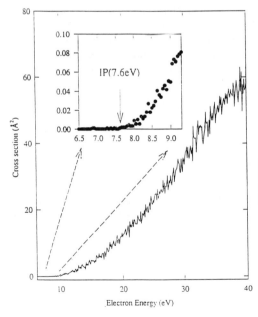

Figure 2. Relative cross section for the Electron impact ionization of C_{60}. Notice the apparent delayed onset is actually a weak cross section at threshold.

The energy scale for the high resolution electron beam (hemispherical sector electron energy analyzer) was calibrated using the onset of argon ions at 15.76 eV. Thus the presence of the plasmon at ~25 eV appears to affect the appearance of the threshold for electron impact ionization of C_{60}.

Thermionic Emission

Multiphoton ionization of fullerenes using long-pulse length (> n sec) lasers occurs mainly through vibrational autoionization. In many cases the laser ionization can be described as thermionic in analogy to the "boiling off" of electrons from a filament. Thermionic emission manifests itself as a "delayed" emission of electrons following pulsed laser excitation. The delayed emission of electrons for microseconds following multiphoton absorption of pulsed laser light has been reported for molecules,[8] metal clusters[9,10] fullerenes,[11-15] and metcars.[16] Klots[17,18] has employed quasiequilbrium theory to calculate rate constants for thermionic emission from fullerenes which seem to quantitatively account for the observed delayed emission times and the measured electron energy distributions.[15] The theory of Klots also accounts for the thermionic emission of C_{60} excited by a low power CW Argon Ion laser.[19] Figure 3 summarizes the mechanism believed responsible for thermionic emission in fullerenes. The triplet state offers a pathway (ISC) to the "heat bath" of the S_0 ground electronic state.

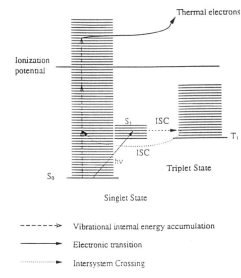

Figure 3. Mechanism leading to thermionic emission in C_{60}.

23

Recently Klots and Compton[20] have reviewed the evidence for thermionic emission from small aggregates where mention was also made of experiments designed to determine the effects of externally applied electric fields on thermionic emission rates. Such effects are well characterized in bulk metals and semiconductors. We have measured the fullerene ion intensity as a function of the applied electric field and normalized this signal to that produced by single photon ionization of an atom in order to correct for all collection efficiency artifacts. Figure 4 shows a Schottky plot for the ion intensity ratio C_{60}^+/Cs^+ produced by XeCl pulsed laser light (308 nm). 308 nm light efficiently ionizes cesium directly into the continuum, thus any variation in the Cs^+ ion signal with electric field should represent a collection efficiency change.

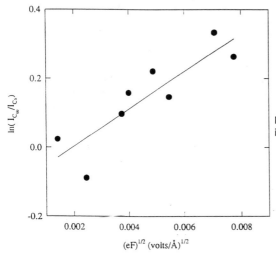

Figure 4. Shottky plot for field enhanced ionization of gas phase C_{60} at 308 nm.

The increase in fullerene ion signal relative to that of Cs^+ is attributed to field enhanced thermionic emission. From the slope of the Schottky plot we obtain a temperature of approximately 1000 K. This temperature is comparable to but smaller than that estimated from measurements of the electron kinetic energies (~1600 - 3700 K). This result for field enhanced thermionic emission is discussed further by Klots and Compton.[21]

Thermionic emission from neutral clusters has long been known for autodetachment from highly excited negative ions.[22] Similarly, electron attachment to C_{60} in the energy range from 8 to 12 eV results in C_{60} anions with lifetimes in the range of microseconds.[23,24] Quasiequilibrium theory (QET) calculations[20] are in reasonable accord with these measurements. More recently we have observed the sequential attachment of two electrons to C_{84} in the gas phase, i.e.,

$$e + C_{84} \rightleftharpoons C_{84}^{-*}$$

Followed by

$$e + C_{84}^{-*} \rightleftharpoons C_{84}^{=**}$$

After mass selecting the metastable $C_{84}^{=**}$ ion, we have observed its autodetachment into the C_{84}^{-} anion. This is the first observation of the decay of a doubly-charged anion.

The observed magnitude of the $C_{84}^{=}$ signal relative to that of C_{84}^{-} implies that the cross section for the attachment of the second electron is as large as that for attachment of the first electron. This is a surprising result since the long range interaction between the incident electron and C_{84}^{-} is repulsive. The capture of electrons into a state of high angular momentum might be facilitated by the excitation of vibrations (phonons) in the C_{84}^{-} anion. The possibility that of the formation of a "Cooper-pair" in the gas phase is intriguing. The fact that alkali atom doped C_{60} (3:1) is a known superconductor strengthens this argument.

The two extra electrons in $C_{84}^{=}$ are calculated to form a bound state (Mark Pederson, private communication). The dissociation of $C_{84}^{=}$ into C_{84}^{-} + e is further inhibited by the presence of a Coulomb barrier (see Ref. 25). The decay of $C_{84}^{=**}$ into C_{84}^{-} over the top of the Coulomb barrier occurs mainly as a result of thermionic emission and from tunneling through the barrier to a much lesser extent. Again, the measured lifetime of $C_{84}^{=**}$ can be accounted for by QET.

ACKNOWLEDGMENTS

Oak Ridge National Laboratory, managed by Lockheed Martin Energy Research Corporation for the U.S. Department of Energy under contract number DE-AC05-96OR22464. Research sponsored by the National Science Foundation CHE9508609 and the Office of Naval Research Molecular Design Institute through the Georgia Institute of Technology. Discussions with Joe Macek and Eph Klots are gratefully acknowledged.

REFERENCES

1. Savina, M. R., Lohr, L. L., Francis, A. H., *Chem. Phys. Lett.* **205**, 200 (1993).
2. Zanasi, R., and Fowler, P. W., *Chem. Phys. Lett.* **238**, 270 (1995).
3. Catalan, et al., J., *Angew. Chem. Int. Ed. Engl.* **34**, 105 (1995).
4. Koudoumas, E., Ruth, A. A., Couris, S., and Leach, S., *Molecular Physics* **88**, 125-133 (1996).
5. Bertsch, G. F., Bulgac, A., Tomanek, D., and Wang, Y., *Phys. Rev. Lett.* **67**, 2690 (1991).
6. Yoo, R. K., Ruscic, B., and Berkowitz, J., *J. Chem. Phys.* **96**, 911 (1992).

7. De Varies, J., Steger, H., Kamke, B., Menzel, C., Weisser, B., Kamke, W., and Hertel, I. V., *Chem. Phys. Lett.* **188**, 159 (1992).
8. P. D. Dao and A. W. Castleman, Jr., *J. Chem. Phys.* **84**, 1434 (1986).
9. Neiman, G. C., Parks, E. K., Richtsmeier, S. C., Liu, K., Pobo, L. G., and Riley, S. R., *High Temp. Science* **22**, 115 (1986).
10. Leisner, T., Athanassenas, K., Echt, O., Kandler, O., Kreisle, D., and Rechnagel, E., *Z. Phys. D.* **24**, 81 (1992).
11. Campbell, E. E. B., Ulmer, G., and Hertel, I. V., *Phys. Rev. Lett.* **67**, 1986 (1991).
12. Wang, L.-S., Concieicao, J., Jin, C., and Smalley, R. E., *Chem. Phys. Lett.* **182**, 5 (1991).
13. Amrein, A., Simpson, R., and Hackett, P., *J. Chem. Phys.* **95**, 1781 (1991).
14. Wurz, P., and Lykke, K. R., *J. Chem. Phys.* **95**, 7008 (1991).
15. Ding, D., Compton, R. N., Haufler, R. E., and Klots, C. E., *J. Phys. Chem.* **97**, 2500 (1993).
16. Cartier, S. F., May, B. D., and Castleman, Jr., A. W., *J. Chem. Phys.* **104**, 3423-3432 (1996).
17. Klots, C. E., *Chem. Phys. Lett.* **186**, 73 (1991).
18. Klots, C. E., *J. Chem. Phys.* **100**, 1035 (1994).
19. Ding, D., Huang, J., Compton, R. N., Klots, C. E., and Haufler, R. E., *Phys. Rev. Lett.* **73**, 1084 (1994).
20. Klots, C. E., and Compton, R. N., *Surface Science Letters and Review* (in press).
21. Klots, C. E., and Compton, R. N., *Phys. Rev. Lett.* **76**, 4092 (1996).
22. Compton, R. N., et al., *J. Chem. Phys.* **45**, 4634 (1966).
23. Jaffke, et al., T., *Chem. Phys. Lett.* **226**, 213 (1994).
24. Huang, J., Carman, H.S., and Compton, R. N., *J. Phys. Chem.* **99**, 1719 (1995).
25. Scheller, M. K., Compton, R. N., and Cederbaum, L. S., *Science* **270**, 1160 (1995).

Abnormal Pressure Effects in H_2 $(B^1\Sigma_u^+ / C^1\Pi_u - X^1\Sigma_g^+)$ Three–Photon Resonant Enhanced Ionization Spectroscopy

G.Sha, G.Zhang, and J.Xu

Dalian Institute of Chemical Physics, Chinese Academy of Sciences, 116023, Dalian, China

D.Proch, W.Knott, and K.L.Kompa

Max-Planck-Institut für Quantenoptik, P.O.Box 1513, 85740 Garching, Germany

Abstract. Pressure-induced abnormal line shift, line broadening and spectral intensity variations have been measured in 3-photon enhanced ionization spectroscopy of H_2 $B^1\Sigma_u^+ - X^1\Sigma_g^+$ (12-0) and $C^1\Pi_u - X^1\Sigma_g^+$ (2-0) rotational transitions. All of these abnormal pressure effects can be well explained by reabsorption of the third harmonic generated in the medium, considering the phase matching conditions that govern Third Harmonic Generation (THG) in a focused beam geometry, anomalous dispersion near any transition, and resonance enhancement of third-order susceptibility. The failure to obtain ion signal from circularly polarized light supports the above explanation. Approximate expressions have been derived which yield line-shifts, -splitting and -intensities at different pressures in agreement with the experimental results.

Since its first observation in Xe (in 1980), the abnormal pressure effect has been a common place in the three-photon resonance enhanced ionization (REMPI) spectroscopy of many atomic and molecular gases. The Xe(6s[3/2], J=1) line disappears once the pressure exceeds 0.4 mbar; this adverse pressure effect is accompanied by the anomalously large blue-shift of the excitation line, line broadening, and THG radiation in the forward direction. Similar phenomena have been found in other atomic (Kr, Ar, Cl), diatomic (Cl_2, H_2, CO) and polyatomic (CH_3I, C_2H_2) molecules with sharp Rydberg transitions. During the past decade, the overwhelmingly accepted explanation[2] has been the interference quenching of 3-photon excitation by THG. For molecular systems, however, some authors[3] have predicted that total suppression will not usually occur. Instead, the authors[3,4] assume that appearance as well as amount of an adverse pressure effect are dominated by the off-resonant reabsorption of the third harmonic which is generated once the phase matching conditions are met. This summary of our H_2 (3+1)-photon REMPI experiments presented here will offer additional and conclusive evidence in support of this mechanism.

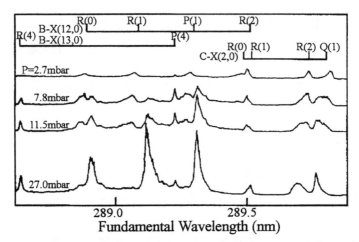

FIGURE 1. Multiphoton ionization spectra of 3-photon resonant H_2 $B^1\Sigma_u^+ / C^1\Pi_u - X^1\Sigma_g^+$ transition under various pressures.

Experiment and Results

An excimer-pumped pulsed dye laser covering the wavelength range 577nm $\leq$ $\lambda \leq$ 584nm with a bandwidth of $\Delta v \approx 0.2cm^{-1}$ is frequency-doubled by a KDP crystal. The linearly polarized UV beam ($E \approx 1mJ$, $\tau_{FWHM} \approx 15ns$) is then focused (f = 100mm) into a temperature-stabilized (T=276K) gas cell equipped with a pair of collection electrodes. In some experiments the focal length was reduced to f = 75 or 50mm. H_2 pressure was varied between 0.1 and 90 mbar. The ion signal picked up by the electrodes while the laser scanned was amplified and fed to a boxcar averager.

A selection of spectra obtained at different pressures is shown in Fig.1, jointly with an identification of the pumped transitions. Most lines appear blue-shifted to a larger or lesser degree, with growing tendency towards higher pressure. At pressure ranging between ~7.8 and ~15mbar, the R(0), R(1), and P(1) lines of B←X(12–0) transition surprisingly each split into one blue and one red-shifted peak. While the latter grows with pressure, the former gradually fades away and eventually disappear altogether (See Fig2). No ion signal could be detected when pumping with circularly polarized light.

Discussion

Assuming the single photon (THG) excitation step to be the MPI rate determining and the medium to be optically dense for the wavelength of the third-harmonic, we may express the ion signal generated by a focused laser beam as:

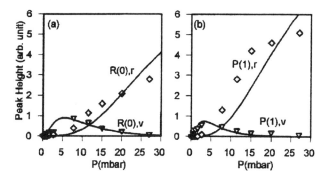

FIGURE 2. Peak height of H_2 B-X(12,0) 3-photon REMPI vs. pressure (a) R(0) rotational line (b) P(1)line,v and r denote blue shifted and red shifted peaks respectively. Symbols ▵ and ◊ are the experimental data and curves are calculated.

$$S_i \propto \left[\chi^{(3)}\right](\Delta k)^2 \exp(b\Delta k) \qquad (1)$$

where $\chi^{(3)}$ represents the third-order nonlinear susceptibility, b denotes the confocal parameter, and $\Delta k = k_3 - 3k_1 = 2\pi/\lambda_{3,0} \times (n_3 - n_1) \approx 2\pi/\lambda_{3,0} \times (n'-1)$, with $n' = n_3$. Phase matching achieves its optimum at $\Delta k = -2/b$[6]. Fig.3 graphs our calculations, at different pressures, of $(n'-1)$ vs. wavelength detuning. The right panel illustrates the Q(2) C←X transition where phase matching can only be achieved to the blue side of the resonance. (Fig.3(a)), however, phase matching is possible to either side of the resonance frequency. Pressure increase moves the "blue" $(n'-1)$ traces away from resonance. As this frequency shift advances, however, $\chi^{(3)}$ drops sharply, resulting in an adverse pressure effect (see Eq.(1)). The "red" $(n'-1)$ curves, on the other hand, draw near resonance upon raising pressure, leading to the prediction of a steadily rising ion signal, due to the rapid growth of $\chi^{(3)}$. Both forecasts are confirmed by the experimental material (see Fig.2).

Close to an excitation frequency the third-order susceptibility is dominated by the resonant contribution. For the off-resonant behavior we may assume that $\chi^{(3)} \propto N_i f_i / (\Delta\omega - i\Gamma)$[7], where $\Delta\omega$ and Γ represent the frequency detuning and the sum of all linewidth contributions, respectively. Introducing the relationship into Eq.(1) and assuming $dS_i/d\omega = 0$ yields an approximate expression for the line shift:

$$\Delta\omega_p = \frac{1}{8} r_e N_i f_i b\left[1 + 4/b\Delta k_{bac} \pm \left(1 + \left(4/b\Delta k_{bac}\right)^2\right)^{1/2}\right] \qquad (2)$$

where $r_e = 2.818 \times 10^{-13}$ cm, N_i the number density of H_2 in state "i", f_i the oscillator strength of i-th transition, $\Delta k_{bac} = 2\pi/\lambda_{i,0} \times (n'_{bac} - 1)$. The background term, n'_{bac}, summarizes all those contributions to the refractive index which are due to non-resonant transitions. For $\Delta k_{bac} \geq 0$, the line shift is positive throughout. $\Delta\omega_p$ is proportional to p at low values, but approaches a limit towards high pressures. For $\Delta k_{bac} < 0$, Eq.(2) indicates the possibility of blue as well as red shifts. Symmetric

29

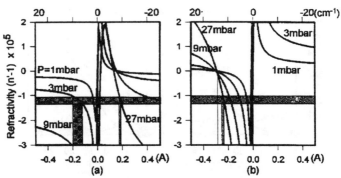

FIGURE 3. Calculated $(n'-1)$ vs. frequency detuning at different H_2 pressures. The bottom abscissa denotes the fundamental wavelength change. The shaded horizontal area highlights the phase matching condition of $-2/b$.

line splitting, i.e. $\left|\Delta\omega_p\right| = \left|-\Delta\omega_p\right|$, is obtained if $4/b\Delta k_{bac} = -1$. The spacing between both lines, $\Delta\omega_{sp}$, then amounts to $\Delta\omega_{sp} = 2\left|\Delta\omega_p\right| = \sqrt{2}/4\, r_0 N_i f_i b$. At p=7.8 mbar we calculated the splitting of the R(0), R(1) and P(1) lines of B←X to be 9.5cm^{-1} (expt: 10.0cm^{-1}), 24.5cm^{-1} (expt: 22.7cm^{-1}), and 17.4cm^{-1} (expt: 17.6cm^{-1}), respectively. Eqs.(1) and (2) permit the calculation of the relative peak intensity, as displayed in Fig.2.

The absence of MPI signal under conditions of circularly polarized excitation is consistent with the presumed mechanism of third harmonic reabsorption since THG is forbidden in this case. An other interpretation of the observations could be based on the assumption of a vanishing weight-3 irreducible tensor component $T^{(3)}$ for the 3-photon excitation of H_2 $B^1\Sigma_u^+$ or $C^1\Pi_u$ states. This hypothesis, however, is in conflict with the well established theory[3]. Some calculations have been performed using the explicit expressions for the 3-photon line strength polarization ratio S^{cir}/S^{Lin}, communicated by [8]. For the R(1), B←X transition we obtain $S^{cir}/S^{Lin} = 0.27$, or 0.54, depending on the symmetry of the virtual intermediate states.

References

1. Miller.J.C, Compton.R.N, Payne.M.G, and Garrett.W.W, *Phys.Rev.Lett.* **45**, 114 (1980)
2. Jackson.D.J and Wynne.J.J, *Phys.Rev.Lett.* **49**, 543 (1982)
3. Li.L, Wu.M, and Johnson.P.M, *J.Chem.Phys.* **86**, 1131 (1987)
4. Jiang,B, Sha.G, Sun.W, Zhang.Ch, He.J, Xu.S and Zhang.C, *J.Chem.Phys.* **97**, 4697 (1992)
5. Xu.S, Sha.G, He.J and Zhang.C, *J.Chem.Phys.* **100**, 1858 (1994)
6. Ward,J.F and New.G.H.C *Phys.Rev.* **185**,57-71 (1969)
7. Friedberg.R, Hartmann.S.R and Manassah.J.T, *J.Phys.B* **24**, 2883-2897 (1991)
8. Maïnos.C, Le Duff. Y and Boursey.E, *Mol.Phys.* **56**,1165-1174 (1985)

Studies of Rare Gas Excimers using (2+1) REMPI/Time-of-flight Mass Spectrometry

R.H. Lipson, S.S. Dimov, X.K. Hu, and D.M. Mao

Department of Chemistry, University of Western Ontario
London, Ontario, Canada, N6A 5B7

Abstract. New single isotopomer spectra for the heteronuclear rare gas excimers (ArKr and NeXe) were recorded using the combined experimental techniques of (2+1) resonantly enhanced multiphoton ionization and time-of-flight mass spectrometry. Vibrational quantum numbering and constants were obtained for several Rydberg states in the region of their $np^5(n+1)p$ and np^5nd asymptotes, while excited state bond lengths were deduced from Franck-Condon factor calculations. Electronic symmetry assignments were made from separate spectra recorded with linearly and circularly polarized light. The interpretation of some of our ArKr results are found to be at odds with conclusions drawn from photofragment imaging experiments. Finally, preliminary dispersive photoelectron spectra for the gerade Rydberg states of Xe_2 are presented.

1. INTRODUCTION

Readily analyzable heteronuclear rare gas dimer electronic spectra leading to well-determined potential energy curves are difficult to obtain because the molecules are only weakly bound by the van der Waals interaction in their ground states. Since the excited states of these particular dimers are also weakly bound, many Rydberg state ← ground state transitions tend to come to the blue of their associated atomic resonance lines. Consequently, the resultant spectra tend to be congested in large part due to the presence of many naturally occurring isotopomers, each with their own unique set of vibrational and rotational term values within each electronic state due to mass effects.

Our group has recently undertaken a systematic examination of the Rydberg states of Xe_2 (1), Kr_2 (2), XeKr (3) and ArXe (4) in the vicinity of the $np^5(n+1)s$, $np^5(n+1)p$, and np^5nd atomic asymptotes, (n = 4 and 5 for Kr and Xe, respectively). Single isotopomer spectra were obtained by exciting jet-cooled dimers by (2+1) resonantly enhanced multiphoton ionization (REMPI), and then analyzing the resultant ions in a time-of-flight (TOF) mass spectrometer. Similar experimental techniques were used here to examine ArKr and NeXe. New dispersive photoelectron spectra (PES) of Xe_2 are also reported.

2. EXPERIMENTAL

The apparatus used to record (2+1) REMPI/TOF spectra of jet-cooled dimers has fully been described elsewhere (1). Two-photon transitions in the wave number ranges between ~ 92380 and 94250 cm^{-1}, and ~ 78000 and 80400 cm^{-1} were examined for ArKr and NeXe, respectively. Single isotopomer ArKr and NeXe spectra were recorded using a linear TOF spectrometer (Comstock, Model TOF-101). Transition wave numbers were calibrated to a precision of ≈ 0.2 cm^{-1} against the optogalvanic spectrum of Ne.

Dispersive PES were obtained for the gerade states of Xe_2 in the same transition wave number range noted above for NeXe. Photoelectron kinetic energies were measured with a double focusing electrostatic analyzer (Comstock, Model AC-901).

3. SPECTROSCOPIC BACKGROUND

Under Hund's case (c) coupling the two-photon electric dipole selection rules limit transitions from the $\Omega = 0^+$ ground states of the heteronuclear dimers, where Ω is the component of total angular momentum along the bond axis, to $\Omega = 0^+$, 1, and 2 excited states. Since the spectra were not rotationally resolved, $\Omega = 0^+$ excited states were distinguished from those with $\Omega = 1$ or 2 by recording separate REMPI / TOF spectra with linearly and circularly polarized light under identical conditions (5). Excited state vibrational constants were computed by least-squares fitting individual sets of isotopic transition wave numbers to an appropriate mass-reduced Dunham expansion employing theoretical ground state parameters (6). Excited state bond lengths were established by Franck-Condon factor calculations (7).

4. RESULTS AND DISCUSSION

Overview (2+1) REMPI/TOF spectra for $^{20}Ne^{132}Xe$ and $40Ar^{84}Kr$ obtained by linearly and circularly polarized two-photon excitation are presented in Figs. 1 and 2, respectively. Since the NeXe spectral analyses are still in progress, the following discussion will focus only on ArKr. A summary of those results are listed in Table I. Consider the higher energy band systems in the region of the $Ar^* 4s[3/2]_1^0$ and $Kr^* 5p[1/2]_0$ atomic lines at 93750.6 cm^{-1} and 94092.9 cm^{-1}, respectively. From our polarization

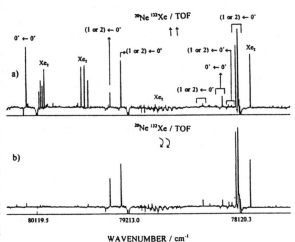

Figure 1. Overview (2+1) REMPI/TOF spectra for $^{20}Ne^{132}Xe$. a) linearly and b) circularly polarized excitation.

measurements the two molecular band systems are assigned to adiabatic $\Omega = 0^+$

excited states formed by a homogeneous interaction between a bound diabatic state dissociating to Ar + Kr* 5p[1/2]$_0$, and a repulsive level which produces Ar* 4s[3/2]$^o_1$ + Kr at dissociation. This contradicts the recent assignments put forward by Chandler and co-workers (8). On the basis of their photofragment imaging experiments where a wavelength dependent production of Ar (but not Kr) was observed, the excited states were assigned Ω = 1 and 2 symmetry, and an Ar* 4s[3/2]$^o_2$ + Kr dissociation limit. However, states with 0$^+$ symmetry are not possible from their proposed dissociation products. The two sets of observations have yet to be reconciled.

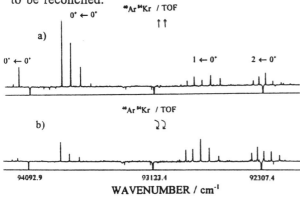

Figure 2. Overview (2+1) REMPI/TOF spectra for ^{40}Ar^{84}Kr. a) linearly and b) circularly polarized excitation.

Since the Rydberg states of an RgRg' dimer (Rg , Rg' = Xe, Kr, Ar, or Ne) can be viewed as an [RgRg']$^+$ molecular ion plus a non-bonding electron, the dissociation energy of a given level should resemble that of the ion-core. Neglecting spin-orbit interactions, four ion states need to be considered. In order of increasing energy, they are A$^2\Sigma^+_{(u)}$, B$^2\Pi_{(g)}$, C$^2\Pi_{(u)}$, and D$^2\Sigma^+_{(g)}$, where the g/u parity subscripts apply only to the homonuclear species. In the molecular orbital picture they result from the removal of antibonding or bonding electrons from the neutral dimer ground state electronic configuration. Consequently, D$_e$(A) >> D$_e$(B) > D$_e$(C) > D$_e$(D). In principle, dispersive PES can provide both experimental proof of the ion-core identity, and evidence of excited state perturbations and predissociation (9). We have applied this method to the Rydberg states of Xe$_2$ which dissociate to Xe*(6p) + Xe, since those levels are now well characterized (1). The results presented in Table II are encouraging, and suggests that similar application to the heteronuclear dimers should be successful.

Table I: Molecular constants (cm^{-1}) for the Rydberg states of ^{40}Ar^{84}Kr.

Asymptote	Ω	T'_e	ω'_e	$\omega'_e x'_e$	D'_e	r'_e [d]
Ar + Kr* 5p[1/2]$_0$ [a] Ar* 4s[3/2]$^o_1$ + Kr	0$^+$	—[b]	44.19 [c]	—	—	<3.89
Ar + Kr* 5p[1/2]$_0$ [a] Ar* 4s[3/2]$^o_1$ + Kr	0$^+$	93498.59(72) [c]	92.53(57)	2.86(10)	<709.8	≤3.37Å
Ar + Kr* 5p[3/2]$_1$	1	92241.2(1.1)	99.69(34)	2.847(24)	838.7(10.8)	3.21(5)
Ar + Kr* 5p[5/2]$_2$	2	91993.01(93)	84.11(44)	3.563(48)	429.9(10.8)	3.25(8)
Ar + Kr(^{1}S$_0$)	0$^+$	0.0	26.748 [f]	1.549 [f]	115.5(10.8)	3.894

a) Perturbed state	c) $\Delta G'_{1/2}$
b) Not determined	d) Equilibrium bond length in Angstrom units

e) Quoted errors are 1σ standard errors
f) Derived ground state constants

Table II: Transitions, photoelectron kinetic energies (KE), and dominant core assignments for the Rydberg states of Xe_2 dissociating to $Xe + Xe^*$ (6p,5d).

Transition	Wavenumber/ cm^{-1}	KE/eV	Dominant core assignment	Transition	Wavenumber /cm^{-1}	KE/eV	Dominant core assignment
$0_g^+ (v'=0) \leftarrow 0_g^+ (v''=0)$	80046	2.85 1.52	$Xe_2^+(C^2\Pi_{3/2u})^{a)}$	$2_g (v'=0) \leftarrow 0_g^+ (v''=0)$	78016	2.47	$Xe_2^+(C^2\Pi_{3/2u})^{a)}$
$0_g^+ (v'=0) \leftarrow 0_g^+ (v''=0)$	79611	2.77 1.50	$Xe_2^+(C^2\Pi_{3/2u})^{a)}$	$0_g^+ (v'=32) \leftarrow 0_g^+ (v''=0)$	77587	3.12	$Xe_2^+(A^2\Sigma_{1/2u}^+)$
$1_g (v'=9) \leftarrow 0_g^+ (v''=0)$	79488	2.62	$Xe^+(^2P_{3/2})^{b)}$	$1_g (v'=9) \leftarrow 0_g^+ (v''=0)$	77017	3.10	$Xe_2^+(A^2\Sigma_{1/2u}^+)^{a)}$
$1_g (v'=7) \leftarrow 0_g^+ (v''=0)$	77378	2.48	$Xe^+(^2P_{3/2})^{c)}$	a) perturbed			
$2_g (v'=4) \leftarrow 0_g^+ (v''=0)$	78978	2.71 2.53	$Xe_2^+(C^2\Pi_{3/2u})$ $Xe_2^+(B^2\Pi_{3/2g})^{d)}$	b) predissociation to produce Xe* 6p[3/2]$_2$ c) predissociation to produce Xe* 6p[5/2]$_2$ d) perturber photoelectron peak			
$1_g (v'=26) \leftarrow 0_g^+ (v''=0)$	78868	2.47	$Xe^+(^2P_{3/2})^{e)}$	e) predissociation to produce Xe* 6p[5/2]$_3$			

ACKNOWLEDGMENTS

Funding from the Natural Sciences and Engineering Research Council of Canada (NSERC), and the Academic Development Fund of the University of Western Ontario is gratefully acknowledged.

REFERENCES

1. Dimov, S.S, Cai, J.Y., and Lipson, R.H. *J. Chem. Phys.* **101**, 10313-10322 (1994); Hu, X.K., Mao, D.M., Dimov, S.S., and Lipson, R.H. *Chem. Phys.* **201**, 557-565 (1995).
2. Lipson, R.H., Dimov, S.S., Cai, J.Y., Wang, P., and Bascal, H.A. *J.Chem.Phys.* **102**, 5881-5889 (1995).
3. Lipson, R.H., Dimov, S.S. Hu, X.K., Mao, D.M. and Cai, J.Y, *J. Chem. Phys.* **103**, 6313-6324 (1995).
4. Dimov, S.S., Hu, X.K., Mao, D.M., Cai, J.Y., and Lipson, R.H., J. Chem. Phys. **104**, 1213-1224 (1996); Mao, D.M., Hu, X.K., Dimov, D.M., and Lipson, R.H., J. Phys. B **29**, L89-L94 (1996).
5. Bray, R.G., and Hochstrasser, R.M., *Mol Phys.* **31**, 1199-1211 (1976); Dimov, S.S., Hu, X.K., Mao, D.M., and Lipson, R.H. *Chem. Phys. Lett.* **239**, 332-338.
6. Bobetic, M.V. and Barker, J.A., *J.Chem. Phys.* **64**, 2367-2369 (1976).
7. Xu, Y. Jäger, W., Djauhari, J., and Gerry, M.C.L. *J.Chem. Phys.* **103**, 2827-2833 (1995).
8. Heck, A.J.R., Neyer, D.W., Zare, R.N., and Chandler, D.W., *J. Phys. Chem.* **99**, 17700-17710 (1995).
9. Dehmer, P.M., Pratt, S.T., and Dehmer, J.L., *J. Phys. Chem.* **91**, 2593-2598 (1987).

SESSION II: FEMTOSECOND RIS

Charge Resonance Enhanced Ionization (CREI) of Molecules in Intense Laser Fields

André D. Bandrauk, Stephan Chelkowski, Tao Zuo, Hengtai Yu

Laboratoire de Chimie Théorique, Faculté des Sciences
Université de Sherbrooke, Que, J1K 2R1, Canada

Abstract. Short ($\tau \leq 100$ fs) intense ($I \geq 10^{14}$ W/cm^2) laser pulses are shown from nonperturbative numerical simulations of the time-dependent Schroedinger equation to induce unusually large ionization rates in molecules often exceeding those of the dissociated fragments. Linear one electron, H_2^+, H_3^{++} and two electron systems H_2, H_3^+ have been studied numerically in order to understand this phenomenon. It will be shown that enhanced ionization is due to the existence of large divergent transition moments in these symmetric molecules due to degeneracies of electronic orbitals upon dissociation. Such degenerate states called charge resonance states in 1939 by Mulliken, give rise to charge resonance oscillations. We show that these charge resonance states are responsible for charge resonance enhanced ionization, CREI. A static field picture of CREI will be shown to explain adequately the enhanced ionization and the critical distances at which it occurs through field induced barrier suppression of the electron-nuclear coulomb potentials in these molecular systems.

INTRODUCTION

Intense laser field ionization of atoms is now reasonably well understood, especially in the high intensity, long or short wavelength regime (1). In the long wavelength limit, atomic ionization can be described as a process of barrier suppression and electron tunneling out of the combined field of the attractive coulomb potential and the classical instantaneous laser electric field to reach the continuum. This model can be described qualitatively by the Keldysh-Faisal-Reiss Theory (2-3) and its extension to long wavelengths as a dc-field tunneling ionization mechanism (4). Similar qualitative ideas have been considered for molecules (5-7). We have performed the first exact numerical calculation of intense field ionization of the one-electron molecules H_2^+ (8-9) and discovered that such a simple system exhibits unusually large ionization rates, exceeding those of neutral H by at least one-order of magnitude (9). This was rationalized as due to laser-induced dynamical localization, i.e., electron tunneling suppression between the protons or equivalently destruction of the chemical bond was shown to occur at critical values of the internuclear distance R_c for particular laser intensities. Following up on the initial barrier suppression ideas of Codling and Frasinski (6), as applied to molecules, recent 3D (9) and 1D (10-11) calculations of one-electron diatomics have shown that barrier suppression of the electron-coulomb potentials also plays a role in creating enhanced ionization. Such an idea is also now being used in explaining unusually large ionization rates of clusters (12-14).

Experimental measurements of intense field multiphoton ionization with ultrashort pulses has lead to unusual discoveries. The group of Normand in France has discovered that the kinetic energy of highly charged atomic fragments from diatoms is well below that expected from direct Franck-Condon transitions (15) so that some form of molecular stabilization at large R has been proposed (16) akin to the earlier dressed state molecular stabilization model of suppressed photodissociation (5), (17-18). Levis has recently measured femtosecond photoionization of aromatic molecules at high intensities and has found little or no fragmentation of benzene at intensities of 4 x 10^{13} W/cm^2, whereas larger aromatics fragment readily (19). The behaviour of small molecules in intense laser fields will be the focus of the present article. We have obtained exact numerical solutions of the time dependent Schroedinger equation, TDSE, for linear one electron systems: H_2^+, H_3^{++} and linear two-electron systems H_2 and H_3^+ in intense ($I \geq 10^{14}$ W/cm^2), ultrashort ($\tau \leq 100$ fs) pulses. In all cases we observe enhanced ionization of these systems at critical distances R_c greater than the equilibrium. Static field calculations are used to try to rationalize these large ionization rates in terms of static field-induced barrier suppression mechanisms proposed earlier (6-11). We will show the importance of charge resonance (CR) states for what we call CREI - Charge Resonance Enhanced Ionization. Full electron-nuclear dynamical calculations beyond the Born-Oppenheimer approximation confirm that the observed low kinetic energies of fragments of diatoms (15-16) are indeed due to CREI (20).

MODELS AND CALCULATIONS

Dissociative intense field ionization of simple molecules such as H_2 was considered as early as 1961 by Hiskes (21) and 1975 by Hanson (22) in static fields. Current experiments are being done using frequency dependent laser fields so that numerical simulations must rely on nonperturbative solutions of the TDSE. We have performed the first 3D ionization calculations of H_2^+ for fixed nuclei (8-9). Such a one-electron system can be treated accurately with a supercomputer for molecular configurations parallel to the laser field. We have now also been able to perform a complete electron-nuclear calculation, i.e. with both electron and proton motions treated simultaneously (20). However, much of the relevant physics can also be obtained from the much simpler static nuclei 1D model of H_2^+ corresponding to the TDSE (10-11),

$$i\frac{\partial \varphi}{\partial t} = -\frac{1}{2}\frac{\partial^2 \varphi}{\partial z^2} + V_c(z, R, q)\varphi - eE(t)z\varphi \quad, \tag{1}$$

where

$$V_c(z, R, q) = -q\left[(z \pm R/2)^2 + a^2\right]^{-1/2}. \tag{2}$$

z is the single electron coordinate, R the internuclear distance and q the nuclear charge. V_C is the electron-nuclear coulomb potential with the two nuclei situated at the positions $\pm$ R/2 respectively. In the case of H_2 (23), one introduces two independent electronic coordinates z_1, z_2 and one adds then the electron-electron repulsion.

$$V_e = +\left\{(z_1 - z_2)^2 + a^2\right\}^{-1/2}. \tag{3}$$

One can readily generalize to linear H_3^{++} and H_3^+ by adding two R distances. We illustrate in Fig. 1 the total electrostatic potential a) $V = V_c + V_{ext}$, b) $V = V_c + V_e + V_{ext}$ with $V_{ext} = -eEz$ or $-eE(z_1 + z_2)$ respectively for H_2^+ and H_2. E is chosen to be the peak field strength of the laser field (e.g. $I = 10^{14}$ W/cm^2, $E = 3 \times 10^8$ V/cm). The factor $a = 1$, serves to remove the coulomb singularities and can be adjusted to give similar ionization potentials as the exact 3-D systems. Having removed the coulomb singularities, one can then proceed with highly accurate split-operator methods to solve the TDSE, equations 1-3, (24). Absorbing boundaries are used to remove ionized

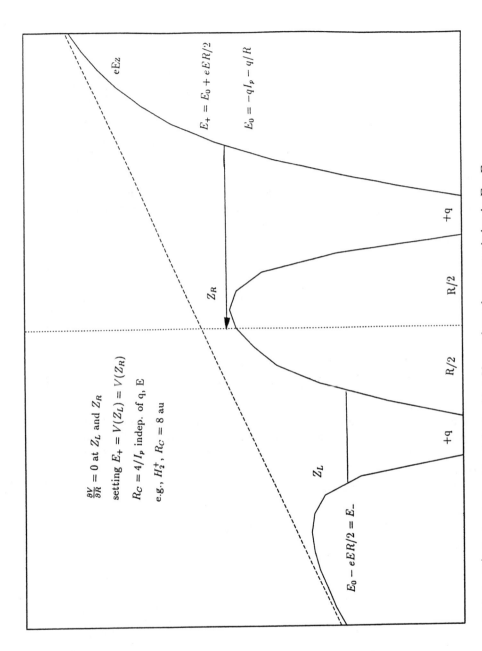

FIG. 1. H_2^+ electrostatic potential $V = V_c + V_{ext}$ and two lowest static levels E_+, E_- .

Labels within figure:

eEz

$E_+ = E_0 + eER/2$

$E_0 = -qI_p - q/R$

$+q$

$R/2$

Z_R

$R/2$

$+q$

Z_L

$E_0 - eER/2 = E_-$

$\frac{\partial V}{\partial R} = 0$ at Z_L and Z_R

setting $E_+ = V(Z_L) = V(Z_R)$

$R_C = 4/I_p$ indep. of q, E

e.g., H_2^+, $R_C = 8$ au

40

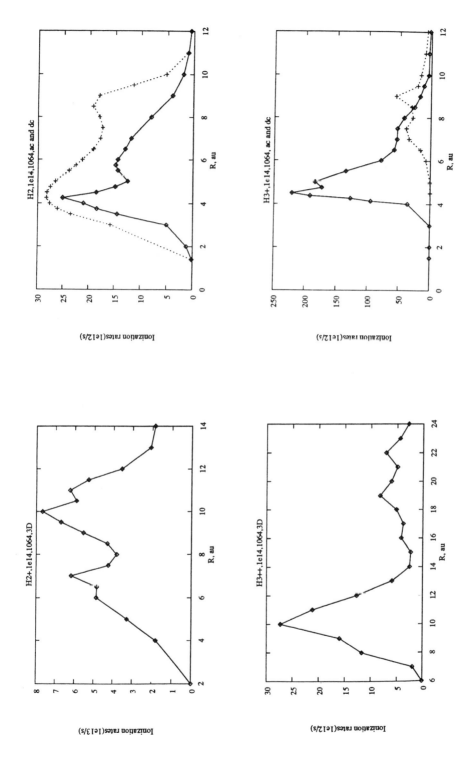

FIG. 2. Ionization rates of H_2^+, H_2, H_3^{++}, H_3^+ for $\lambda = 1064$ nm, $I = 10^{14}$ W/cm^2

41

electrons at the edges of the large numerical grid (dimension ~ 500 a.u.). Ionization rates $\Gamma(s^{-1})$ are obtained from the decaying norm of the wavefunction $N(t) = e^{-\Gamma t}$.

Typical results for the ionization rates $\Gamma(s^{-1})$ as a function of interproton distance R are shown in Fig. 2 at intensity 10^{14} W/cm^2, wavelength λ - 1064 nm. The rise time of the pulse is taken to be 5 cycles (1 cycle - 3.55 fs) and taking the logarithm of the norm gives a linear relation ln $N(t) = - \Gamma t$ after the rise time. In all cases one observes peaks in the ionization rates: a) at R = 6 and 8 a.u. for H_2^+; b) at R = 4 and 6 a.u. for H_2; c) at R = 5 a.u. for H_3^{++}; d) at R = 4.5 a.u. for H_3^+. Thus for the one-electron systems, H_2^+ and H_3^{++}, the peaks occur at somewhat larger distances than the two-electron systems H_2 and H_3^+. The ionization rates at the peaks are seen to exceed the asymptotic rates of the fragments, H or H_2^+, by about one-order of magnitude.

The maxima in the ionization rates can be rationalized in terms of two mechanisms, i) a laser-induced dynamic localization of the electron and ii) barrier suppression (9C,10). Both mechanisms occur whenever there are two essential states, e.g. the charge resonance states $1\sigma_g$ and $1\sigma_u$ in H_2^+ which are strongly coupled in the field. The energy separation of such two states in a strong field is eER, i.e. the potential energy difference between two protons at distance R for an electron in a field of amplitude E. This is seen in the static field picture, Fig. 1. The field E distorts the symmetric double well by the electrostatic energy eEz, but at the same time the $1\sigma_u$ level is shifted up by the energy eER/2, the radiative coupling energy $<1\sigma_g/eEz/1\sigma_u>$. Calculating the critical distance R_C at which the $1\sigma_u$ level touches the barrier which is now suppressed by the field, yields the result in atomic units (au) (10).

$$R_C = 4/I_p$$

where I_p is the ionization potential of the <u>neutral</u> atom fragment. Thus for H_2^+, $I_p(H) = 0.5$ au so that $R_C \simeq 7$-10 au. The splitting of the single peak predicted by the barrier suppression model is due to the dynamic localization effect of the laser (9). We see that the double peak structure remains in the case of H_2. In the case of the large systems, H_3^+ and H_3^{++}, only one peak now occurs since there is now more than one charge resonance transition (23). Coulomb explosion experiments (15) and numerical simulations (20) now confirm CREI as being the main mechanism for dissociative ionization of diatomic molecules in which unusually low kinetic energy fragments are produced.

REFERENCES

1. Gavrila, M., *Atoms in Laser Fields* (Academic Press, NY 1992)
2. Keldysh, L.V., *Sov. Phys. JETP* **20**, 1307 (1965); Faisal, F.H.M., *J. Phys.* **B6**, L89 (1973): Reiss, H. **A22**, 1786 (1980).
3. Ammosov, M.Y., Delone, N.B., Krainov, V.P., *Sov. Phys. JETP* **64**, 1191 (1986).
4. Corkum, P.B., Burnett, N.H., Brunel, F., *Phys. Rev. Lett.* **62**, 1259 (1989).
5. Bandrauk, A.D., *Molecules in Laser Fields* (M. Dekker, N.Y. 1993).
6. Codling, K., Frasinski, L., Hatherly, P., *J. Phys.* **B22**, L321 (1989); **26**, 783 (1993).
7. Chin, S.L., Liang, Y., Decker, J., Ilkov, F.A., Ammosov, M.V., *J. Phys.* **B25**, L249 (1992).
8. Chelkowski, S., Zuo, T., Bandrauk, A.D., *Phys. Rev.* **A46**, 5342 (1992).
9. Zuo, T., Bandrauk, A.D., a) *Phys. Rev.* **A48**, 3837 (1993); b) **A49**, 3943 (1994); c) **A52**, 2511 (1995).
10. Chelkowski, S., Bandrauk, A.D., *J. Phys.* **B28**, L723 (1995).
11. Seideman, T., Ivanov, M.Y., Corkum, P.B., *Phys. Rev. Lett.* **75**, 2819 (1995).
12. Rose-Petruck, C., Schafer, K.J., Barty, C.P.J., SPIE Proc., *Applications of Laser Plasma Radiation*, **2523**, 272 (1995).
13. Villeneuve, D.M., Ivanov, M.Y., Corkum, P.B., *Phys. Rev. A* (to appear).
14. Snyder, E.M., Buzza, S.A., Castleman, A.W., 1996 (preprint).
15. Schmidt, M., Normand, D., Cornaggia, C., *Phys. Rev.* **A50**, 5037 (1994).
16. Schmidt, M., Normand, D., Lewenstein, M., D'Oliveira, P. (preprint).
17. Bandrauk, A.D., Sink, M.L., *J. Chem. Phys.* **74**, 1110 (1981).
18. Aubanel, E., Gauthier, J.M., Bandrauk, A.D., *Phys. Rev.* **A48**, 2145 (1993); *Phys. Rev. A* **48**, R4011 (1993).
19. DeWitt, M.J., Levis, R.J., *J. Chem. Phys.* **102**, 8670 (1995).
20. Chelkowski, S., Zuo, T., Atabek, O., Bandrauk, A.D., *Phys. Rev.* **A52**, 2977 (1995)
21. Hiskes, J.R., *Phys. Rev.* **122**, 1207 (1961).
22. Hanson, G.R., *J. Chem. Phys.* **62**, 1161 (1975).
23. Yu, H., Bandrauk, A.D., *Phys. Rev. A* (submitted).
24. Shen, H., Bandrauk, A.D., *J. Chem. Phys.* **99**, 1185 (1993).

Photoionization of Polyatomic Molecules Using Intense, Near-Infrared Radiation of Femtosecond Duration

Robert J. Levis* and Merrick J. DeWitt

Department of Chemistry
Wayne State University
Detroit, MI 48202

Abstract. The relative crossections for the photoionization and the photodissociation of a series of aromatic compounds is presented in this paper. The molecules are ionized using an intense pulse of 780 nm radiation of duration 170 fs. The photo-induced processes are measured using time-of-flight mass spectrometry. We find that the photoionization rate scales as $e^{-\text{ionization potential}}$. We propose that the reason for the exponential dependence has to do with the three dimensional structure of the molecule under investigation.

INTRODUCTION

The photoionization of complex molecular species using intense radiation of femtosecond duration represents a qualitatively new field of light-matter interaction. This is true for two reasons. The first reason is that at high laser powers, the resonance condition for excitation is relaxed because the electric field of the laser approaches the binding energy of the valance electron to the nuclei. This leads to large Stark shifts as well as tunnel ionization phenomena. Thus it is likely that virtually any molecule can be ionized using any wavelength with high enough powers and short enough pulses. The second reason that femtosecond photoionization is of interest is that with pulse lengths of the order of molecular vibrations, the degree of nuclear excitation is markedly diminished. This allows the observation of the parent molecular ion which may have important analytical applications. For example, this laboratory has observed the parent molecular ion for molecules as diverse as ferrocene, dicholorodifluorocarbon, benzene, hexane and rhodamine 6G using only 780 nm excitation. The mechanism of the coupling of the radiation with the molecules remains unclear. In addition, the reason why intact molecular ions are observed is also unknown. Hence the focus of this paper is the mechanism of molecular ionization using intense pulses of near-infrared radiation of femtosecond duration.

At the intensities employed in this investigation, essentially two schemes for photoionization must be considered. The first scheme is multiphoton ionization wherein multiple photons are absorbed simultaneously by the molecule. At the intensities employed in this investigation, 10^{13-14} W cm^{-2}, highly nonlinear absorption process are possible. The other ionization mechanism that must be considered is a tunneling phenomena wherein the Coulombic potential of the molecule is perturbed to such an extent that it is possible for valence electrons to directly tunnel into the vacuum. For atomic species, the delineation between these two regimes is rather clear both theoretically[1] and experimentally[2]. In the near-infrared region only atoms with sufficiently high ionization potential, He and Ne, can tunnel ionize. The experimental signature for tunnel ionization is a continuum in the photoelectron spectrum extending to high kinetic energy. The noble gas atoms Xe, Kr and Ar have sufficiently low IP that multiphoton ionization is observed. The signature for MPI is a series of well-defined peaks separated by the photon energy, hv, in the photoelectron spectrum. For the case of molecules with extended molecular orbitals and low IP, neither the theory nor the current experimental measurements are sufficient to predict which mechanism will prevail at 780 nm. At 10 μm and for diatomic and small polyatomic molecules it is clear that the tunneling mechanism accurately predicts the physics of the ionization process[3]. In this case, the tunneling rate is accounted for by the theory proposed by Ammosov, Delone and Krainov[4]:

$$w = \left(\frac{3e}{\pi}\right)^{3/2} \frac{Z^2}{3n^{*3}} \frac{2l+1}{2n^*-1} \left[\frac{4eZ^3}{(2n^*-1)n^{*3}F}\right]^{2n^*-3/2} \exp\left[\frac{-2Z^3}{3n^{*3}F}\right] \tag{1}$$

In this theory, it is the ionization potential of the molecule, E_0, where $n^* = Z/\sqrt{2E_0}$, and the field strength of the laser, F, that exclusively govern the photoionization rate. However, more complex molecules do not have the simple Coulombic potentials required for the ADK theory. We might then expect that factors other than the ionization potential influence the ionization probability σ780, rec. Four such factors include the permanent dipole moment, the polarizability, the hyperpolarizability and the structure of the molecular system.

EXPERIMENTAL

To investigate the mechanisms of near infrared femtosecond photoionization of polyatomic molecules we have coupled a regeneratively amplified Ti:Sapphire laser[5] to a time-of-flight mass spectrometer.[6] The laser system delivers a maximum of 500 μJ of 780 nm radiation with a pulse duration of 170 fs. In this system a 150 fs oscillator pulse is stretched to 100 ps using two gratings, amplified in a YAG pumped regenerative amplifier and recompressed to 170 fs using a second pair of diffraction gratings. The linearly polarized laser pulse is then focused to a ~50 μm beam spot in the extraction region of a linear time-of-flight mass spectrometer. The resulting mass spectra are signal averaged for 1,000 laser shots and the average spectrum is stored on an external computer for subsequent analysis.

RESULTS AND DISCUSSION

A number of molecules have been studied with this system and we focus here on two series: benzene, naphthalene, phenanthrene and anthracene; and benzene toluene, ethyl benzene and propyl benzene. The relative ionization probability for each of these molecules has been measured and calibrated to that for benzene under identical conditions. The ionization potential, polarizability, the relative ionization probability at 780 nm and ionization order for these molecules are listed in the table shown below. The relative ionization probability is defined as the integrated sum under all of the mass peaks in the spectrum. The ionization order is obtained by plotting the logarithm of the relative ionization probability as a function of the logarithm of the laser power.

TABLE
Properties of Aromatic Molecules

	IP	Polarizability	$\sigma_{780, rel}$	Order
Benzene	9.24	10.32	1	8.1
Toluene	8.82	12.26	1.2	8.8
Ethylbenzene	8.77	14.2	1.4	8.2
Propylbenzene	8.72	16.0	0.35	7.2
Naphthalene	8.13	16.5	20	8.5
Phenanthrene	7.89	36.8	40	6.9
Anthracene	7.44	25.4	200	8.0

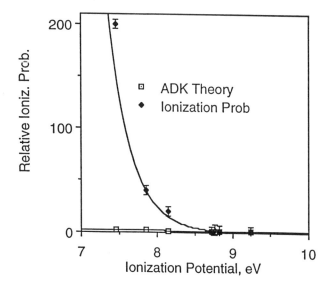

Figure 1. A plot of the relative ionization probability at 3.8×10^{13} W cm^{-2}, 780 nm, 170 fs duration as a function of ionization potential for the molecule, anthracene, phenanthrene, naphthalene propylbenzene, ethylbenzene, toluene and benzene (filled diamonds). The line through the open squares represents the ionization frequency calculated using Eq. 1 under our experimental conditions and normalized to benzene.

An interesting trend in the data is obtained by plotting the ionization probability as a function of the ionization potential for the molecules. As can be seen from Fig. 1, the correlation between the relative ionization probability and the ionization potential appears to be an inverse exponential. A plot of the relative ionization probability as a function of ionization potential for a simple tunneling model normalized to benzene is also shown in the figure as the line through the unfilled squares. The tunnel model employed is based on a structureless atom as set out by Eq. 1. As can be seen the simple tunneling model is essentially linear over the range of ionization potential employed. We propose that the bulk of the exponential increase in the ionization probability with decreasing ionization potential is due to the increasing size of the molecular orbitals as the series progresses from benzene to anthracene. To determine whether molecular orbital structure plays an important role we calculated the electrostatic potential cube of the molecule using *ab initio* methods. An electric field was then superimposed on various one-dimensional slices through the molecule so that a tunneling calculation could be performed. A simple model for the enhancement of the ionization probability is shown in Fig. 2 where we compare the tunnel ionization barrier for the case of benzene and anthracene with the application of a 1.7 V Å^{-1} field. As can be seen, the anthracene has essentially zero barrier for ionization while the benzene has an appreciable barrier to tunnel ionization. The tunnel calculations including molecular structure demonstrate an exponential increase in ionization rate for the series benzene, naphthalene and anthracene.

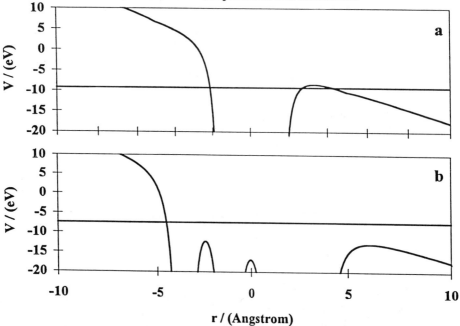

Figure 2. The electrostatic potential of a) benzene and b) anthracene (along the long axis for anthracene) with an electric field of 1.7 V Å^{-1} superimposed. The electrostatic potentials were calculated using *ab initio* methods at the 6-31g* level.

Including molecular structure into the tunnel model may also account for the significant fragmentation observed for anthracene in comparison with the limited fragmentation observed for benzene. In fact, under our laser conditions, the laser intensity can not be lowered to the point where the anthracene parent ion is the largest peak in the mass spectrum. As seen in Figure 2, the electric field of the laser operates on the anthracene molecule for three times the length of the benzene molecule. This means that the anthracene molecule rapidly enters the barrier suppression regime as the laser intensity is increased. In addition the polarizability of anthracene is larger than benzene, thus the localization of electron density will be much larger than that for benzene. It is likely that the enhanced interaction of the highest occupied molecular orbitals of anthracene with the electric field of the laser results in the increased fragmentation rate.

ACKNOWLEDGEMENT

We acknowledge the support of the NSF and the NIH through a Young Investigator Award and grant HG 485, respectively. RJL is a Camille Dreyfus Teacher Scholar and a Sloan Fellow.

REFERENCES

1. Keldysh, L.V.; 1965 Sov. Phys. JETP, 20, 1307-14.
2. Mevel E.; Breger, P.; Trainham, R.; Petite G.; Agostini, P.; Migus, A.; Chambaret, J.P.; Antonetti, A.; 1993 Phys. Rev. Lett., 70, 406-9.
3. Squier J.; Salin, F., Mourou, G., Harter, D. 1991 Opt. Lett., 16, 324-6.
4. Ammosov, M.V.; Delone, N.B.; Krainov, V.P.; Sov. Phys. JETP, 91, 1191-4.
5. DeWitt, M.J.; Levis, R.J. 1995, J. Chem. Phys., 102, 8670-3.
6. Schilke, D.; Levis, R.J.; 1994 Rev. Sci. Instrum., 65, 1903-11.

Time Evolution of the Lowest Singlet States of Z-1,3,5-Hexatriene and 1,3-Cyclohexadiene Studied by Femtosecond Photoionization

W. Fuß, T. Schikarski, W. E. Schmid, S. A. Trushin, K. L. Kompa

Max-Planck-Institut für Quantenoptik, D-85740 Garching, Germany

Abstract. Ultrafast dynamics in the lowest singlet states of Z-1,3,5-hexatriene (ZHT) and 1,3-cyclohexadiene (CHD) after excitation in the range 260–295 nm has been monitored by delayed photoionization in the region of 280–415 nm. While in ZHT a decay of the $2A_1$ state with an excess energy dependent rate constant of 0.4–1.4 ps^{-1} is directly observed by measuring the delayed total ion yield, in the case of CHD the appearance of vibrationally hot products after internal conversion/isomerization with a rate constant of 1.7 ps^{-1} can only be seen by detection of the ion fragment $C_6H_7^+$.

The study of the low lying excited singlet states of cyclohexadiene (CHD) and hexatriene (HT) is of great importance for the understanding of the fundamental photochemical reactions such as pericyclic ring-opening or Z-E (cis-trans) isomerization (1). Some of them play a crucial role in photobiology, e. g., in the visual system and in the photochemistry of vitamin D. It is commonly accepted that for most of the polyenes, after excitation in the intense $S_2 \leftarrow S_0$ ($1^1B_2 \leftarrow 1^1A_1$) band, the main products are formed only after relaxation to the dark S_1 (2^1A_1) state. ZHT is a product of the photochemical ring opening of CHD. We measured the decay kinetics of these states by using femtosecond time-resolved pump-probe two-color photoionization.

The species under study were excited in the range 260–291 nm by ~300 fs pulses. The subsequent decay of these excited states was monitored by ionization of the excited molecules in the range 280–415 nm with variable delay. The experimental system is described in detail elsewhere (2–4). The tunable pump and probe radiation was supplied by an optical parametric generator pumped with the 800 nm output of a Ti-sapphire laser and subsequent two-fold frequency mixing. Measurements were performed in gas phase with mass selective detection of ions.

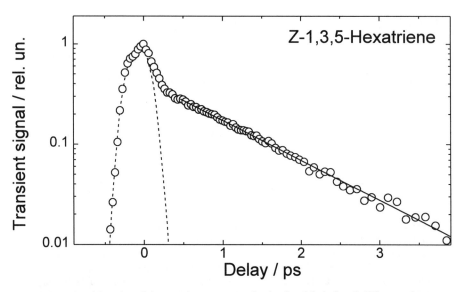

FIGURE 1. Semi-log plot of the transient mass-unselective ion (circles) and difference frequency (dashed curve) signals *vs.* delay time for ZHT pumped by one photon at 268 nm and probed by two-photon ionization at 400 nm at magic angle polarization. The solid line is a linear fit to the data in the region 0.5–4 ps giving a decay rate of 0.90 ± 0.02 ps^{-1}.

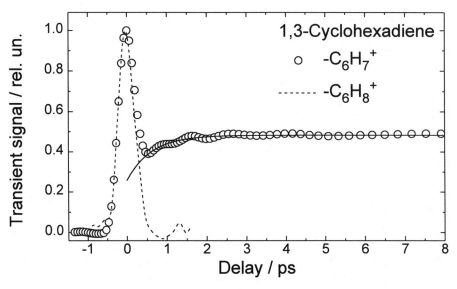

FIGURE 2. Linear plot of the transient mass-selective signals for the ions $C_6H_7^+$ (mass 79, open circles) and $C_6H_8^+$ (mass 80, dashed curve) for CHD pumped by one photon at 268 nm and probed by two photon ionization at 290 nm at magic angle polarization. The solid line shows the result of a single-exponential fit to the data in the region 0.5–8 ps giving a rise rate of 1.7 ± 0.1 ps^{-1}.

RESULTS AND DISCUSSION

In the pump wavelength range 260-291 nm, ZHT has a weak absorption due to the $S_1 \leftarrow S_0$ ($2^1A_1 \leftarrow 1^1A_1$, origin at ca. 291 nm) transition (its analog in the E isomer is symmetry-forbidden). In the case of ZHT, the kinetics measured via the total ion yield coincides with the kinetics measured via the parent ion ($C_6H_8^+$) at mass 80. When the pump and probe beams are polarized at the magic angle, a typical transient ion signal *vs.* pump-probe delay consists of an initial spike followed by a single-exponential tail (see Fig. 1). The shape of the initial spike practically coincides with the pump-probe cross-correlation function, which has been recorded separately by difference-frequency mixing. At parallel and orthogonal polarization the transient ion signal shows a polarization dependent additional process with a time constant of ~250 fs. This orientational anisotropy is due to free rotation of the ZHT molecule around the long molecular axis. The fact that after excitation in the $S_1 \leftarrow S_0$ band the polarization anisotropy is mainly sensitive to the rotation around this axis, implies that this band has a dipole transition along the C_2 axis, perpendicular to the molecular axis. The rate constant derived from the exponential decay depends on the excitation wavelength and increases from 0.4 to 1.4 ps^{-1} when the pump wavelength changes from 291 to 260 nm (i. e., 0–4000 cm^{-1} above the origin of the 2^1A_1 state). This rate reflects the decay of the 2^1A_1 state into the ground state 1^1A_1 due to internal conversion and/or isomerization. From the temperature dependence we infer a barrier of about 170 cm^{-1}. Our direct measurements of the lifetime of the 2^1A_1 state of ZHT are consistent with the linewidth of the 0–0 $2^1A_1 \leftarrow 1^1A_1$ origin of about 1 cm^{-1} (5), suggesting a lifetime of ≥5 ps, and the existence of fluorescence only in a narrow region (200 cm^{-1}) above the origin (6). Moreover, recent measurements of the decay after excitation of the 1^1B_2 band found a lifetime of 730 fs (7). Since the 1^1B_2 state is rapidly converted to the 2^1A_1 state, this time certainly corresponds to the lifetime of 2^1A_1 with the corresponding excess energy of 5000 cm^{-1}.

Our experimental findings are consistent with recent quantum chemical calculations (8), which show that for ZHT there are two pathways with different (but small) activation energies, going down from 2^1A_1 via two different conical intersections, where they branch to the ground state of either the educt (tZt conformer) or the products (the E isomer and the tZc conformer).

For CHD the pump wavelength of 268 nm was near the origin of the allowed $1^1B_2 \leftarrow 1^1A_1$ transition. In contrast to ZHT, the transient ion signal at mass 80 corresponding to the parent ion ($C_6H_8^+$) shows no delayed part within our present time resolution when probed in the wavelength region 280-415 nm. However, when probed in the region 280-290 nm and in the limit of weak pump fluence, the fragment at mass 79 ($C_6H_7^+$) falls down after the initial spike and then shows a rising feature (see Fig. 2) with an appearance rate constant of 1.7±0.2 ps^{-1}. This delayed signal, we believe, is due to resonant two-photon ionization of the ground state of the products CHD/HT, which are produced with high vibrational excess energy of ~4 eV, and subsequent dissociation of the vibrationally hot ion $C_6H_8^+$ into $C_6H_7^+$ and H. Therefore we interpret the 1.7 ps^{-1} as the rate constant of the photochemical ring opening of CHD. This rate is an order of magnitude higher than the value deduced from Raman spectroscopical measurements, where a rate constant of 0.17 ps^{-1} has been found for the formation of ground state product (9). However, recent transient absorption studies of the photochemical ring-opening of CHD in solution suggest that its rate could be faster than 1 ps^{-1} (10, 11).

The rate determining step of the photochemical ring opening of CHD is the transition from the 2^1A_1 state via a conical intersection to the ground state surface of the educt and the product (12, 13). In this mechanism, the 2^1A_1 surface acts as a common intermediate for CHD and cZc-HT. The measured rate of 1.7 ps^{-1} for photochemical ring-opening in CHD is most probably the decay rate of the 2^1A_1 state with several thousand cm^{-1} of vibrational excess energy. It should be compared with the decay rate of 1.4 ps^{-1} for tZt-HT at high excess energy reported in the first part of this work. So these two values are very similar for the two conformers.

REFERENCES

1. Orlandi, G., Zerbetto, F., and Zgierski, M. Z., *Chem. Rev.* **91**, 867–891 (1991).
2. Fuß, W., Höfer, T., Hering, P., Kompa, K. L., Lochbrunner, S., Schikarski, T., and Schmid, W. E., *J. Phys. Chem.* **100**, 921–927 (1996).
3. Fuß, W., Schikarski, T., Schmid, W. E., Trushin, S. A., Hering, P., and Kompa, K. L., *J. Chem. Phys.*, submitted.
4. Fuß, W., Schikarski, T., Schmid, W. E., Trushin, S. A., and Kompa, K. L., *Chem. Phys. Lett.*, submitted.
5. Buma, W. J., Kohler, B. E., and Song, K., *J. Chem. Phys.* **94**, 6367–6376 (1991).
6. Petek, H., Bell, A. J., Christensen, R. L., and Yoshihara, K., *J. Chem. Phys.* **96**, 2412–2415 (1992).
7. Hayden, C. C., and Chandler, D. W., *J. Phys. Chem.* **99**, 7897–7903 (1995).
8. Olivucci, M., Bernardi, F., Celani, P., Ragazos, I., and Robb, M. A., *J. Am. Chem. Soc.* **116**, 1077–1085 (1994).
9. Reid, P. J., Lawless, M. K., Wickham, S. D., and Mathies, R. A., *J. Phys. Chem.* **98**, 5597–5606 (1994).
10. Pullen, S., Walker II, L. A., Donovan, B., and Sension, R. J., *Chem. Phys. Lett.* **242**, 415–420 (1995).
11. Fuß, W., Lochbrunner, S., Schmid, W. E., and Kompa, K. L., "Comparison of the photochemical ring opening of cyclohexadiene and dehydrocholesterol in solution, studied by ultrafast absorption spectroscopy," in *1996 Technical Digest Series Vol. 8, The 10th Int. Symp. on Ultrafast Phenomena,* San Diego, California, USA, May 27–June 1, 1996, pp. 339–340.
12. Celani, P., Ottani, S., Olivucci, M., Bernardi, F., and Robb, M. A., *J. Am. Chem. Soc.* **116**, 10141–10151 (1994).
13. Celani, P., Bernardi, F., Robb, M. A., and Olivucci, M., *J. Am. Chem. Soc.* **118**, in press (1996).

Formation, Stability and Fragmentation of Biomolecular Clusters in a Supersonic Jet Investigated with Nano- and Femtosecond Laser Pulses

Anja Meffert and Jürgen Grotemeyer

Institut für Physikalische Chemie der Universität Würzburg,
Marcusstrasse 9-11, 97070 Würzburg

Abstract. Within the scope of our work the cluster formation of small amino acids with organic molecules typically employed as matrix substances in Matrix assisted Laser Desorption / Ionization (MALDI) (1) has been investigated. The aim of our study was to find whether heterogeneous clusters consisting of matrix and sample molecules may play a role in the MALDI ionization process which is not completely understood until now.
The investigation was performed in a time-of-flight mass spectrometer via laser desorption of the sample mixtures into a supersonic beam of argon, followed by Multiphoton Ionization of the neutral molecules and clusters either with nano- or femtosecond laser pulses for comparison. The main results are that clusters could only be detected in the case of femtosecond ionization and that an intermolecular proton transfer within the heterogeneous dimers from matrix molecule to the amino acids was observed.

INTRODUCTION

The formation of molecular clusters in the gas phase has become a subject of growing interest throughout the last years, especially clusters of biomolecules. The reason is that weak bound molecular clusters can be regarded as model systems for biologically relevant processes and that understanding of the structure of such clusters provides information about fundamental physical proceedings between molecules. On the other hand Matrix Assisted Laser Desorption / Ionization (MALDI) has become a central tool for the investigation of large biomolecules especially for peptides, but details about the MALDI ionization process are still unknown. In the last years some indications for the participation of gas phase reactions in the ionization process were found (2). Therefore the cluster formation between small amino acids and MALDI matrix molecules in the gas phase was studied by means of laser desorption time-of-flight mass spectrometry in order to find out if heterogeneous clusters are formed and if they might play a role in the MALDI ionization process.

EXPERIMENTAL

The experiments were performed in a reflectron time-of-flight mass spectrometer (Bruker TOF1). A pulsed CO_2 laser (10.6 μm) was used for the desorption of the sample mixtures out of a matrix (Polyethylene) into a supersonic beam of argon. In most experiments MALDI matrix and amino acid were mixed in equal molar parts. The neutral sample molecules and formed clusters were transported through a skimmer into the ion source of the RETOF-MS where they were postionized with multiphoton ionization (MUPI). MUPI was performed either with nanosecond or femtosecond laser pulses for comparison. The fs laser pulses were generated with a dye laser pumped by a XeCl excimer laser (λ~260 nm, pulse length 500 fs, energy/pulse 5-10 μJ). For nanosecond ionization the frequency doubled output of a Nd:YAG pumped dye laser was used (λ = 260 nm, pulse length 8 ns, energy/pulse ~ 500 μJ).

The MALDI matrices under investigation are derivatives of benzoic and cinnamic acid (ferulic acid, sinapic acid, 2,5-dihydroxybenzoic acid and vanillic acid) and the employed amino acids are such without an aromatic chromophor (glycine, leucine, glutamic acid and lysine).

RESULTS AND DISCUSSION

Investigation of the pure MALDI Matrices

First experiments were performed with pure matrix substances as the aromatic chromophor of the MALDI matrices dominates the spectroscopy of the molecular clusters. Multiphoton ionization was performed with nano- and femtosecond laser pulses at λ = 260 nm.

In the case of nanosecond ionization only weak molecular ions of the matrix molecules are detected but strong fragmentation resulting from loss of the carboxyl group is observed. No formation of any homogeneous clusters can be recognized here. In contrast in the case of femtosecond ionization the molecular ions represent the base peak of the respective spectrum and much less fragmentation is observed. In addition homogeneous clusters of the matrix substances M_n up to n = 5 for ferulic acid are detected.

For illustration the mass spectrum of ferulic acid obtained with femtosecond multiphoton ionization is presented in figure 1.

Obviously the duration of the laser pulses used for ionization is important for the detection of molecular matrix clusters . This leads to the conclusion that the excited S_1 states of either the matrix molecules and the clusters have only a short lifetime and that competitive processes like intersystem crossing ISC to the triplet state T_1, relaxation to the ground state or neutral fragmentation lead to a loss of the excited molecules before absorption of a second photon which would lead to ionization.

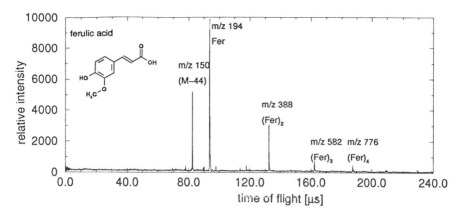

FIGURE 1. Time of flight mass spectrum of ferulic acid obtained with femtosecond multiphoton ionization at λ = 260 nm (pulse length 500 fs, energy/pulse ~ 20 μJ).

Experiments concerning Cluster Formation with NH_3

After investigating the spectroscopic properties of the pure MALDI matrices the question of interest is: *What happens in the case of cluster formation with small polar molecules?* The topics of interest are whether MALDI matrices form heterogeneous clusters with polar molecules and whether intracluster reactions can be observed which might play a role in the MALDI ionization process. First clustering experiments were performed with ammonia. The argon jet was enriched with NH_3 in front of the desorption chamber and multiphoton ionization was performed with femtosecond laser pulses again. Figure 2 shows the mass spectrum obtained for ferulic acid and ammonia as representative example.

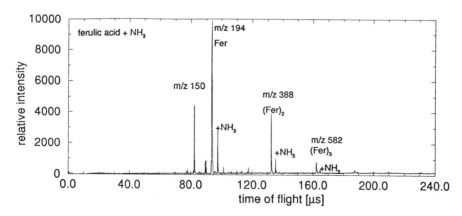

FIGURE 2. Mass spectrum of ferulic acid + ammonia. Multiphoton ionization performed with femtosecond laser pulses (λ = 260 nm).

Besides the molecular ion the spectrum shows again the homogeneous clusters of ferulic acid $(Fer)_n$. In addition to each signal of ferulic acid a further peak seventeen mass units higher is observed resulting from clustering of one molecule NH_3 to free ferulic acid as well as to its homogeneous clusters. But no "neat" ammonia or ammonium ion NH_4^+ is observed. Analogous results were obtained for the other MALDI matrices under investigation. Only in the case of 2,5-dihydroxybenzoic acid no clustering between matrix and ammonia is observed.

Experiments concerning clustering with amino acids

The results obtained with ammonia led to similar experiments with amino acids as small polar biomolecules in order to study the clustering behavior of typical MALDI matrices and the components of proteins. In first experiments glycine and leucine, small non aromatic amino acids, were employed.The investigated matrix substances were mixed with the respective amino acid in the molar ratio 1:1 and co-desorbed into the supersonic beam of argon. Postionization was again performed with femtosecond laser pulses at $\lambda = 260$ nm. Before discussing the results it must be emphasized that ionization of the free amino acids is not possible under the chosen ionization conditions as they do not dispose of an adequate chromophor.

In all measurements with **glycine** a heterogeneous dimer [matrix+glycine] of only little intensity is observed. Obviously the formation of homogeneous clusters is the dominant process as intensive signals of homogeneous clusterions of the respective matrix are to be noticed. No addition of glycine to these homogeneous clusters is detected. In the experiments with ferulic acid a very small signal at m/z 76 which corresponds to protonated glycine appears in the mass spectrum. In the spectra obtained with the other matrices no such signal is detected.

More interesting are the results obtained with **leucine**. The mass spectrum of ferulic acid and leucine is presented in figure 3.

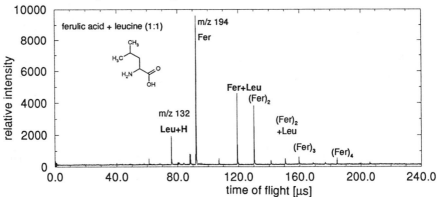

FIGURE 3. Cluster formation of ferulic acid with leucine.

All matrices form heterogeneous dimers [matrix+leucine] and the signals are more intensive as those of the homogeneous clusterions in contrast to the results obtained with glycine. Even more important is the fact that in all spectra a peak at m/z 132 is observed indicating protonated leucine. In the case of 2,5-dihydroxybenzoic acid the signal of the heterogeneous dimer as well as that of the protonated leucine is of relative low intensity compared to the spectra obtained with ferulic, vanillic or sinapic acid.

As no direct ionization of the amino acid is possible under the chosen conditions for MUPI it seems likely that a proton transfer from matrix molecule to amino acid occurs within the heterogeneous dimer after ionization. Subsequently the dimer dissociates releasing the protonated amino acid.

$$[M+A]^{+ \cdot} \rightarrow [M-H]^{\cdot} + [A+H]^{+}$$

Such a process is known as *dissociative proton transfer* and studied for cluster systems of benzene derivatives and small solvens molecules like toluene $+H_2O$ or NH_3 (3).

Is their any evidence for this assumed mechanism for the formation of protonated amino acids ? One indication is the peak profile observed for the protonated amino acids. The signals show a metastable widening which indicates that the protonated molecules result from the fragmentation of larger molecular aggregates.

Further support is obtained from the investigations with functionalized non aromatic amino acids. Analogous experiments were performed with glutamic acid and lysine, which is a basic amino acid possessing a second amino group. All spectra obtained in the case of **lysine** show only a small amount of heterogeneous dimer but very intensive [lysine+H]-signals. These results suggest that the formed heterogeneous dimer decays rapidly after ionization but that a proton transfer between matrix substance and lysine has taken place to a large extent in the cluster. Figure 4 presents the spectrum for ferulic acid as example.

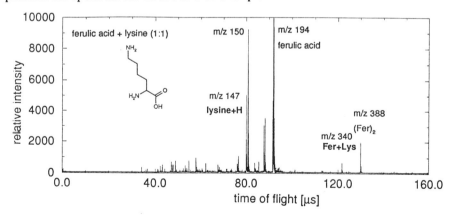

FIGURE 4. Cluster formation between ferulic acid and lysine.

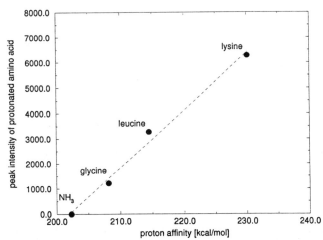

FIGURE 5. Correlation of the signal intensities of protonated amino acids with their proton affinities. Results for ferulic acid as employed MALDI matrix.

Comparing the spectra obtained with glycine, leucine and lysine an increase of the signal intensity of protonated amino acid is noticed. This result suggests a correlation with the proton affinities of the investigated amino acids which is reasonable for the assumed intracluster proton transfer. Indeed exact evaluation shows that the amount of detected protonated amino acid can be correlated to the proton affinities (4) as well as the heat of reaction ΔH_R for the dissociation of the ionized heterogeneous cluster releasing the protonated amino acid which were estimated with the help of thermochemical data. This correlation is presented in figure 5.

From our results it can be **concluded** that cluster formation in the gas phase is not likely to play a role in the MALDI ionization process.

ACKNOWLEDGMENTS

This work was supported by the Deutsche Forschungsgemeinschaft. We thank the Fonds der Chemischen Industrie for support.

REFERENCES

1. Ehring, H., Karas, M. and Hillenkamp, F., *Org. Mass Spectrom.* **27**, 472-480 (1992).
   Karas, M., Bahr, U., Ingendoh, A., Nordhoff, E., Stahl, B., Strupat, K. And Hillenkamp, F., *Anal. Chim. Acta* **241**, 175-185 (1990).
2. Wang, B.H., Dreisewerd, K., Bahr, U., Karas, M. And Hillenkamp, F., *J. Am. Soc. Mass. Spectrom.* **4**, 393-398 (1993).
3. Brutschy, B., *Chem. Rev.* **92**, 1567-1587 (1992).
4. Meot-Ner (Mautner), M. Hunter, E.P. and Field, F.H., *J. Am. Chem. Soc.* **101**, 686-689 (1979).

Photoionization Mechanisms of Metal Carbonyls with High Power Femtosecond Laser Pulses

K. F. Willey, C. L. Brummel, and N. Winograd

Department of Chemistry, The Pennsylvania State University
184 Materials Research Institute Building, University Park, PA 16802 USA

Abstract. High intensity pulses are utilized to explore the photoionization mechanisms involved during femtosecond excitation. A molecular ion signal is observed for both $Fe(CO)_5$ and $Cr(CO)_6$ at all wavelengths studied (i.e. 800, 400, and 266 nm). Two distinct ionization mechanisms are proposed. MPI dominates when the multiphoton cross section is large. If the neutral molecules, however, are exposed to sufficient laser intensity barrier suppression ionization occurs. At our highest obtainable powers for 800 nm fs excitation, intact doubly and triply charged molecular ions are observed.

INTRODUCTION

The advent of high power ultra-short pulse laser systems has made it possible to investigate previously unattainable photoionization mechanisms. In "traditional" MPI experiments incorporating ns pulses, a main source of photofragmentation is deactivation of the excited neutral through processes such as internal conversion, intersystem crossing, and electronic or vibrational predissociation. Metal carbonyls have proven to be especially challenging systems when employing ns excitation due to fragmentation prior to ionization through these deactivation channels (1). At all wavelengths studied, the bare metal ion is the dominate species observed in multiphoton ionization experiments (2). It is believed that the main mechanism for photoionization is excitation leading to the stripping of all CO ligands followed by ionization of the central metal. Femtosecond laser pulses, however, are capable of generating absorption rates which may outrun the fragmentation processes which are prevalent in ns excitation of metal carbonyls. In this paper we present results on the photoionization of $Fe(CO)_5$ and $Cr(CO)_6$ with femtosecond pulses of 800, 400, and 266 nm light. The molecular ion is observed at all wavelengths for both compounds. Furthermore, intact doubly and triply charged molecular ions are detected with 800 nm light above 2×10^{14} W/cm^2.

EXPERIMENTAL

Photoionization experiments are performed in a reflectron based TOF-MS. Metal carbonyl samples are introduced into the chamber by way of a leak valve. The leak rate is controlled to maintain an operating pressure no greater than 1×10^6

torr to assure collision free measurements. The focused laser beam interacts with the gas-phase molecules in the extraction region of the mass spectrometer. The ionized species are accelerated through a field-free region, travel through the reflecting mirror, traverse a second field-free region, and are finally detected at a dual channel-plate assembly. The reflectron compensates for the initial energy spread introduced by the spatial distribution of the laser beam in the extraction field such that mass resolution of 9000 can be obtained at 200 m/z (3). The laser system used is a Ti:sapphire based regenerative amplified system manufactured by Clark-MXR, Inc. At the fundamental (800 nm), 1.5 mJ 150 fs pulses are generated at 1 kHz repetition rate. The 800 nm light can in turn be doubled and tripled with a LBO and BBO crystal, respectively.

RESULTS AND DISCUSSION

At the high peak powers achieved by focusing the fs laser beam, photoionization mechanisms other than resonant or non-resonant multiphoton ionization become operative. The electric field generated by the laser pulse begins to distort the potential energy surface such that the barrier to ionization is lowered. In turn the probability of tunneling through the coulombic barrier becomes significant. The Keldysh parameter (4,5):

$$\gamma = \omega(2m_eIP)^{1/2}/eE_0, \tag{1}$$

in SI units (where ω is the angular frequency of the laser, IP is the ionization potential, E_0 is the electric field strength, and m_e and e are the mass and charge of the electron, respectively), aids in distinguishing between a MPI and tunneling ionization regime. If $\gamma \gg 1$ then MPI occurs, while at conditions where $\gamma \ll 1$ tunneling ionization is believed to dominate. At even higher laser intensities the coulombic barrier is suppressed such that the electron is no longer bound. The onset for barrier suppression ionization is at the laser intensity (6)

$$I_{(th)} = 4.0 \times 10^9 \ IP^4(eV)/ Z^2, \tag{2}$$

where Z is equal to the charge state produced.

As stated above, ns excitation of metal carbonyls leads to the detection of mainly bare metal ions. Whereas with 800 nm fs excitation of $Fe(CO)_5$ and $Cr(CO)_6$, shown in Figure 1, intact molecular ions are observed. M refers to the metal core and L denotes the CO ligands. The spectra were taken under identical conditions at an average laser intensity of 8.0×10^{12} W/cm^2, 200 fs pulses, and at a pressure of 1×10^{-6} torr. The $Fe(CO)_5$ spectrum exhibits photofragmentation corresponding to consecutive loss of CO ligands, but the molecular ion remains as a prominent peak. The molecular ion, however, is the dominant peak in the $Cr(CO)_6$ spectrum where very little photofragmentation is observed. The ionization potential for $Fe(CO)_5$ and $Cr(CO)_6$ is near 8 eV which corresponds to $\gamma = 2.9$ at these laser conditions. This value suggests that ionization is not through a tunneling mechanism. The threshold value for barrier suppression is 1.6×10^{13} W/cm^2 for both $Fe(CO)_5$ and $Cr(CO)_6$. The reported laser intensities are average powers, therefore barrier suppression ionization may occur at the peak of the pulse. Possible causes for the differing degree of photofragmentation in the two spectra are ionization through different mechanisms and the lifetime of the neutral intermediate states.

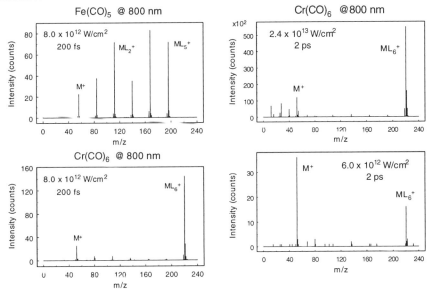

FIGURE 1. Photoionization spectra of Fe(CO)$_5$ and Cr(CO)$_6$.

FIGURE 2. Photoionization spectra of Cr(CO)$_6$ at 800 nm with ps pulses.

To explore the role the lifetime of the neutral excited states plays, we performed photoionization experiments with 2 ps pulses. In the Fe(CO)$_5$ studies the dominant channel observed at all laser conditions is Fe$^+$. Channels corresponding to Fe(CO)$_n^+$, where n≥1, comprise less than 25% of the observed peaks, with no species larger than Fe(CO)$_3^+$ detected. As shown in Figure 2, however, excitation of Cr(CO)$_6$ with 800 nm ps pulses exhibits very different behavior. At our highest obtainable peak powers with 2 ps pulses (i.e. 2.4x10^{13} W/cm^2), the spectrum closely resembles that of fs excitation experiments where the molecular ion is the dominant peak. When the laser intensity is attenuated by a factor of four, an increase in the degree of fragmentation occurs. Photoionization of Cr(CO)$_6$ with 800 nm light requires six photons. A power dependence study, where the log of the ion counts is measured as a function of the log of the laser intensity over the range 4.0x10^{12} - 2.4x10^{13} W/cm^2, reveals a slope of 3.8 for the bare metal ion and a slope of 5.4 for the molecular ion channel. This suggests that ionization occurs through two distinct mechanisms. At low laser intensities, the prevailing mechanism is a weak MPI process that leads to the prominent Cr$^+$ fragment. As the laser intensity is increased, especially above the threshold value for barrier suppression (i.e. 1.6x10^{13} W/cm^2), the molecular ion channel begins to dominate. Our data suggest that Cr(CO)$_6^+$ arises due to a barrier suppression ionization mechanism. Generation of the molecular ion signal through a MPI process can not be ruled out, but one would predict a measured photon order closer to that of Cr^+ instead of a totally non-resonant process.

Femtosecond excitation experiments have also been performed at 400 and 266 nm for $Fe(CO)_5$ and $Cr(CO)_6$. At both wavelengths for $Fe(CO)_5$, Figure 3, the molecular ion is the base peak in the spectrum. Photofragmentation channels correspond to consecutive loss of CO ligands. Femtosecond excitation of $Cr(CO)_6$ at 400 and 266 nm, not shown, exhibits similar results. As the laser intensity increases, for both systems and at both wavelengths, the degree of fragmentation increases. This trend is observed with 800 nm fs excitation also, but an interesting result occurs at our highest obtainable laser intensities. Intact multiply charged molecular ions are detected.

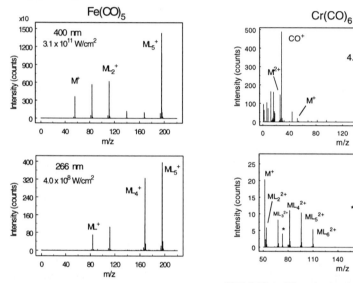

FIGURE 3. Photoionization spectra of $Fe(CO)_5$ at 400 and 266 nm.

FIGURE 4. Photoionization spectrum of $Cr(CO)_6$ at peak power of fs excitation.

As can be seen in Figure 4, even at these powers the molecular ion channel is observed for photoionization of $Cr(CO)_6$. The lower trace in Figure 4 is an expansion of the range from 50 to 240 m/z. The ML_6^{2+} channel is clearly visible with photofragment channels corresponding to the loss of CO ligands. The triply charged molecular ion is also detected and is labeled with an asterisk. One would predict multiply charged molecules of this size to be unstable and undergo Coulomb explosion if the charges are localized on different ligands and the metal. No peak broadening is observed in the mass spectrum as the laser intensity is increased which would rule out an explosion process following ionization. Bonding in the metal carbonyl systems is governed by electrostatics upon ionization. Therefore, if the charge builds on the metal core the CO ligands become more strongly bound to the metal ion due to ion-dipole and ion induced-dipole interactions.

CONCLUSIONS

Unlike ns MPI experiments, we observe molecular ion signal with fs excitation at all wavelengths studied. Two distinct photoionization mechanisms are proposed. For the $Fe(CO)_5$ system where the MPI cross section is high, ionization occurs before the neutral molecules can be exposed to the peak power of the laser pulse. However, the multiphoton cross section for 800 nm excitation of $Cr(CO)_6$ is low, thus molecules can be photoionized through a barrier suppression mechanism. With fs excitation at 800 nm intact doubly and triply charged molecular ions are detected above 2×10^{14} W/cm^2. This observation suggests that the multiple charges may be localized on the metal core. We are currently exploring other systems to determine if this observation is specific for metal carbonyls or is a general phenomena seen for all metal-ligand systems.

ACKNOWLEDGEMENTS

The authors graciously acknowledge the financial support of the National Institute of Health and the National Science Foundation..

REFERENCES

1. Duncan, M. A., Dietz, T. G., and Smalley, R. A., *Chem. Phys.* **44**, 415-419 (1979).
2. Gobeli, D. A., Yang, J. J., and El-Sayed, M. A., *Chem. Rev.* **85**, 529-554 (1985).
3. Brummel, C. L., Willey, K. F., Vickerman, J. C., and Winograd, N., *Int. J. Mass Spectrom. Ion Processes* **143**, 257-270 (1995).
4. Keldysh, L. V., *Sov. Phys. JETP* **20**, 1307-1314 (1965).
5. Lompre, L. A., Mainfray, G., Manus, C., Repoux, S., and Thebault, J., *Phys. Rev. Lett.* **36**, 949-952 (1976).
6. Augst, A., Meyerhofer, D. D., Strickland, D., and Chin, S. L., *J. Opt. Soc. Am. B* **8**, 858-867 (1991).

SESSION III: ATOMIC RIS I

Quantum Control of Dynamics

Moshe Shapiro* and Paul Brumer†

* *Chemical Physics Department, The Weizmann Institute of Science, Rehovot, Israel 76100,* and
† *Chemical Physics Theory Group Department of Chemistry, University of Toronto Toronto, Canada M5S 1A1*

Abstract. Coherent Control provides a quantum interference based method for controlling molecular dynamics. This theory is reviewed and applications to a variety of processes including photodissociation, and asymmetric synthesis, are discussed. Control scenarios and computations on the control of IBr, H_2O, DOH and Na_2 photodissociation reactions are discussed. We show that a wide range of yield control is possible under suitable laboratory conditions. Recent experiments on the control photocurrent directionality in semiconductors and of the Na_2 photodissociation to yield Na atoms in different excited states are shown to confirm the theory.

I. INTRODUCTION

Selectivity is at the heart of Chemistry and the control of reactions using lasers has been a goal for decades. Recently, we[1]-[17] and other groups[18]-[27] have demonstrated theoretically that one can achieve this goal by using quantum interference phenomena. We showed that phases acquired by a quantum systems while excited by lasers enable one to control quantum interferences, and hence the outcome, of many dynamical processes. Experimental tests[28]-[38] of our approach, termed Coherent Control (CC), have confirmed many of the theoretical predictions and proven the viability of the method.

The purpose of this article is to provide an introduction to the concepts[39] underlying CC and to discuss its current status in both Chemistry and Physics.

I.A Aspects of Scattering Theory and Reaction Dynamics

The processes we wish to control include branching "half" collisions,

$$AB + C \leftarrow ABC \rightarrow A + BC \qquad (1)$$

and "full" collisions,

$$AB(m'') + C \leftarrow A + BC(m) \rightarrow A + BC(m') \qquad (2)$$

In the above, A, B, C are either atoms, groups of atoms, electrons, or photons. m, m' denote the internal (vibrational, rotational, photon occupation) quantum numbers of the reactants or products.

Given $\Psi(t = 0)$, the system wavefunction at an initial time, the evolution of the system is determined by the time dependent material Schrödinger Equation,

$$H_M \Psi(t) = i\hbar \partial \Psi(t)/\partial t. \tag{3}$$

where H_M is the system Hamiltonian. The long time behavior of $\Psi(t)$ is intimately connected with the nature of the time independent continuum energy eigenstates. For every continuum energy value E, each of the possible outcomes observed in the product region is represented by an independent wavefunction. This "boundary" condition is expressed more precisely by denoting the different possible chemical products of the breakup of ABC in Eq. (1) by an index q (e.g. $q = 1$ denotes the $A + BC$ products), and all additional identifying state labels by m. The set of continuum eigenfunctions of the material Hamiltonian,

$$H_M | E, m, q^- \rangle = E | E, m, q^- \rangle, \tag{4}$$

is now defined via the requirement that asymptotically every $| E, m, q^- \rangle$ state goes over to a state of the separated products, denoted $| E, m, q^0 \rangle[40]$, which is of energy E, chemical identity q and remaining quantum numbers m.

The description of the system in terms of $| E, m, q^- \rangle$ has an important advantage: Expressing the state of the system in the present in terms of these states, i.e., writing and initial continuum state as

$$\Psi(t = 0) = \sum_{q,m} \int dE c_{q,m}(E) | E, m, q^- \rangle, \tag{5}$$

means that we know the fate of the system in the future. Since each of the $| E, m, q^- \rangle$ states correlates with a *single* product state, the probability of observing each $| E, m, q^0 \rangle$ product state is simply given by $|c_{q,m}(E)|^2$ - the *preparation* probabilities. The probability of producing a chemical product q in the future is therefore given as

$$P_q = \sum_m \int dE |c_{q,m}(E)|^2. \tag{6}$$

Below we demonstrate that the key to laser control is to change one $c_{q,m}(E)$ coefficient relative to another $c_{q',m}(E)$ coefficient *at the same energy*. In order to understand how this can be done we discuss now the process of preparation.

I.B Perturbation Theory, System Preparation and Coherence

Consider the effect of an electric field $\epsilon(t)$ on an initially bound eigenstate $| E_g \rangle$ of the radiation-free Hamiltonian, H_M. The overall Hamiltonian is then given by:

$$H = H_M - \mathsf{d}[\bar{\epsilon}(t) + \bar{\epsilon}^*(t)] \tag{7}$$

where d is the component of the dipole moment along the electric field.

If the impinging photon is energetic enough to dissociate the molecule, it is then necessary to expand $|\Psi(t)\rangle$ in the bound and scattering eigenstates of the radiation-free Hamiltonian,

$$|\Psi(t)\rangle = \sum_i c_i(t)|E_i\rangle \exp(-iE_it/\hbar) + \sum_{m,q} \int dE c_{E,m,q}(t)|E, m, q^-\rangle \exp(-iEt/\hbar). \tag{8}$$

Insertion of Eq. (8) into the time dependent Schrödinger Equation results in a set of first-order differential equations for the $c_\nu(t)$ coefficients, where ν represents either the bound (i) or scattering (E, m, q) indices.

For weak fields use of first order perturbation theory gives, for the post-pulse preparation coefficient,

$$c_{E,m,q}(t \gg \Gamma) = (\sqrt{2\pi}/i\hbar)\epsilon(\omega_{E,E_g})\langle E, m, q^-|\mathsf{d}|E_g\rangle, \tag{9}$$

where Γ is the pulse duration and

$$\epsilon(\omega) = (1/\sqrt{2\pi})\int_{-\infty}^{\infty} \exp(i\omega t)\,\bar{\epsilon}(t)\,dt. \tag{10}$$

It follows from Eq. (6) and Eq. (9) that the probability $P(E, q)$ of forming asymptotic product in arrangement q is,

$$P(E, q) = \sum_m |c_{E,m,q}(t \gg \Gamma)|^2 = (2\pi/\hbar^2)\sum_m |\epsilon(\omega_{E,E_g})\langle E_g|\mathsf{d}|E, m, q^-\rangle|^2 \tag{11}$$

and that the branching ratio $R(1, 2; E)$ between the $q = 1$ products and the $q = 2$ products at energy E is given as,

$$R(1, 2; E) = \frac{\sum_m |\langle E_g|\mathsf{d}|E, m, 1^-\rangle|^2}{\sum_m |\langle E_g|\mathsf{d}|E, m, 2^-\rangle|^2} \tag{12}$$

I.C Coherent Control of Chemical Reactions

We now address the issue of how to alter the above yield ratio $R(1, 2; E)$ in a *systematic* fashion. Equation (12) makes clear that (at least in the weak field regime) this can not be achieved by altering the laser intensity, since the field strength cancels out in the expression for R. Quantum interference phenomena can, however, alter the numerator or denominator of R in an independent and controlled way. This can be achieved by accessing the final continuum state via two or more interfering pathways. One of the first examples which we studied [1], involves preparing a molecule in a superposition $c_1|\phi_1\rangle + c_2|\phi_2\rangle$

71

state and exciting the two components to the same final continuum energy E by using two CW sources. The field employed is of the form,

$$\bar{\epsilon}(t) = \epsilon_1 e^{-i\omega_1 t + i\chi_1} + \epsilon_2 e^{-i\omega_2 t + i\chi_2} \tag{13}$$

where $\hbar\omega_i = E - E_i$. A straightforward computation[1] yields that,

$$R(1, 2; E) = \frac{\sum_m |\langle \tilde{\epsilon}_1 c_1 \phi_1 + \tilde{\epsilon}_2 c_2 \phi_2 | \mathbf{d} | E, m, 1^- \rangle|^2}{\sum_m |\langle \tilde{\epsilon}_1 c_1 \phi_1 + \tilde{\epsilon}_2 c_2 \phi_2 | \mathbf{d} | E, m, 2^- \rangle|^2} \tag{14}$$

where $\tilde{\epsilon}_i = \epsilon_i \exp(i\chi_i)$. Expanding the square gives:

$$R(1, 2; E) =$$

$$\frac{\sum_m [|\tilde{\epsilon}_1 c_1 \langle \phi_1 | \mathbf{d} | E, m, 1^- \rangle|^2 + |\tilde{\epsilon}_2 c_2 \langle \phi_2 | \mathbf{d} | E, m, 1^- \rangle|^2 + 2Re[c_1 c_2^* \tilde{\epsilon}_1 \tilde{\epsilon}_2^* \langle \phi_1 | \mathbf{d} | E, m, 1^- \rangle]}{\sum_m [|\tilde{\epsilon}_1 c_1 \langle \phi_1 | \mathbf{d} | E, m, 2^- \rangle|^2 + |\tilde{\epsilon}_2 c_2 \langle \phi_2 | \mathbf{d} | E, m, 2^- \rangle|^2 + 2Re[c_1 c_2^* \tilde{\epsilon}_1 \tilde{\epsilon}_2^* \langle \phi_1 | \mathbf{d} | E, m, 2^- \rangle]} \tag{15}$$

The structure of the numerator and denominator of Eq. (15) is of the type desired, i.e. each has a term associated with the excitation of the $|\phi_1 >$ state, a term associated with the excitation of the $|\phi_2 >$ state, and a term corresponding to the interference between the two excitation routes. The interference term, which can either be constructive or destructive, is in general different for the two product channels. What makes Eq. (15) so important *in practice* is that the interference term has coefficients whose magnitude and sign depend upon *experimentally controllable* parameters. In the case of Eq. (15) the experimental parameters which alter the yield[1] are contained in the complex quantity $A = \tilde{\epsilon}_2 c_2 / \tilde{\epsilon}_1 c_1$. Both $x \equiv |A|$ and $\theta_1 - \theta_2 \equiv arg(A)$ can be controlled separately in the experiment.

The "real time" analogue of the above two CW frequencies scenario, in which the superposition state preparation is affected by a single broad-band pulse and the dissociation by a second pulse, is discussed in detail in Section II.B below.

II. REPRESENTATIVE CONTROL SCENARIOS

The two step approach is but one particular implementation of coherent control; numerous other scenarios may be designed. They all rely upon the same "coherent control principle", that *in order to achieve control one must drive a state through multiple independent optical excitation routes to the same final state.*

It would seem that laser incoherence would lead to loss of control since incoherence implies that the phases of $\tilde{\epsilon}_1$ and $\tilde{\epsilon}_2$ in Eq. (15) are random. An ensemble average of these phases is expected to lead to the disappearance of the interference term. This is true, however, only in the fully chaotic limit. Control can persist in the presence of some laser incoherence[16] or when the initial state is described by a *mixed*, as distinct from *pure*, state[7]. Most surprising is the fact, described below, that by utilizing strong laser fields one can attain quantum interference control with completely *incoherent* sources[41].

We now describe in more detail three control scenarios.

II.A 1-Photon 3-Photons Interferences

So far, we exploited quantum interference phenomena by dissociating a superposition of several energy eigenstates with a single type (one photon absorption) process. It is possible instead to start with a *single* energy eigenstate and employ interference between optical routes of *different* types. Such is the interference between two multiphoton processes of different multiplicities. In order to satisfy the coherent control principle, which requires that we reach the same final energy E, we must use photons of commensurate frequencies, i.e. frequencies which satisfy an $m\omega_1 = n\omega_2$ relation, with integer m and n. Selection rules dictate the acceptable n, m pairs.

As the simplest example, we examine a one photon process interfering with a three photon process ("3+1" control). Let H_g and H_e be the nuclear Hamiltonians for a ground and excited electronic states. H_g is assumed to have a discrete spectrum and H_e to possess a continuous spectrum. The molecule, initially in an eigenstate $|E_i\rangle$ of H_g is subjected to two electric fields given by

$$\epsilon(t) = \epsilon_1 \cos(\omega_1 l + \mathbf{k}_1 \cdot \mathbf{R} + \theta_1) + \epsilon_3 \cos(\omega_3 t + \mathbf{k}_3 \cdot \mathbf{R} + \theta_3) , \qquad (16)$$

Here $\omega_3 = 3\omega_1$, $\epsilon_l = \epsilon_l \hat{\epsilon}_l$, $l = 1, 3$; ϵ_l is the magnitude and $\hat{\epsilon}_l$ is the polarization of the electric fields. The two fields are chosen parallel, with $\mathbf{k}_3 = 3\mathbf{k}_1$.

The probability $P(E, q; E_i)$ of producing product with energy E in arrangement q from a state $|E_i\rangle$ is given by

$$P(E, q; E_i) = P_3(E, q; E_i) + P_{13}(E, q; E_i) + P_1(E, q; E_i) . \qquad (17)$$

where $P_1(E, q; E_i)$ and $P_3(E, q; E_i)$ are the probabilities of dissociation due to the ω_1 and ω_3 excitation, and $P_{13}(E, q; E_i)$ is the term due to interference between the two excitation routes.

In the weak field limit, $P_3(E, q; E_i)$ is given by

$$P_3(E, q; E_i) = (\frac{\pi}{\hbar})^2 \epsilon_3^2 F_3^{(q)} \qquad (18)$$

where

$$F_3^{(q)} = \sum_n |\langle E, n, q^- | (\hat{\epsilon}_3 \cdot \mathbf{d})_{e,g} | E_i \rangle|^2 . \qquad (19)$$

d is the electric dipole operator, and

$$(\hat{\epsilon}_3 \cdot \mathbf{d})_{e,g} = \langle e | \hat{\epsilon}_3 \cdot \mathbf{d} | g \rangle , \qquad (20)$$

with $|g\rangle$ and $|e\rangle$ denoting the ground and excited electronic states, respectively. $P_1(E, q; E_i)$ is given in third order perturbation theory by[6]

$$P_1(E, q; E_i) = (\frac{\pi}{\hbar})^2 \epsilon_1^6 F_1^{(q)}, \qquad (21)$$

73

where,

$$F_1^{(q)} = \sum_n |\langle E, n, q^- | T | E_i \rangle|^2 , \tag{22}$$

with

$$T = (\hat{\epsilon}_1 \cdot \mathbf{d})_{e,g} (E_i - H_g + 2\hbar\omega_1)^{-1} (\hat{\epsilon}_1 \cdot \mathbf{d})_{g,e} (E_i - H_e + \hbar\omega_1)^{-1} (\hat{\epsilon}_1 \cdot \mathbf{d})_{e,g} . \tag{23}$$

We assumed that $E_i + 2\hbar\omega_1$ is below the dissociation threshold and that dissociation occurs from the excited electronic state only.

A similar derivation[6] gives the cross term in Eq. (17) as

$$P_{13}(E, q; E_i) = -2(\frac{\pi}{\hbar})^2 \epsilon_3 \epsilon_1^3 \cos(\theta_3 - 3\theta_1 + \delta_{13}^{(q)}) |F_{13}^{(q)}| \tag{24}$$

with the amplitude $|F_{13}^{(q)}|$ and phase $\delta_{13}^{(q)}$ defined by

$$|F_{13}^{(q)}| \exp(i\delta_{13}^{(q)}) = \sum_n \langle E_i | T | E, n, q^- \rangle \langle E, n, q^- | (\hat{\epsilon}_3 \cdot \mathbf{d})_{e,g} | E_i \rangle . \tag{25}$$

The branching ratio $R_{qq'}$ between the q and q' products can then be written as

$$R_{qq'} = \frac{F_3^{(q)} - 2x\cos(\theta_3 - 3\theta_1 + \delta_{13}^{(q)})|F_{13}^{(q)}| + x^2 F_1^{(q)}}{F_3^{(q')} - 2x\cos(\theta_3 - 3\theta_1 + \delta_{13}^{(q')})|F_{13}^{(q')}| + x^2 F_1^{(q')}} . \tag{26}$$

where x is defined as,

$$x = \epsilon_1^3/\epsilon_3. \tag{27}$$

The numerator and denominator of Eq. (26) contain contributions from two independent routes and an interference term. Since the interference term is controllable through variation of laboratory parameters, so too is the product ratio $R_{qq'}$. Thus the principle upon which this control scenario is based is the same as in the first example above, although the interference is introduced in an entirely different way.

Experimental control over $R_{qq'}$ is obtained by varying the difference ($\theta_3 - 3\theta_1$) and the parameter x. The former is the phase difference between the ω_3 and the ω_1 laser fields and the latter, via Eq. (27), incorporates the ratio of the two lasers amplitudes. Experimentally one envisions using "tripling" to produce ω_3 from ω_1, the subsequent variation of the phase of one of these beams provides a straightforward method of altering $\theta_3 - 3\theta_1$. Indeed, generating ω_3 from ω_1 allows for compensation of any phase jumps in the two laser sources. Thus the relative phase $\omega_3 - 3\omega$, is well defined.

As pointed out above, "3+1" is not necessarily the only viable control scenario in the "$n + m$" family. It has the advantage that one may generate one of the frequencies (the tripled photon) from the other. This is indeed the reason why the "3+1" route was the first control scenario to be implemented experimentally. (see discussion below).

Control of *integral* (in contrast to *differential*) cross-sections requires that the $| E, n, q^- \rangle$ continuum states be made up of equal parity $| J, M \rangle$ angular momentum states. This means that in the "$m + n$" control scheme, the integer n must have the same parity as the integer m. Thus, studies of a "2+2" scheme for the control of the Na_2 photodissociation[15, 42] (discussed in detail in Section II.C) and of a "2+4" scenario for the control of the Cl_2 photodissociation[43], have been published. In addition, studies of "3+1" control with strong fields, have also appeared[44, 45, 46]. These studies and others[47] have verified that "$n + m$" control is viable even when strong fields are used, although the dependence on the x amplitude, and the $\theta_n - 3\theta_m$ phase factors is no longer as transparent as in the weak field case, discussed above.

The weak field "3+1" scenario has now been experimentally implemented in part in REMPI type experiments. The experiments demonstrated control of the total ionization rate, first in Hg[28], and then in HCl and CO[29] In the case of HCl[29], the molecule was excited to an intermediate $^3\Sigma^-(\Omega^+)$ vibrotational resonance, using a combination of three ω_1 ($\lambda_1 = 336$ nm) photons and one ω_3 ($\lambda_3 = 112$ nm) photon. The ω_3 beam was generated from an ω_1 beam by tripling in a Kr gas cell. Ionization of the intermediate state takes place by absorption of one additional ω_1 photon.

The relative phase of the light fields was varied by passing the ω_1 and ω_2 beams through a second Ar or H_2 ("tuning") gas cell of variable pressure. The HCl REMPI experiments verified the prediction of a sinusoidal dependence of the ionization rates on the relative phase of the two exciting lasers of Eq. (26). The HCl experiment also verified the prediction of Eq. (26) of the dependence of the strength of the sinusoidal modulation of the ionization current on the x amplitude factor. More recently control over branching processes such as dissociation vs. ionization in HI was demonstrated by Gordon et al.[30].

If one is content with controlling angular distributions, one can lift the equal-parity restriction. Absorption of two photons of perpendicular polarizations[5, 8], or of two photons interfering with their second-harmonic photon ("2+1" scenario)[8, 33, 34], result in states of different parities. Though such processes do not lead to control of integral quantities, they do allow for control of differential cross-sections. The "1+2" scenario (discussed in Section IV) has been implemented experimentally for the control of photo-current directionality in semiconductors, using no bias voltage[33, 35] and for the control of the orientation of the HD^+ photodissociation[36].

II.B The Pump-Dump Scheme

A useful extension of the scenario outlined in Section I.C is a "pump-dump" scheme[18, 19], in which an initial superposition of bound states is prepared with one laser pulse and subsequently dissociated with another. The pump and dump steps are assumed to be temporally separated by a time delay τ. The analysis below shows that under these circumstances the control parameters are the central frequency of the pump pulse, and the time delay between the two pulses.

Consider a molecule, initially ($t = 0$) in eigenstate $|E_g\rangle$ of Hamiltonian H_M, subjected to two transform limited light pulses. The field $\bar{\epsilon}(t)$ consists of two temporally separated pulses $\bar{\epsilon}(t) = \bar{\epsilon}_x(t) + \bar{\epsilon}_d(t)$, with the Fourier transform of $\bar{\epsilon}_x(t)$ denoted $\epsilon_x(\omega)$, etc. For convenience, we have chosen Gaussian pulses peaking at $t = t_x$ and t_d respectively. As discussed in Section I.B, the $\bar{\epsilon}_x(t)$ pulse induces a transition to a linear combination of two excited bound electronic state with nuclear eigenfunctions $|E_1\rangle$ and $|E_2\rangle$, and the $\bar{\epsilon}_d(t)$ pulse dissociates the molecule by further exciting it to the continuous part of the spectrum. Both fields are chosen sufficiently weak for perturbation theory to be valid[48].

The superposition state prepared by the $\bar{\epsilon}_x(t)$ pulse, whose width is chosen to encompass just the two E_1 and E_2 levels, is given in first order perturbation theory as,

$$|\phi(t)\rangle = |E_g\rangle e^{-iE_g t/\hbar} + c_1|E_1\rangle e^{-iE_1 t/\hbar} + c_2|E_2\rangle e^{-iE_2 t/\hbar}, \qquad (28)$$

where,

$$c_k = (\sqrt{2\pi}/i\hbar)\langle E_k|d|E_g\rangle\epsilon_x(\omega_{kg}), \quad k = 1, 2, \qquad (29)$$

with $\omega_{kg} \equiv (E_k - E_g)/\hbar$.

After a delay time of $\tau \equiv t_d - t_x$ the system is subjected to the $\bar{\epsilon}_d(t)$ pulse. It follows from Eq. (28) that after this delay time each preparation coefficient has picked up an extra phase factor of $e^{-iE_k\tau/\hbar}$, $k = 1, 2$. Hence, the phase of c_1 relative to c_2 at that time increases by $[-(E_1 - E_2)\tau/\hbar = \omega_{2,1}\tau]$ Thus the natural two-state time evolution replaces the relative laser phase of the two-frequency control scenario of Section I.C.

After the decay of the $\bar{\epsilon}_d(t)$ pulse the system wavefunction is given as,

$$|\psi(t)\rangle = |\phi(t)\rangle + \sum_{n,q}\int dE B(E, n, q|t)|E, n, q^-\rangle e^{-iEt/\hbar}. \qquad (30)$$

The probability of observing the q fragments at total energy E in the remote future is therefore given as,

$$P(E, q) = \sum_n |B(E, n, q|t = \infty)|^2$$

$$= (2\pi/\hbar^2)\sum_n |\sum_{k=1,2} c_k\langle E, n, q^-|d|E_k\rangle\epsilon_d(\omega_{EE_k})|^2 \qquad (31)$$

where $\omega_{EE_k} = (E - E_k)/\hbar$, c_k is given by Eq. (29).

Expanding the square and using the Gaussian pulse shape gives:

$$P(E, q) = (2\pi/\hbar^2)[|c_1|^2 d_{1,1}^{(q)}\epsilon_1^2 + |c_2|^2 d_{2,2}^{(q)}\epsilon_2^2 + 2|c_1 c_2^* \epsilon_1 \epsilon_2 d_{1,2}^{(q)}|\cos(\omega_{2,1}(t_d - t_x) + \alpha_{1,2}^{(q)}(E) + \phi)] \qquad (32)$$

where $\epsilon_i = |\epsilon_d(\omega_{EE_i})|$, $\omega_{2,1} = (E_2 - E_1)/\hbar$ and the phases ϕ, $\alpha_{1,2}^{(q)}(E)$ are defined by

$$\langle E_1 |\mathrm{d}| E_g \rangle\langle E_g |\mathrm{d}| E_2 \rangle \equiv |\langle E_1 |\mathrm{d}| E_g \rangle\langle E_g |\mathrm{d}| E_2 \rangle| e^{i\phi}$$

$$d_{i,k}^{(q)}(E) \equiv |d_{i,k}^{(q)}(E)| e^{i\alpha_{i,k}^{(q)}(E)} = \sum_n \langle E, n, q^- |\mathrm{d}| E_i \rangle\langle E_k |\mathrm{d}| E, n, q^- \rangle \qquad (33)$$

Integrating over E to encompass the full width of the second pulse, and forming the ratio, $Y = P(q)/[\sum_q P(q)]$, gives the ratio of products in each of the two arrangement channels, i.e. the quantity we wish to control. Once again it is the sum of two direct photodissociation contributions, plus an interference term.

Examination of Eq. (32) makes clear that the product ratio Y can be varied by changing the delay time $\tau = (t_d - t_x)$ or ratio $x = |c_1/c_2|$; the latter is most conveniently done by detuning the initial excitation pulse.

An example of this type of pump-dump control[11] is provided by the example of IBr photodissociation. Specifically, we showed that it is possible to control the Br* vs. Br yield in this process, using two conveniently chosen picosecond pulses. The first pulse was chosen to prepare a linear superposition of two bound states which arise from mixing of the X and A states. A subsequent pulse pumps this superposition to dissociation where the relative yields of Br and Br* are examined. The results show the vast range of control which is possible with this relatively simple experimental setup.

Theoretical work on similar pump dump scenarios for the control of the

$$\mathrm{D} + \mathrm{OH} \leftarrow \mathrm{HOD} \rightarrow \mathrm{H} + \mathrm{OD}$$

dissociation via the B-state[49] of HOD and the A-state[50] of HOD have recently been published.

II.C Resonantly Enhanced "2+2" Control of a Thermal Ensemble

In practice there are a number of sources of incoherence which tend to diminish control. Prominent amongst these are effects due to an initial thermal distribution of states and effects due to partial coherence of the laser source. Below we describe one approach, based upon a resonant "2+2" scenario, which deals effectively with both problems. An alternative method in which coherence is retained in the presence of collisions is discussed elsewhere[7].

The specific scheme we advocate is the particular case of Na$_2$ photodissociation. Here the molecule is lifted from an initial bound state $|E_i, J_i, M_i\rangle$ to energy E via two independent two photon routes. To introduce notation, first consider a single such two photon route. Absorption of the first photon of frequency ω_1 lifts the system to a region close to an intermediate bound state $|E_m J_m M_m\rangle$, and a second photon of frequency ω_2 carries the system to the dissociating states $|E, \hat{\mathbf{k}}, q^-\rangle$, where the scattering angles are specified by $\hat{\mathbf{k}} = (\theta_k, \phi_k)$. Here the J's are the angular momentum, M's are their projection along the z-axis, and the values of energy, E_i and E_m, include specification of the vibrational quantum numbers. Specifically, if we denote the phases of the coherent states by ϕ_1 and ϕ_2, the wavevectors by $\mathbf{k}_1$ and $\mathbf{k}_2$ with overall phases $\theta_i = \mathbf{k}_i \cdot \mathbf{R} + \phi_i$ $(i = 1, 2)$ and the electric field amplitudes by ϵ_1 and ϵ_2, then the probability amplitude for resonant two photon $(\omega_1 + \omega_2)$ photodissociation is given[15, 42] by

$$T_{\hat{\mathbf{k}}q,i}(E, E_i J_i M_i, \omega_2, \omega_1) =$$

$$\sum_{E_m, J_m} \frac{\langle E, \hat{\mathbf{k}}, q^- | d_2 \epsilon_2 | E_m J_m M_i \rangle \langle E_m J_m M_i | d_1 \epsilon_1 | E_i J_i M_i \rangle}{\omega_1 - (E_m + \delta_m - E_i) + i\Gamma_m} \exp[i(\theta_1 + \theta_2)] =$$

$$\frac{\sqrt{2mk_q}}{h} \sum_{J,p,\lambda \geq 0} \sum_{E_m, J_m} \sqrt{2J+1} \begin{pmatrix} J & 1 & J_m \\ -M_i & 0 & M_i \end{pmatrix} \begin{pmatrix} J_m & 1 & J_i \\ -M_i & 0 & M_i \end{pmatrix}$$

$$\cdot D_{\lambda,M_i}^{Jp}(\theta_k, \phi_k, 0) t(E, E_i J_i, \omega_2, \omega_1, q | J p \lambda, E_m J_m) \exp[i(\theta_1 + \theta_2)] \qquad (34)$$

Here d_i is the component of the dipole moment along the electric-field vector of the ith laser mode, $E = E_i + (\omega_1 + \omega_2)$, δ_m and Γ_m are respectively the radiative shift and width of the intermediate state, m - the reduced mass, and k_q is the relative momentum of the dissociated product in q-channel. The D_{λ,M_i}^{Jp} is the parity adapted rotation matrix[51] with λ the magnitude of the projection on the internuclear axis of the electronic angular momentum and $(-1)^J p$ the parity of the rotation matrix. We have set $\hbar \equiv 1$, and assumed for simplicity lasers which are linearly-polarized and with parallel electric-field vectors. Note that the T-matrix element in Eq. (34) is a complex quantity, whose phase is the sum of the laser phase $\theta_1 + \theta_2$ and the molecular phase, i.e. the phase of $\mathbf{t}$.

Because the t-matrix element contains a factor of $[\omega_1 - (E_m + \delta_m - E_i) + i\Gamma_m]^{-1}$ the probability is greatly enhanced by the approximate inverse square of the detuning $\Delta = \omega_1 - (E_m + \delta_m - E_i)$ as long as the line width Γ_m is less than Δ. Hence only the levels closest to the resonance $\Delta = 0$ contribute significantly to the dissociation probability. *This allows us to photodissociate only a select number of states (preferably one) from a thermal bath.*

Consider then the following coherent control scenario. A molecule is irradiated with three interrelated frequencies, $\omega_0, \omega_+, \omega_-$ where photodissociation occurs at $E = E_i + 2\omega_0 = E_i + (\omega_+ + \omega_-)$ and where ω_0 and ω_+ are chosen resonant with intermediate bound state levels. The probability of photodissociation at energy E into arrangement channel q is then given by the square of the sum of the T matrix elements from pathway "a" $(\omega_0 + \omega_0)$ and pathway "b" $(\omega_+ + \omega_-)$. That is, the probability into channel q

$$P_q(E, E_i J_i M_i; \omega_0, \omega_+, \omega_-) \equiv \int d\hat{\mathbf{k}} \left| T_{\hat{\mathbf{k}}q,i}(E, E_i J_i M_i, \omega_0, \omega_0) + T_{\hat{\mathbf{k}}q,i}(E, E_i J_i M_i, \omega_+, \omega_-) \right|^2$$

$$\equiv P^{(q)}(a) + P^{(q)}(b) + P^{(q)}(ab) \qquad (35)$$

Here $P^{(q)}(a)$ and $P^{(q)}(b)$ are the independent photodissociation probabilities associated with routes a and b respectively and $P^{(q)}(ab)$ is the interference term between them, discussed below. Note that the two T matrix elements in Eq. (35) are associated with different lasers and as such contain different laser phases. Specifically, the overall phase of the three laser fields are $\theta_0 = \mathbf{k}_0 \cdot \mathbf{R} + \phi_0$, $\theta_+ = \mathbf{k}_+ \cdot \mathbf{R} + \phi_+$ and $\theta_- = \mathbf{k}_- \cdot \mathbf{R} + \phi_-$, where ϕ_0, ϕ_+ and ϕ_- are the photon phases, and $\mathbf{k}_0$, $\mathbf{k}_+$, and $\mathbf{k}_-$ are the wavevectors of the laser modes ω_0, ω_+ and ω_-, whose electric field strengths are $\epsilon_0, \epsilon_+, \epsilon_-$ and intensities I_0, I_+, I_-.

The optical path-path interference term $P^{(q)}(ab)$ is given by

$$P^{(q)}(ab) = 2\epsilon_0^2 \epsilon_+ \epsilon_- |\mu_{ab}^{(q)}| \exp[i(\delta_a^q - \delta_b^q)] \cos(\alpha_a^q - \alpha_b^q) \qquad (36)$$

with relative phase

$$\alpha_a^q - \alpha_b^q = (\delta_a^q - \delta_b^q) + (2\theta_0 - \theta_+ - \theta_-). \qquad (37)$$

where the amplitude $|\mu_{ab}^{(q)}|$ and the molecular phase difference $(\delta_a^q - \delta_b^q)$ are defined by

$$\epsilon_0^2 \epsilon_+ \epsilon_- |\mu_{ab}^{(q)}| \exp[i(\delta_a^q - \delta_b^q)] = \frac{8\pi \mathrm{m} k_q}{h^2} \sum_{J,p,\lambda \geq 0} \sum_{E_m, J_m; E'_m, J'_m}$$

$$\begin{pmatrix} J & 1 & J_m \\ -M_i & 0 & M_i \end{pmatrix} \begin{pmatrix} J_m & 1 & J_i \\ -M_i & 0 & M_i \end{pmatrix} \begin{pmatrix} J & 1 & J'_m \\ -M_i & 0 & M_i \end{pmatrix} \begin{pmatrix} J'_m & 1 & J_i \\ -M_i & 0 & M_i \end{pmatrix}$$

$$\cdot t(E, E_i J_i, \omega_0, \omega_0, q | Jp\lambda, E_m J_m) t^*(E, E_i J_i, \omega_-, \omega_+, q | Jp\lambda, E'_m J'_m).$$

$$(38)$$

Consider now the quantity of interest $R_{qq'}$, the branching ratio of the product in q-channel to that in q'-channel. Noting that in the weak field case $P^{(q)}(a)$ is proportional to ϵ_0^4, $P^{(q)}(b)$ to $\epsilon_+^2 \epsilon_-^2$, and $P^{(q)}(ab)$ to $\epsilon_0^2 \epsilon_+ \epsilon_-$ we can write

$$R_{qq'} = \frac{\mu_{aa}^{(q)} + x^2 \mu_{bb}^{(q)} + 2x|\mu_{ab}^{(q)}| \cos(\alpha_a^q - \alpha_b^q) + (B^{(q)}/\epsilon_0^4)}{\mu_{aa}^{(q')} + x^2 \mu_{bb}^{(q')} + 2x|\mu_{ab}^{(q')}| \cos(\alpha_a^{q'} - \alpha_b^{q'}) + (B^{(q')}/\epsilon_0^4)} \qquad (39)$$

where $\mu_{aa}^{(q)} = P^{(q)}(a)/\epsilon_0^4$, $\mu_{bb}^{(q)} = P^{(q)}(b)/(\epsilon_+^2 \epsilon_-^2)$ and $x = \epsilon_+ \epsilon_-/\epsilon_0^2 = \sqrt{I_+ I_-}/I_0$. The terms with $B^{(q)}, B^{(q')}$, described below, correspond to resonant photodissociation routes to energies other than $E = E_i + 2\hbar\omega_0$ and hence[4] to terms which do not coherently interfere with the a and b pathways. Minimization of these terms, due to absorption of $(\omega_0 + \omega_-)$, $(\omega_0 + \omega_+)$, $(\omega_+ + \omega_0)$ or $(\omega_+ + \omega_+)$, is discussed elsewhere[15, 42]. Here we just emphasize that the product ratio in Eq. (39) depends upon both the laser intensities and relative laser phase. Hence manipulating these laboratory parameters allows for control over the relative cross section between channels.

The proposed scenario, embodied in Eq. (39), also provides a means by which control can be improved by eliminating effects due to laser jitter. Specifically, the term $2\phi_0 - \phi_+ - \phi_-$ contained in the relative phase $\alpha_a^q - \alpha_b^q$ can be subject to the phase fluctuations arising from laser instabilities. If such fluctuations are sufficiently large then the interference term in Eq. (39), and hence control, disappears[16]. We can eliminate this problem by generating ω_+ as $\omega_+ = 2\omega_0 - \omega_-$ via frequency doubling of ω_0 and frequency differencing of ω_- from the resulting beam. It is easy to see that in this case the quantity $2\phi_0 - \phi_+ - \phi_-$ vanishes, irrespective of the phase jitter and drift of either source! Control is then attained by introducing an extra, perfectly controlled phase, χ, through the addition of a delay line in, say, the ω_- beam.

Typical results for Na_2 are provided elsewhere[15, 42]. Note also that control is not limited to two-product channels, such as those discussed above. Recent computations[42] on higher energy Na_2 photodissociation, where more product arrangement channels are available, show equally large ranges of control for the three channel case.

III. CONTROL OF SYMMETRY BREAKING

Weak field phase interference has one remarkable property; it can lead to controlled *symmetry breaking*[12]. Below we show that the pump-dump scheme described above (section II.B) can lead to symmetry breaking and to the generation of chirality. Other mechanisms for collinear symmetry breaking in *strong* fields have recently been proposed[52, 53]. There, it was shown that one can generate *even* high harmonics by exciting a symmetric double quantum-well.

In general, symmetry breaking occurs whenever a system executes a transition to a *non-symmetric* eigenstate of the Hamiltonian. Strictly speaking, non-symmetric eigenstates (i.e., states which do not belong to any of the symmetry-group representations) can occur if there exist several degenerate eigenstates, each belonging to a different irreducible representation. Any linear combination of such eigenstates is non-symmetric.

Non-symmetric eigenstates of a symmetric Hamiltonian also occur in the continuous spectrum of a BAB type molecule. It is clear that the $| E, n, R^- \rangle$ state, which correlates asymptotically with the dissociation of the right B group, must be degenerate with the $| E, n, L^- \rangle$ state, giving rise to the departure of the left B group. It is also possible to form a *symmetric* $| E, n, s^- \rangle$ and an *anti-symmetric* $| E, n, a^- \rangle$ eigenstates of the same Hamiltonian by taking the $\pm$ combination of these states. There is an important physical distinction between the non-symmetric states and states which are symmetric/antisymmetric: Any experiment performed in the asymptotic $B - AB$ or $BA - B$ regions must, by necessity, measure the probability of populating a non-symmetric state. This follows because when the $B - AB$ distance or the $BA - B$ distance is large, a given group B is either far away from *or* close to group A. Thus symmetric and antisymmetric states are not directly observable in the asymptotic regime.

We conclude that the very act of observation of the dissociated molecule entails the collapse of the system to one of the non-symmetric states. As long as the probability of collapse to the $| E, n, R^- \rangle$ state is equal to the probability of collapse to the $| E, n, L^- \rangle$ state, the collapse to a non-symmetric state does not lead to a preference of R over L in an *ensemble* of molecules. This is the case when the above collapse ("symmetry breaking") is *spontaneous*, i.e., occurring due to some (random) factors not in our control. Coherent control techniques allow us to influence these probabilities. In this case symmetry breaking is stimulated rather than spontaneous. This has far reaching physical and practical significance.

One of the most important cases of symmetry breaking arises when the two B groups (now denoted as B and B') are not identical, but are enantiomers of one another. (Two groups of atoms are said to be enantiomers of one another if one is the mirror image of the other. If these groups are also "chiral", i.e., they lack a center of inversion symmetry, then the two enantiomers are distinguishable and can be detected through the distinctive direction of rotation of linearly polarized light).

The existence and role of enantiomers is recognized as one of the fundamental broken symmetries in nature[54]. It has motivated a longstanding interest in asymmetric synthesis, i.e. a process which preferentially produces a specific chiral species. Contrary to the prevailing belief[55] that asymmetric synthesis must necessarily involve either chiral reactants, or chiral external system conditions such as chiral crystalline surfaces, we show below that preferential production of a chiral photofragment can occur even though the parent molecule is not chiral. In particular two results are demonstrated: (1) ordinary photodissociation, using linearly polarized light, of a BAB' "pro-chiral" molecule may yield different cross sections for the production of right-handed (B) and left-handed (B') products, when the direction of the angular momentum (m_j) of the products is selected; and (2) that this natural symmetry breaking may be enhanced and controlled using coherent lasers.

To treat this problem we return to the formulation of the pump-dump scenario described above. Considering the dissociation of BAB' into right (R) and left (L) hand products we have:

$$Y = P(L)/[P(L) + P(R)], \tag{40}$$

As above, the product ratio Y is a function of the delay time $\tau = (t_d - t_x)$ and ratio $x = |c_1/c_2|$, the latter by detuning the initial excitation pulse. Active control over the products $B \mid AB'$ vs. $B' + AB$, i.e. a variation of Y with τ and x, and hence control over left vs. right handed products, will result only if $P(R)$ and $P(L)$ have different functional dependences on x and τ.

We now show that $P(R)$ may be different from $P(L)$ for the $B'AB$ case. We first note that this molecule belongs to the C_s point group which is a group possessing only one symmetry plane. This plane, denoted as σ, is defined as the collection of the C_{2v} points, i.e., points satisfying the $B - A = A - B'$

81

condition, where $B - A$ designates the distance between the B and A groups. In order to do that we choose the intermediate state $|E_1\rangle$ to be *symmetric* and the state $|E_2\rangle$ to be *antisymmetric* with respect to reflection in the σ plane. Furthermore, we shall focus upon transitions between electronic states of the same representations, e.g. A' to A' or A'' to A'' (where A' denotes the symmetric representation and A'' the antisymmetric representation of the C_s group). We further assume that the ground vibronic state belongs to the A' representation.

The first thing to demonstrate is that it is possible to excite simultaneously, by optical means, both the symmetric $|E_1\rangle$ and antisymmetric $|E_2\rangle$ states. Using Eq. (29) we see that this requires the existence of both a symmetric d component, denoted as d_s, and an antisymmetric d component, denoted d_a, because, by the symmetry properties of $|E_1\rangle$ and $|E_2\rangle$,

$$\langle E_1 |d| E_g\rangle = \langle E_1 |d_s| E_g\rangle, \quad \langle E_2 |d| E_g\rangle = \langle E_2 |d_a| E_g\rangle. \tag{41}$$

The existence of both dipole-moment components occurs in $A' \to A'$ electronic transitions whenever a bent $B' - A - B$ molecule deviates considerably from the equi-distance C_{2v} geometries (where $d_a = 0$). The effect is non Franck-Condon in nature, because we no longer assume that the dipole-moment does not vary with the nuclear configurations. (In the theory of vibronic-transitions terminology the existence of both d_s and d_a is due to a Herzberg-Teller intensity borrowing[56] mechanism).

We conclude that the excitation pulse *can* create a $|E_1\rangle, |E_2\rangle$ superposition consisting of two states of different reflection symmetry, which is therefore non-symmetric. We now wish to show that this non-symmetry established by exciting *non-degenerate bound* states translates to a non-symmetry in the probability of populating the two *degenerate* $|E, n, R^-\rangle, |E, n, L^-\rangle$ *continuum* states. We proceed by examining the properties of the bound-free transition matrix elements of Eq. (33) entering the probability expression of Eq. (32).

Although the continuum states of interest $|E, n, q^-\rangle$ are non-symmetric, they satisfy a closure relation, since, $\sigma|E, n, R^-\rangle = |E, n, L^-\rangle$ and vice-versa. Working with the symmetric and antisymmetric continuum eigenfunctions,

$$|E, n, R^-\rangle \equiv (|E, n, s^-\rangle + |E, n, a^-\rangle)/\sqrt{2}, \tag{42}$$

$$|E, n, L^-\rangle \equiv (|E, n, s^-\rangle - |E, n, a^-\rangle)/\sqrt{2}, \tag{43}$$

using the fact that $|E_1\rangle$ is symmetric and $|E_2\rangle$ antisymmetric, and adopting the notation $A_{s2} \equiv \langle E, n, s^- |d_a| E_2\rangle$, $S_{a1} \equiv \langle E, n, a^- |d_s| E_1\rangle$, etc. we have,

$$d_{11}^{(q)} = \sum{}' [|S_{s1}|^2 + |A_{a1}|^2 \pm 2Re(A_{a1}S_{s1}^*)] \tag{44}$$

$$d_{22}^{(q)} = \sum{}' [|A_{s2}|^2 + |S_{a2}|^2 \pm 2Re(A_{s2}S_{a2}^*)] \tag{45}$$

82

$$\mathrm{d}_{12}^{(q)} = \sum' \left[S_{s1} A_{s2}^* + A_{a1} S_{a2}^* \pm S_{s1} S_{a2}^* \pm A_{a1} A_{s2}^* \right] \tag{46}$$

where the plus sign applies for $q = R$ and the minus sign for $q = L$.

We have demonstrated the extent of expected control, by considering a model symmetry breaking in the HOH photodissociation in three dimensions, where the two hydrogens are assumed distinguishable[12]. The computation was done using the formulation and computational methodology of Segev et al.[57].

IV. CONTROL WITH INTENSE LASER FIELDS

We now discuss some extensions of CC to strong laser fields. Parallel work involving other strong field scenarios has been done by Bandrauk et al.[43], Corkum et al.[47], Bardsley et al.[44] Lambropoulos et al.[58] and Guisti-Suzor et al.[45]. Here we concentrate on a strong field control scenario in which the dependence on the relative phase between the two laser beams, hence on laser coherence, disappears. As a result, coherence plays no role in this scenario (save for being intimately linked with the existence of the narrow-band laser sources needed for its execution). Although the unimportance of coherence means that we lose phase control, the effect still depends on quantum interference phenomena. The scenario is therefore called Interference Control.

To illustrate interference control we look at the control of the electronic states of Na atoms generated by the photodissociation of Na_2, a process treated in the context of weak field CC in Section IIC. We envision a scenario, in which we employ two laser sources: One laser, (not necessarily intense) with center frequency ω_1 is used to excite a molecule from an initially populated bound state $| E_i \rangle$ to a dissociative state $| E, m, q^- \rangle$. A second laser, with frequency ω_2, is used to couple ("dress") the continuum with some (initially unpopulated) bound states $| E_j \rangle$. With both lasers on, dissociation to $| E, m, q^- \rangle$ occurs via one direct, $| E_i \rangle \to | E, m, q^- \rangle$, and a multitude of indirect, e.g., $| E_i \rangle \to | E, m, q^- \rangle \to | E_j \rangle \to | E, m, q^- \rangle$, pathways. The interference between these pathways to form a given channel q at product energy E can be either constructive or destructive. As we show below, varying the frequencies and intensities of the two excitation lasers strongly affects this interference term, providing a means of controlling the photodissociation line shape and the branching ratio into different products.

With this scenario in mind we now briefly discuss the methodology of dealing with strong laser fields and the extension of CC ideas to this domain. We consider the photodissociation of a molecule with Hamiltonian H_M in the presence of a radiation field with Hamiltonian H_R, whose eigenstates are the Fock states $|n_k\rangle$ with energy $n_k \hbar \omega_k$. (In the case of several frequencies the repeated index in $n_k \omega_k$ implies the sum over the modes.)

Strong field dynamics is completely embodied[59] in the fully interacting eigenstates of the total Hamiltonian H, $H = H_M + H_R + V$, where V is the light-matter interaction, denoted $|(E, m, q^-), n_k^-\rangle$

$$H|(E, m, q^-), n_k^-\rangle = (E + n_k \hbar\omega_k)|(E, m, q^-), n_k^-\rangle . \qquad (47)$$

The minus superscript on n_k is used in exactly the same way as in the weak field domain: it is a reminder that each $|(E, m, q^-), n_k^-\rangle$ state correlates to a non-interacting $|(E, m, q^-), n_k\rangle \equiv |E, m, q^-\rangle|n_k\rangle$ state when the light-matter interaction V is switched off.

If the system is initially in the $|E_i, n_i\rangle \equiv |E_i\rangle|n_i\rangle$ state and we suddenly switch on V, the photodissociation amplitude to form in the future the product state $|E, m, q^-\rangle|n_k\rangle$ is simply given[59] as the overlap between the initial and fully interacting state $\langle(E, m, q^-), n_k^-|E_i, n_i\rangle$. This overlap assumes the convenient form

$$\langle(E, m, q^-), n_k^-|E_i, n_i\rangle = \langle(E, m, q^-), n_k|VG(E^+ + n_k\hbar\omega_k)|E_i, n_i\rangle, \qquad (48)$$

by using the Lippmann-Schwinger equation

$$\langle(E, m, q^-), n_k^-| = \langle(E, m, q^-), n_k| + \langle(E, m, q^-), n_k|VG(E^+ + n_k\hbar\omega_k). \qquad (49)$$

Here $G(\mathcal{E}) = 1/(\mathcal{E} - H)$ and $E^+ = E + i\delta$, with $\delta \to 0^+$ at the end of the computation. Equation (48) is exact and provides a connection between the photodissociation amplitude and the VG matrix element. It is the latter which we compute exactly using a high field extension of the artificial channel method[60, 61].

Two quantities are of interest: the channel specific line shape,

$$A(E, q, n_k|E_i, n_i) = \int d\hat{\mathbf{k}} \, |\langle(E, \hat{\mathbf{k}}, q^-), n_k^-|E_i, n_i\rangle|^2, \qquad (50)$$

and the total dissociation probability to channel q

$$P(q) = \sum_{n_k} \int dE \, A(E, q, n_k|E_i, n_i). \qquad (51)$$

In Eq. (51) the sum is over photons that excite the molecule above the dissociation threshold. In writing Eq. (50) diatomic dissociation is assumed, so that $m = \hat{\mathbf{k}}$.

Consider for example the photodissociation of Na_2 from the $|E_i\rangle = |v = 19, {}^3\Pi_u\rangle$ initial state, where v denotes the vibrational quantum number in the ${}^3\Pi_u$ electronic potential[62] $|E_i\rangle$ is assumed to have been prepared by previous excitation from the ground electronic state. Excitations from $|E_i\rangle$ by ω_1 and mixing of the initially unpopulated $|E_j\rangle$ by ω_2 to the dissociating continua produce $Na(3s)+Na(3p)$ and $Na(3s)+Na(4s)$. Computations were done with ω_1 chosen within the range 15,430 cm^{-1}< ω_1 < 15,700 cm^{-1}with intensity $I_1 \sim 10^{10}$ W/cm^2, which is sufficiently energetic to dissociate levels of the ${}^3\Pi_u$ state with

84

$v \geq 19$ to both Na(3s) + Na(3p) and Na(3s) + Na(4s). The second laser has fixed frequency $\omega_2 = 13,964$ cm^{-1}and intensity I$_2$=3.2×10^{11} W/cm^2 and can dissociate levels with $v \geq 26$ to both products. Under these circumstances the contribution of above threshold dissociation is found to be negligible. However cognizance must be taken of the possibility of dissociation of $| E_i \rangle$ by ω_2 and of $| E_j \rangle$ by ω_1. These processes do not interfere and can not be controlled. Hence we must find the range of parameters that minimizes them.

V. CONCLUSIONS

Our discussion makes clear that the characteristic features which we invoke in order to control chemical reactions are purely quantum in nature. There is, for example, little classical about the time dependent picture where the ultimate outcome of the deexcitation, i.e. product H + HD or H$_2$ + D depends entirely upon the phase and amplitude characteristics of the wavefunction. Indeed, as repeatedly emphasized above, if, e.g. collisional effects are sufficiently strong so as to randomize the phases then reaction control is lost. Hence reaction dynamics is intimately linked to the wavefunction phases which are controllable through coherent optical phase excitation.

These results must be viewed in light of the history of molecular reaction dynamics over the past two decades. Possibly the most useful result of the reaction dynamics research effort has been the recognition that the vast majority of qualitatively important phenomena in reaction dynamics are well described by classical mechanics. Quantum and semiclassical mechanics were viewed as necessary only insofar as they correct quantitative failures of classical mechanics for unusual circumstances and/or for the dynamics of very light particles. Considering reaction dynamics in traditional chemistry to be essentially classical in character therefore appeared to be essentially correct for the vast majority of naturally occurring molecular processes. Coherence played no role. The approach which we have introduced above makes clear, however, that coherence phenomena have great potential for application. The quantum phase is always present and can be used to our advantage, even though it is irrelevant to traditional chemistry. By calling attention to the extreme importance of coherence phenomena to controlled chemistry we herald the introduction of a new focus in atomic and molecular science, i.e. introducing coherence in controlled environments to modify molecular processes, thus defining the area of coherence chemistry.

Acknowledgments: We acknowledge support for this research by the U.S. Office of Naval Research, grant no. N00014-96-1-0433.

References

[1] Brumer P., and Shapiro M., *Chem. Phys. Lett.* **126**, 541 (1986).

[2] Shapiro M., and Brumer P., *J. Chem. Phys.* **84**, 4103 (1986).

[3] Brumer P., and Shapiro M., Faraday Disc. Chem. Soc. **82**, 177 (1986).

[4] Shapiro M., and Brumer P., *J. Chem. Phys.* **84**, 4103 (1986).

[5] Asaro C., Brumer P., and Shapiro M., *Phys. Rev. Lett.* **60**, 1634 (1988).

[6] Shapiro M., Hepburn J., and Brumer P., *Chem. Phys. Lett.* **149**, 451 (1988).

[7] Brumer P., and Shapiro M., *J. Chem. Phys.* **90**, 6179 (1989).

[8] Kurizki G., Shapiro M., and Brumer P., *Phys. Rev. B* **39**, 3435 (1989).

[9] Seideman T., Shapiro M., and Brumer P., *J. Chem. Phys.* **90**, 7136 (1989).

[10] Krause J., Shapiro M., and Brumer P., *J. Chem. Phys.* **92**, 1126 (1990).

[11] Levy I., Shapiro M., and Brumer P., *J. Chem. Phys.* **93**, 2493 (1990).

[12] Shapiro M., and Brumer P., *J. Chem. Phys.* **95** 8658 (1991).

[13] Brumer P., and Shapiro M., *Ann. Rev. Phys. Chem.* **43**, 257 (1992).

[14] Shapiro M., and Brumer P., *J. Chem. Phys.* **97** 6259 (1992).

[15] Chen Z., Brumer P., and Shapiro M., *Chem. Phys. Lett.* **198** 498 (1992).

[16] Jiang X-P., Brumer P., and Shapiro M., *J. Chem. Phys.* **104** 607 (1996).

[17] Dods J., Brumer P., and Shapiro M., *Can. J. Chem.* **72**, (Polanyi Honor Issue) 958 (1994)

[18] Tannor D. J. , and Rice S. A. , , *J. Chem. Phys.* **83**, 5013 (1985); Tannor D. J. , Kosloff R., and Rice S. A., *J. Chem. Phys.* **85**, 5805 (1986).

[19] Rice S.A., Tannor D.J., and Kosloff R., *J. Chem. Soc. Faraday Trans.* **82** 2423 (1986);

[20] Tannor D.J., and Rice S.A., *Adv. Chem. Phys.* **70** 441 (1988).

[21] Kosloff R., Rice S.A., Gaspard P., Tersigni S., and Tannor D.J., *Chem. Phys.* **139** 201 (1989).

[22] Tersigni S., Gaspard P. and Rice S.A., *J. Chem. Phys.* **93** 1670 (1990).

[23] Shi S., Woody A., and Rabitz H., *J. Chem. Phys.* **88** 6870 (1988); Shi S., and Rabitz H., *Chem. Phys.* **139** 185 (1989).

[24] Peirce A.P., Dahleh M., and Rabitz H., *Phys. Rev. A.* **37** 4950 (1988).

[25] Shi S., and Rabitz H., *J. Chem. Phys.* **92** 364 (1990).

[26] Krause J.L., Whitnell R.M., Wilson K.R., Yan Y., and Mukamel S., *J. Chem. Phys.* **99** 6562 (1993).

[27] Jakubetz W., Just B., Manz J., and Schreier H.-J., *J. Phys. Chem.* **94** 2294 (1990).

[28] Chen C., Yin Y-Y., and Elliott D.S., *Phys. Rev. Lett.* **64** 507 (1990); *ibid* **65** 1737 (1990).

[29] Park S.M., Lu S-P., and Gordon R.J., *J. Chem. Phys.* **94** 8622 (1991); Lu S-P., Park S.M., Xie Y., and Gordon R.J., *J. Chem. Phys.* **96** 6613 (1992).

[30] Zhu L., Kleiman V., Li X., Lu S.P., Trentelman K., and Gordon R.J., *Science* **270**, 77 (1995).

[31] Scherer N.F., Ruggiero A.J., Du M., Fleming G.R., *J. Chem. Phys.* **93** 856 (1990).

[32] Boller K.J., Imamoglu A., and Harris S.E., *Phys. Rev. Lett.* **66** 2593 (1991).

[33] Baranova B.A., Chudinov A.N., and Zel'dovitch B. Ya, *Opt. Comm.* **79** 116 (1990).

[34] Yin Y-Y., Chen C., Elliott D.S., and Smith A.V., *Phys. Rev. Lett.* **69** 2353 (1992)

[35] E. Dupont, P.B. Corkum, H.C. Liu, M. Buchanan and Z.R. Wasilewski, Phys. Rev. Letters **74**, 3596 (1995)

[36] B. Sheeny, B. Walker and L.F. Dimauro, Phys. Rev. Lett. **74**, 4799 (1995)

[37] Y.-Y. Yin, R. Shehadeh, D. Elliott, and E. Grant, Chem. Phys. Lett. **241**, 591 (1995).

[38] Shnitman A., Sofer I., Golub I., Yogev A., Shapiro M., Chen Z., and Brumer P., *Phys. Rev. Lett.* **76** 2886 (1996).

[39] For a discussion see, e.g., Macomber J.D., *The Dynamics of Spectroscopic Transitions*, Wiley, N.Y., 1976.

[40] This is the asymptotic condition of scattering theory [see, Taylor J.R., *Scattering Theory*, J. Wiley, N.Y., 1972].

[41] Chen Z., Shapiro M., and Brumer P., *Chem. Phys. Lett.* **228** 289 (1994).

[42] Chen Z., Brumer P., and Shapiro M., *J. Chem. Phys.* **98** 6843 (1993).

[43] Chelkowski S., and Bandrauk A.D., *Chem. Phys. Lett.* **186** 284 (1991); Bandrauk A.D., Gauthier J.M., McCann J.F., *Chem. Phys. Lett.* **200** 399 (1992).

[44] Szöke A., Kulander K.C., and Bardsley J.N., *J. Phys. B* **24**, 3165 (1991); Potvliege R.M., and Smith P.H.G., *J. Phys. B* **25**, 2501 (1992).

[45] Charron E., Guisti-Suzor A., and Mies F.H., *Phys. Rev. Lett.* **71** 692 (1993).

[46] Blank R., and Shapiro M., *Phys. Rev.* **A52** 4278 (1995).

[47] Chelkowski S., Bandrauk A.D., and Corkum P.B., *Phys. Rev. Lett.* **65** 2355 (1990).

[48] The use of perturbation theory does not necessarily imply small total yields. Computational results (Brumer P., and Shapiro M. - to be published) indicate that perturbation theory is quantitatively correct for dissociation probabilities as large as 0.25.

[49] Shapiro M., and Brumer P., *J. Chem. Phys.* **98** 201 (1993).

[50] Henriksen N.E., and Amstrup B., *Chem. Phys. Lett.* **213** 65 (1993); *J. Chem. Phys.* **97** 8285 (1993)

[51] Levy I., and Shapiro M., *J. Chem. Phys.* **89** 2900 (1988).

[52] Bavli R., and Metiu H., *Phys. Rev. Lett.* **69** 1986 (1992).

[53] Ivanov M. Yu, Corkum P.B., and Dietrich P., *Laser Physics*, **3**, 375 (1993).

[54] Barron L.D., *Molecular Light Scattering and Optical Activity*, Cambridge Univ. Press, Cambridge, 1982; Woolley R.G., *Adv. Phys.* **25**, 27 (1975); "Origins of Optical Activity in Nature", ed. Walker D.C., (Elsevier, Amsterdam, 1979).

[55] For a discussion see, Barron L.D., *Chem. Soc. Rev.* **15**, 189 (1986).

[56] Hollas J.M., *High Resolution Spectroscopy*, Butterworths, London, 1982.

[57] Segev E., and Shapiro M., *J. Chem. Phys.* **77**, 5604 (1982).

[58] Nakajima T., and Lambropoulos P., *Phys. Rev. Lett.* **70** 1081 (1993).

[59] Brumer P., and Shapiro M., *Adv. Chem. Phys.* **60** 371, Lawley K. P., Ed., (Wiley-Interscience Pub., 1986).

[60] Shapiro M., and Bony H., *J. Chem. Phys.* **83**, 1588 (1985); Balint-Kurti G. G., and Shapiro M., *Adv. Chem. Phys.* **60** 403, Lawley K. P., Ed., (Wiley-Interscience Pub., 1986).

[61] Bandrauk A. D., and Atabek O., *Adv. Chem. Phys.* **73**, 823 (1989)

[62] The potential curves and the relevant electronic dipole moments are taken from: Schmidt I., Ph.D. Thesis, Kaiserslautern University, 1987.

Dynamic Stark shift in a 3-level atomic system

Do-Young Jeong, K.S. Lee, A.S. Choe, Jongmin Lee,
and Bum Ku Rhee*

*Laboratory for Quantum Optics, Korea Atomic Energy Research Institute,
P.O. Box 105, Yusong, Taejon, 305-600, Korea
*Permanent Address : Department of Physics, Sogang University,
1 Sinsoo-dong, Mapo-ku, Seoul, 121-742, Korea*

Abstract. Dynamic Stark shift of a two-photon excited state has been investigated theoretically and experimently in a 2-photon resonant 3-photon ionization scheme in a 3-level atomic system, where the resonance frequencies of the first and second transitions are in resonance within a few GHz. Detailed analysis of the experiment is given using density matrix formalism, accounting for the phase fluctuations of the laser field. A good agreement between theory and experiment has been obtained.

Introduction

The fluctuations of the electric field and the associated bandwidth can make significant effects on near-resonant laser-atom interactions. This is because the partial coherence of the field introduces extra time-scale relationship into the interaction dynamics. Field fluctuation rates can be comparable with several other rates, those related to spontaneous decay, Rabi oscillation, detuning, etc. Competition among these time scales make the theory of light absorption more complicated. Exact theories exist only for some simple model problems, although methods for attacking more realistic models have been discussed recently[1,2].

There has been of interest in developing theoretical models to describe laser fields. Two particular models, the phase diffusion (PD) model where the phase is stochastic and the amplitude is constant, and chaotic field (CF) model where only the amplitude is fluctuating and the phase constant, have attracted great interest. The PD model corresponds to a well-stabilized cw laser with a diffusing phase and the CF model describes thermal light or light from a multimode laser with a large number of independent modes.

In this report, we present the experimental and theoretical results for the dynamic Stark effect in a 3-level atomic system. In the theoretical work, exponentially correlated phase fluctuations is introduced into the electric field, and the amplitude of the field is treated as a Gaussian variable. For numerical simulation of exponentially correlated colored noise the integral algorithm developed by Fox *et al*[3] is employed in this work.

Theoretical Background

We consider an atomic system with a ground state $|1\rangle$ and two excited states $|2\rangle$, $|3\rangle$ with respective energies $\hbar\omega_i$, $i=1,2,3$, where the resonance frequencies of the first and second transitions are in resonance within a few *GHz*. The atom is assumed to interact with a single nonmonochromatic field written as

$$\vec{E}(\omega, t) \ = \ \frac{1}{2}\vec{E_0}(t)\exp[i\omega t + i\phi(t)] \ + \ c.c. \tag{1}$$

where the stochastic frequency $\varepsilon(t) \ = \ d\phi(t)/dt$ is assumed to be a Gaussian random variable with the correlation function of the form :

$$\langle\,\varepsilon(t)\varepsilon(t+\tau)\,\rangle \ = \ b\beta\exp(-\beta|\tau|) \tag{2}$$

where $b\beta$ is the variance and $1/\beta$ is the correlation time of the frequency fluctuations. In Eq. (1) the optical frequency of the laser field is $\omega + \varepsilon(t)$. We approximate the intensity profile of a laser pulse with a Gaussian shape such as

$$I(t) \ = \ \frac{\varepsilon_0 c}{2}|\vec{E}(t)|^2 \ = \ \frac{I_0}{\sqrt{2\pi(\varDelta t)^2}}\exp(-\frac{t^2}{2(\varDelta t)^2}) \tag{3}$$

where ε_0, c, I_0 and $\varDelta t$ are the dielectric constant of free space, light velocity, pulse energy and the characteristic time of a laser pulse, respectively.

The Fourier transform of the electric field gives the intensity spectrum arising from the phase fluctuation of the laser field assumed to have a constant amplitude E_0 [4] :

$$P_E(\omega) \ = \ \frac{E_0^2}{2}\int_{-\infty}^{\infty}\exp(-i(\omega-\omega_0)\tau)\exp\left[-b|\tau| - b\frac{\exp(-\beta|\tau|)-1}{\beta}\right]d\tau \tag{4}$$

Its shape changes from Lorentzian of FWHM equal to $2b$ if $\beta \gg b$ to Gaussian of FWHM $\sqrt{8\ln(2)b\beta}$ if $b \gg \beta$ [4,5]. The time evolution of the density matrix can be described by the Liouville equation :

$$i\hbar\left(\frac{d}{dt}+\hat{\varGamma}\right)\hat{\rho} \ = \ [\hat{H}+\hat{V}, \hat{\rho}]$$
$$\langle\,\hat{\rho}_{ij}\,\rangle \ = \ \langle\,\varPsi\,|\,i\rangle\langle j\,|\,\varPsi\,\rangle \tag{5}$$

where $\hat{\varGamma}$ is the operator of the system relaxation, $\hat{H}$ and $\hat{V}$ are the Hamiltonians of the unperturbed atomic system and of the laser-atom interaction, respectively.

Results and Discussion

The intermediate state of the atomic system has the red detuning of 3.5 GHz with respect to the frequency of the unperturbed 2-photon transition. The cross-sections of the first and second transitions are 6.2×10^{-13} and 2.5×10^{-14} cm^{-1}, respectively.

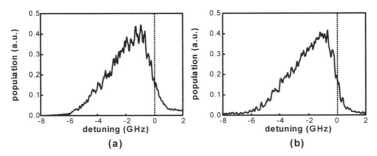

FIGURE 1. Theoretical and experimental two-photon excitation spectra : (a) theoretical result, when the peak value of rabi frequency of the first transition is 20.0 GHz and $b = 0.5$ GHz, $\beta = 1.2$ GHz (b) experimental result, when the dye laser has temporal pulse width of 10 ns, energy density of 3.2 mJ/cm^2 and spectral linewidth of about 1 GHz with multimodes.

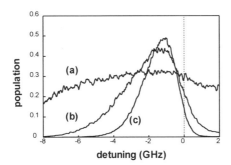

FIGURE 2. Variation of the two-photon excitation spectrum (numerical result) as a function of the correlation time at the same bandwidth, where the peak value of the first rabi frequency is 20.0 GHz : (a) $b = 0.45$ GHz, $\beta = 9.0$ GHz (b) $b = 0.5$ GHz, $\beta = 1.2$ GHz (c) $b = 1.0$ GHz, $\beta = 0.2$ GHz

The theoretical and experimental ionization profiles are shown in Fig. 1(a) and 1(b), respectively. There is a good agreement between the figures, in particular for the width and asymmetry of the profiles. The typical result shown in Fig. 1 has several noticeable features. (1) The line shape is asymmetric with a pronounced red wing. (2) The peak of the spectrum is shifted to the red. (3) The width of the spectrum significantly exceeds the laser bandwidth of 1 GHz.

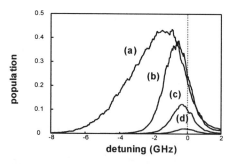

FIGURE 3. Variation of the two-photon excitation spectra (numerical result) as a function of peak values of the rabi frequency at the same laser spectrum of $b = 0.5$ GHz, $\beta = 1.2$ GHz : (a) 18.0 GHz (b) 5.0 GHz (c) 3.0 GHz (d) 1.0 GHz, where the population is amplified by 10.

Fig. 2 shows the 2-photon excitation spectrum depending on the type of phase fluctuation of the laser field. The numerical simulations reveal a significant dependence of the asymmetry of the excitation spectrum on the correlation time $1/\beta$, even if the bandwidth of the laser is kept constant. Fig. 3 presents some ionization profiles of the 2-photon excited state at four different values of the peak rabi. It shows dynamic Stark shift as a function of laser pulse energy.

Conclusion

We have measured and calculated the line shape of the two-photon resonant three-photon ionization spectrum. In the numerical calculation, the integral algorithm for exponentially correlated colored noise is adapted and the calculational results agree well with experimental ones, in spite that the electric field of the real multimode laser is treated as a single mode nonmonochromatic field with the same bandwidth.

The numerical simulations show a significant dependence of the asymmetry of the excitation spectrum on the correlation time $1/\beta$, even if the bandwidth of the laser is kept constant.

References

1. S.N. Dixit, P. Zoller, and P. Lambropoulos, Phys. Rev. A 21, 1289 (1980)
2. P. Zoller, J. Phys. B 15, 2911 (1982)
3. R.F. Fox, I.R. Gatland, R. Roy, and G. Vemuri, Phys. Rev. A 38, 5938 (1988)
4. M.H. Anderson, R.D. Jones, J. Cooper, S.J. Smith, D.S. Elliott, H. Ritsch, and P. Zoller, Phys. Rev. A 42, 6690 (1990)
5. G. Vemuri and R. Roy, Opt. Comm. 77, 318 (1990)

Effects of interference among two-photon coherences on the population dynamics of a Four-level Λ system

A.S.Choe, Yong-joo Rhee, and Jongmin Lee

Laboratory for Quantum Optics
Korea Atomic Energy Research Institute
P.O.Box 105 Yusong, Taejon 305-600 Korea

Abstract. We show that *optimal detuning method* describing population transfer passage in a three-level system has an essential difference from the *adiabatic following passage* by counterintuititive sequence of two coherent laser pulses, by examining the dressed eigenstates and the Bloch vector precessing around a torque vector. In a four-level Λ system, population dynamics is studied for the case where, on the optimal detuning condition, two-photon coherences causing adiabtic transfer interfere each other. For a scheme of three degenerate levels, we show a possibility of complete population inversion by the optimal detuning method.

INTRODUCTION

Nowadays, adiabatic following passages of complete population transfer into a specific level of multilevel system by laser radiations were achieved by the counterintuitive sequence or the pulse chirping methods.(1) Counterintuitive sequence has been of great attraction in the experimental realization because of its possibility of broad applications.(2) Recently, Choe(3) reported a new *optimal detuning method* in which the optimal control of laser frequencies near two-photon resonance caused complete population inversion in a three level system, which can be comparable with the two methods mentioned above. In this paper, we present some results of studies relevant to the optimal detuning method.

TWO ADIABATIC PASSAGES

We start with the quantized three-level system interacting with a classical harmonic field, as shown in Figure 1 where both the three-level and the four-level systems are illustrated. The electric dipole approximation is used. All the relaxations are neglected. Within the rotating wave approximation, we can write the perturbation Hamiltonian of the system by

$$H = \frac{\hbar}{2} \begin{bmatrix} 0 & V_1(t) & 0 \\ V_1(t) & 2\Omega & V_2(t) \\ 0 & V_2(t) & 0 \end{bmatrix} \tag{1}$$

on the two-photon resonant condition, where $V_{1,2}(t)$ is the Rabi frequencies, and Ω is the one-photon detuning. We focus our attention on adiabatic inversion of population by means of slowly varying lasers of the Gaussian shape pulses.

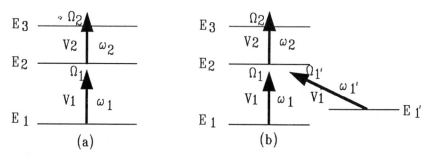

FIGURE 1. (a) Three-level atomic system; (b) Four-level atomic system

From the eigenvalue equation defined by $Hu = \lambda u$, we obtain the solution as

$$u_o = \frac{(V_2(t), 0, -V_1(t))}{\sqrt{V_1(t)^2 + V_2(t)^2}} \quad \text{for } \lambda_o = 0, \qquad (2)$$

$$u_\pm = \frac{(V_1(t), \lambda_\pm, V_2(t))}{\sqrt{V_1(t)^2 + V_2(t)^2 + \lambda_\pm^2}} \quad \text{for } \lambda_\pm = \Omega \pm \sqrt{\Omega^2 + V_1(t)^2 + V_2(t)^2}. \qquad (3)$$

As studied by Kuklinski[4] u_o is the adiabatic solution corresponding to the adiabatic following passage by counterintuitive sequence. In this case, $u_o(t)=(1,0,0)$ and $(0,0,-1)$ before and after the interaction, respectively. If, furthermore, the interaction process is adiabatic, the system is confined in the trajectory evolved from the initial dressed eigenstate, leading to the following of the inversion according to u_o state.

Meanwhile, u_o and u_- are the adiabatic solutions for the case of the optimal detuning method as mentioned by Choe. [5] The reason is simple. The necessary conditions of the optimal detunings are $\Omega \gg V_{1,2}$ and ω_1 and ω_2 being in near two-photon resonance. Under the condition that a quasi-steady state is maintained during the interaction with the lasers, the state of the system is a superposition of the two dressed states described by u_o and u_- when the initial population are assumed to be only in the first level. In this case, the phase evolution of the non-zero eigenvalue state plays an essential role for complete population transfer. In Figure 2, adiabatic passages of the eigenstate components of the probability amplitudes, $C_1(t)$ and $C_3(t)$ are shown and a significant difference between the counterintuitive and the optimal detuning methods are found. Now, we compare the traces of the system vector precessing around a torque vector. As in the two-level system we define the Bloch vector $\vec{\rho}$ by $(\text{Re}(2\rho_{13}), \text{Im}(2\rho_{13}), \rho_{33} - \rho_{11})$, where $\rho_{ij} = C_i^* C_j$ and the torque vector $\vec{\Gamma}$ by $(2V_1V_2/\Omega, 0, (V_2^2 - V_1^2)/\Omega)$, respectively. From the difference components of complex-valued two-photon coherence, ρ_{13}, we can find that in the counterintuitive method the Bloch vector (ρ_{cis}) adiabatically follows the torque vector while in the optimal detuning method the Bloch vector (ρ_{od}) moves in the perpendicular plane to the torque vector as shown in Figure 2 (c), where the Bloch vector (ρ_{int}) of intermediate situation is also drawn. More detailed description will follow soon in reference (6)

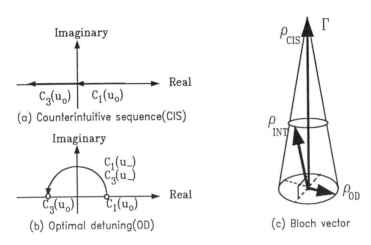

(a) Counterintuitive sequence(CIS)

(b) Optimal detuning(OD)

(c) Bloch vector

FIGURE 2. Adiabatic passages of probability amplitudes and Bloch vector

EFFECTS OF INTERFERENCE

In Figure 1 (b), three adiabatic passages, denoted by the paths of transitions by $1 \to 2 \to 3$, $1' \to 2 \to 3$, and $1 \to 2 \to 1'$, are found. When all the paths are under the condition of the adiabatic inversion, what are the atomic-level populations going to be with the initial condition like $\rho_{11}(0) = \rho_{1'1'}(0)=0.5$? This speculation leads to finding of the effects of interference among two-photon coherences represented by $\rho_{13}, \rho_{1'3}$, and $\rho_{11'}$ on the population dynamics, in the theoretical point of view. In the view point of practical application, information concerning the laser parameters to enhance the ionization efficiency in the selective photoionization process can be drawn.

We examine the dynamics for the case where $V_1(t) = V_{1'}(t) = V_2(t)$ denoted by $V(t)$ and $\Omega_1 = \Omega_1' = -\Omega_2$ denoted by Ω. In our previous work(7) we found that the analytic solutions within an effective two-level approach are not adequate in quantitative description of the inversion behavior. Now we examine the detailed dynamics by numerical solutions of the Schrödinger equation and Figure 3 shows the results. Interference effects of two-photon coherences significantly reduces the final inversion ratio and optimal inversion occurs when the Rabi frequencies are reduced by 20 % while maintaining the optimal detunings. Remarkable is the fact that the optimal detuning formula of the three-level system are equally applicable in describing the adiabatic inversion process of the four-level Λ system.

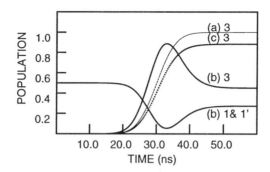

FIGURE 3. Populations of the four-level Λ system interacting with Gaussian shape laser pulses; (a) without any interference, (b) with interference effects, and (c) optimized with correction of Rabi frequencies

THREEFOLD DEGENERATE LEVEL SYSTEM

We consider a laser-light scheme, such as J=1-0-1, consisting of three degenerate levels to examine the coupling effects of two-photon coherences on adiabatic inversion. Two linearly, right-circularly, and left-circularly polarized laser pulses are applied so as to achieve two-photon coherent inversion condition for the different sets of laser polarization. We solved the Schrödinger equation analytically within an effective two-level approach and found the solutions to be in good agreement with the numerical results. As results, the coupling effect, arising from the common intermediate level, of the two-photon coherences causes no change in the inversion amount, but significant change takes place in the effective Rabi frequency. The optimal inversion for the degenerate system occurs when the Rabi frequency is reduced by a factor of $1/\sqrt{3}$ of that for the inversion condition of non-degenerate system.

REFERENCES

1. Oreg, J., Hioe, F. T., and Eberly, J. H., Phys. Rev. A**29**, 690-697 (1984)
2. (a) Schiemann, S., Kuhn, A., Steuerwald, S., and Bergmann, K., Phys. Rev. Lett. **29**, 637-3640 (1993); (b) Lawell, J. and Prentiss, M., Phys. Rev. Lett. **72**, 993-996 (1994)
3. (a) Choe, A. S., Rhee, Y., Lee, J., Han, P. S., Borisov, S. K., Kuzmina, M. A., and Mishin, V. A., Phys. Rev. A **52**, 382-386 (1995); (b) Choe, A. S., Rhee, Y., Lee, J., and Kuzmina, M. A., and Mishin, V. A., J. Phys. B: At. Mol. Optics **28**, 3821-3829 (1995); (c) Choe, A. S., Rhee, Y., Yoo, B., Kim, S., and Lee, J., J. Korean Phys. Soc. **29** 162-169 (1996)
4. Kuklinski, J. R., Gaubatz, U., Hioe, F. T., and Bergmann, K., Phys. Rev. A **11**, 6741-6744 (1989)
5. Choe, A. S., Lee, J., Phys. Rev. A (submitted)
6. Kim, J. B., Lee, J., Choe, A. S., and Rhee, Y., Phys. Rev. A(submitted).
7. Choe, A. S., Jung, E. C., Yi, J. H., Yoo, B., Rhee, Y., and Lee, J., Sae Mulli (meaning of *New Physic* in Korean), **36** No. 3, (1996)

Observation of Doubly Excited States in Negative Ions Using Resonance Ionization Spectroscopy

D. Hanstorp*, U. Ljungblad*, U. Berzinsh† and D. J. Pegg§

*Department of Physics, Göteborg University and Chalmers University of Technology,
S-412 96 Göteborg, Sweden
†Department of Spectroscopy, University of Latvia, LV 1586 Riga, Latvia
§Department of Physics, University of Tennessee, Knoxville, TN 37 996 USA

Abstract. Negative ions are especially sensitive to electron correlation since in these loosely bound systems the electron-electron interaction can become comparable to or dominate the electron-core interaction. This statement is particularly true in the case of double excitation where two electrons are simultaneously excited to large distances from the residual core. In this paper it is demonstrated that RIS is a powerful tool for investigations of doubly excited states of negative ions under the simultaneous condition of high resolution and sensitivity.

A proof-of-principle experiment is presented, where doubly excited states of Li⁻ situated just below the Li(4p) are investigated. Such autodetaching states manifest themselves as resonances in the photodetachment cross section. We have measured the photon energy dependence of the partial cross section for leaving the residual Li atom in the 3s state. Fano profiles were fitted to the individual resonances in order to extract the energies and widths of the doubly excited states.

INTRODUCTION

Double excitation was first investigated by Madden and Codling (1) in an experiment where they studied the helium atom using synchrotron radiation. In this system, the long range Coulomb field due to the He⁺ core results in an infinite Rydberg series of doubly excited states below each hydrogenic He⁺ threshold. In order to explain the observed spectra it was necessary to abandon the independent particle model and instead describe the two excited electron as two particle performing a correlated motion about the nucleus (2).

Another prototype system for investigation of electron correlation is a negative ion. In this case, the lack of a long range force between the neutral atom and the outermost electron leads to a binding dominated by the polarization potential. As a matter of fact, the independent particle model predicts unstable negative ions for many elements that experimentally are observed to be stable. Consequently, doubly excited states of negative ions are very good systems for testing theoretical models describing electron correlation. Such investigations, both experimental (3) and theoretical (4), have been performed on the H^- ion during the past two decades.

In this paper we present the results of a study of $^1P^o$ doubly excited states of Li^-. The Li^- ion is of particular interest since, with the exception of H^-, it has the simplest structure of any negative ion and is therefore tractable to theory. In the "frozen core" approximation it become an effective two-electron system. Further, the small ionization potential and electron affinity of Li, which in total is only 6 eV, makes it possible to investigate this ion with conventional narrow bandwidth lasers.

Previously (5), a very broad resonance was observed between the Li(3s) and Li(3p) in a photodetachment experiment in which the total yield of neutral atoms was measured. This method does, however, not enable us to detect high lying doubly excited states. With increasing excitation, the number of decay channels increases dramatically. Doubly excited states predominantly decay to the highest lying states of the residual atom that are energetically possible, whereas the direct photodetachment process tend to leave the atom its lower states. Therefore the resonance structure tends to "wash out" as the degree of excitation increases. The aim of this paper is to show that resonance ionization spectroscopy, RIS, can be effectively used for a state selective detection thereby allowing us to isolate a particular decay channel of the doubly excited state which produces a well modulated resonance structure. This method has previously been used in our laboratory to accurately measure the electron affinity of the lithium atom (6).

METHOD

The concept of the experiment is illustrated in Fig. 1a. A laser of frequency $\omega 1$ was used to produce doubly excited states of Li^- of $^1P^o$ symmetry which lie just below the Li(4p) photodetachment threshold. Due to an interference with the direct photodetachment process, these autodetaching states manifest themselves as resonances in the photodetachment cross section. The tunable ultraviolet radiation $\omega 1$ was generated by frequency doubling the output of an excimer-pumped dye laser. A second laser of frequency $\omega 2$ was used to selectively detect, via a resonance ionization process, residual Li atoms left in the 3s state following photodetachment via the 3sεp channel. The visible radiation ω2 was produced by a second dye laser pumped by the same excimer laser. In this case the frequency

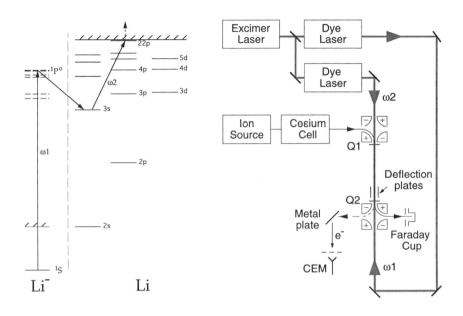

FIGURE 1. a: Energy level diagram for the Li⁻/Li systems.
b: A schematic diagram of the collinear beam apparatus.

was tuned to photoexcite the excited residual Li atoms from the 3s state to the 22p state. These highly excited Rydberg atoms were then efficiently field ionized and the resulting Li^+ ions were detected. The spectra were produced by monitoring the number of Li^+ ions thus produced as a function of the tunable laser frequency ω1.

A schematic of the interacting beam apparatus used in the present measurement is shown in Fig. 1b. A collinear geometry was used to enhance both the sensitivity and energy resolution of the measurement. Li^+ ions were extracted from a plasma-type ion source and accelerated to 4 keV. A beam of Li⁻ ions was produced by passing the Li^+ ions through a Cs vapor charge exchange cell. The Li⁻ beam was merged with the two laser beams ω1 and ω2 in the interaction region, which is defined by two 3 mm apertures placed 0.5 m apart. A pair of electrostatic quadrupole deflectors were used to direct the ion beam into and out of the path of the laser beams. In the interaction region, the Li⁻ ion beam and the laser beam ω1 interacted in an anti-parallel configuration while the ion beam and laser beam ω2 interacted in a parallel configuration. The second quadrupole also served to field ionize those atoms excited to Rydberg states by laser ω2. The Li⁻ ions were collected for normalization purposes at one side of the quadrupole deflector and the Li⁺ ions were detected on the opposite side. The Li⁺ ions were monitored by detecting the secondary electrons produced when the ions impacted on a metal plate. A channel electron multiplier (CEM) was used for this purpose. The major source of background arose from Li⁺ ions produced by the double detachment of Li⁻ by collisions with the residual gas in the interaction chamber. This contribution was minimized by the use of a low pressure (10^{-9} mbar) in the interaction chamber and by a pair of deflection plates positioned just prior to the entrance of the second quadrupole. The latter served to sweep out of the beam any Li⁺ ions produced upstream in collisional events.

RESULTS

Figure 2 shows the observed resonant structure in the relative cross sections for photodetachment of Li⁻ via the 3sεp channel below the 4p threshold. In the figure, the present measurements are compared with the result of a recent eigenchannel R-matrix calculation of partial cross section by Pan et al (7). The experimental energy resolution, which is estimated to be approximately 25 μeV, is sufficiently high compared to typical resonance widths that one can make a direct comparison with theory without resorting to deconvolution procedures. The measured photon energy scale is absolute, being established by calibration techniques based on atomic reference lines of Ne and Ar. Our cross section measurements are, however, relative. They were normalized to the theoretical results of Pan et al (7) by multiplying by a factor that is the ratio of the areas under the experimental and theoretical spectra between the same energy limits.

In the spectra there are three window resonances in the region just below the Li(4p) threshold. Both the energies and strengths of the resonances are well predicted by theory. The observed resonances are associated with the autodetaching decay of doubly excited states of the Li⁻ ion. A double Fano equation (8) of the form

$$\sigma(\omega) = \sigma_0 + \sum_{n=1}^{2} \sigma_n \frac{(q_n + \varepsilon_n)^2}{1 + \varepsilon_n^2} \tag{1}$$

was fitted to our measured cross sections, where $\varepsilon_n = (\hbar\omega - E_m)/(\Gamma_n/2)$, and σ_0 the non-resonant cross section, is assumed to be constant. Three resonance parameters: the resonance energy, E_m, the resonance width, Γ_n, and the profile

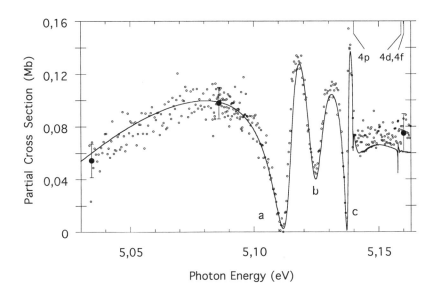

Figure 2. The partial cross section for photodetachment of Li- via the 3sεp decay channel below the Li(4p) threshold. The measured data(circles) are compared to the calculation (solid line) of reference (7).

index, q_n, are then obtained for each resonance from the fit. The two resonances, labeled a and b, yielded resonance energies of 5.1132(4) eV and 5.1234(4) eV, respectively. Their widths are 0.0074(5) eV and 0.0076(11) eV. These values are in good agreement with a recent calculation based on the complex rotation method by Lindroth (9). We also attempted, unsuccessfully, to include the resonance just below the 4p threshold, labeled c, in the fitting procedure. In this case it appears, however, that the resonance is prematurely terminated by the opening of the 4pεs channel.

DISCUSSION

This work represents an investigation of resonance structure in partial cross sections using a state selective detection technique based on resonance ionization. The high sensitivity of this detection method and the large interaction volume provided by the collinearly merged laser and ion beams helps to compensate for the inherently low values of the cross section for photoexcitation of high lying doubly excited states. This experiment demonstrates that RIS is a powerful tool that will likely be used extensively in negative ion spectroscopy in the future.

We plan to continue these experiments to even higher levels of excitation where the high sensitivity and energy resolution of the present apparatus will be of extreme importance. Of particular interest is the energy region where adjacent series of resonances begin to overlap. It has been suggested that chaotic behavior may develop under such conditions. Further we intend to apply the RIS method to study doubly excited states in the heavier alkali negative ions as well as the He⁻ ion. In the case of the alkali ions we will be able to systematically investigate how an extended core will influence the nature of the doubly excited states. The case of He⁻ is theoretically interesting since it has an open core consisting of a single electron.

ACKNOWLEDGMENT

We thank A. F. Starace, C.-N. Liu and E. Lindroth for fruitful discussion and for providing us with their data. Financial support for this research has been obtained from the Swedish Natural Science Research Council (NFR). DJP acknowledge support from the Swedish Institute and the US Department of Energy, Office of Basic Energy Sciences, Division of Chemical Sciences and UB acknowledge personal support from Chalmers University of Technology.

REFERENCES

1. R. P. Madden and K.Codling, Phys. Rev. Lett. **10**, 516 (1963).
2. For example, J. Macek J. Phys B 7, 831 (1968); S. Salomonson et al. Phys. Rev. A **39**, 5111 (1989); Y.K. Ho Phys. Reports **99**, 1 (1983); E. Lindroth, Phys. Rev. A **49**, 4473 (1994); J.-Z. Tang, S. Watanabe and M. Matsuzawa Phys. Rev. Lett. **69**, 1633 (1992);
3. For example, P. G. Harris et al. Phys. Rev. Lett. 65, 309 (1990); H. C. Bryant et al., Phys. Rev. Lett. 38, 228 (1977); M. E. Hamm et al., Phys. Rev. Lett. **43**, 1715 (1979); M. Halka et al., Phys. Rev. A **44,** 6127 (1991).
4. For example, H.R. Sadeghpour and C. H. Greene Phys. Rev. Lett. **65**, 313 (1990); H.R.Sadeghpour and C. H. Greene M. Cavagnero, Phys. Rev. A **45**, 1587 (1992); J. -Z. Tang et al., Phys. Rev. A **49,** 1021 (1994).
5. U. Berzinsh et al. Phys. Rev.Lett. **74,** 4795 (1995).
6. G. Haeffler et al. Phys. Rev. A **53,** 4127 (1996).
7. C. Pan, A. F. Starace, and C. H. Greene, Phys Rev A **53**, 840 (1996).
8. U. Fano, Phys.Rev **124**, 1866 (1961)
9. E. Lindroth, Phys. Rev. A **52**, 2737 (1995).

Resonance Ionization Spectroscopy of Negative Ions

V.V.Petrunin*, H.H.Andersen, P.Balling, P.Kristensen
and T.Andersen

*Institute of Physics and Astronomy, University of Aarhus,
DK-8000 Aarhus C, Denmark*

Abstract. A new technique based on state-selective, resonant-ionization detection of atoms, produced by laser detachment or resonant excess-photon- detachment processes, has made it possible to obtain significant progress in the studies of structural and dynamic properties of negative ions and of nonlinear interaction of an autoionizing state with a near-resonant laser field. New methods are briefly described and illustrated by examples from stable and autoionizing negative ions formed by elements with a closed s-electron subshell: He⁻, Be⁻, Ca⁻, Sr⁻, Ba⁻.

Studies of the structure and dynamics of negative ions are of fundamental interest for atomic physics. Negative ions exhibit properties markedly different from neutral atoms due to the extra electron being bound in a short-range potential and the important role played by electron correlations. The existence of only one truly stable state in most of the atomic negative ions limits the application of resonant techniques which are known to be very efficient for other atomic systems. Smooth threshold behavior of the photodetachment cross section prohibits accurate measurements of binding energies using detachment channels with $l > 0$.

A recently developed technique based on ultra-sensitive detection of atoms produced in a specific state as a result of a laser-photodetachment process has allowed us to investigate selectively even very weak detachment and excess-photon-detachment channels and thus to study structural and dynamic properties of stable and autoionizing negative-ion states, as well as resonant transitions between them, with a sensitivity and accuracy previously unaccessible.

The detection technique is based on the collinear resonant ionization method[1] which was developed and applied previously for detection of rare atomic isotopes, and it combines ultra-high selectivity and sensitivity with high spectral resolution[1,2]. A charge-state-separated negative ion beam, produced as a result of double-electron capture by fast positive ions in alkali-metal vapor, is overlapped collinearly with two or more pulsed laser beams ($\tau_L \sim 8$ ns). The first tunable dye laser is usually applied to perform detachment or resonant excitation of the negative ions in the vicinity of the investigated threshold or doubly-excited negative-ion state. The second dye laser is applied with a time delay relative to the first laser to perform resonant excitation from the specific atomic state to a Rydberg level. The Rydberg atoms are subsequently detected using state-selective field ionization in a non-uniform electric field at the end of the interaction region. The state-selective field ionization is very important in negative-ion studies, not only because it leads to a reduction in collisional-noise signal[2], but also because it eliminates the laser background originating from ionization of neutral atoms by the first powerful pulse used for negative-ion detachment.

The new technique has been applied to clarify the nature and state composition of alkaline-earth, negative-ion beams[3,4,5]. The binding energies of the stable and long-lived, doubly-excited states have been determined with sub-wavenumber accuracy using weak, previously unobservable, s-electron threshold photodetachment channels. Figure 1 shows the s-electron threshold for detachment of $Ca^-(4s^2 4p \ ^2P)$ ions to the $Ca(4s5s \ ^3P_1)$ level[4]. The channel represents only a ~0.1% fraction in the total photodetachment cross section. The threshold positions for the two statistically populated Ca^- fine-structure components are evident. The alkaline-earth ground-state fine-structure components of the negative ions have been observed and studied

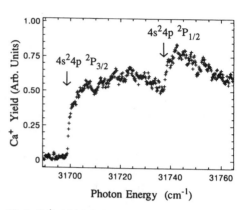

Fig.1. Ca^+ yield follwing the photodetachment of the $Ca^-(4s^2 4p \ ^2P_{1/2,3/2})$ ground state to the $Ca(4s5s \ ^3S_1)$ level, which is subsequently monitored by resonant ionization via the $Ca(4s15p \ ^3P_2)$ Rydberg level.

selectively for the first time in the experiments described (see Table 1). Figure 2 shows the result of the laser scan in the vicinity of the detachment threshold for the He^- $(1s2s2p \ ^4P)$ to the $He(1s3s \ ^3S)$ state and represents an example of a metastable negative-ion study. The electron affinity of the helium $1s2s \ ^3S$ state was determined with an accuracy of about 1 GHz.

Table 1.

Negative ion state	Binding energy (meV)	Parent state
Stable negative ions:		
$Ca^-(4s^2 4p \ ^2P_{3/2})$	19.73(10)	$Ca(4s^2 \ ^1S)$
$(4s^2 4p \ ^2P_{1/2})$	24.55(10)	$Ca(4s^2 \ ^1S)$
$Sr^-(5s^2 5p \ ^2P_{3/2})$	32.236(20)*	$Sr(5s^2 \ ^1S)$
$(5s^2 5p \ ^2P_{1/2})$	52.12(5)*	$Sr(5s^2 \ ^1S)$
$Ba^-(6s^2 6p \ ^2P_{3/2})$	89.60(6)	$Ba(6s^2 \ ^1S)$
$(6s^2 6p \ ^2P_{1/2})$	144.62(6)	$Ba(6s^2 \ ^1S)$
Metastable negative ions:		
$He^-(1s2s2p \ ^2P_{5/2})$	77.524(5)*	$He(1s2s \ ^3S)$
$Be^-(2s2p^2 \ ^4P_{3/2})$	290.99(10)	$Be(2s2p \ ^3P)$
$Be^-(2p^3 \ ^4S_{3/2})$	295.72(11)	$Be(2p^2 \ ^3P)$
$Ba^-(5d6s6p \ ^4F_{9/2})$	144.2(5)	$Ba(5d6s \ ^3D_3)$

*preliminary

The sensitivity of the detection technique has allowed us to disprove previous claims of observation of a long-lived $4s4p^2$ ^4P component in Ca⁻, to discover the long-lived unpredicted metastable Ba⁻(5d6s6p $^4F_{9/2}$) ion, to measure its binding energy, and to establish a lower limit for its lifetime of 50 µs. The measured binding energies of stable and metastable negative ions are summarized in Table 1. An improvement of about two orders of magnitude in the accuracy of the binding energies has made it possible to perform critical tests of theoretical predictions.

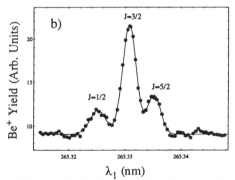

Fig.2. He⁺ yield following the photodetachment of the He⁻(1s2s2p ^4P) state to the He(1s3s ^3S) level which is subsequently monitored by resonant ionization via the He (1s14p ^3P) Rydberg level. One wavelength step corresponds to 7µeV.

The presence of two long-lived, doubly-excited states in the negative beryllium ion (see Fig.3a), which are connected by an optical transition, offers a unique possibility for gaining new information about the dynamic properties of these states and for the investigation of resonant

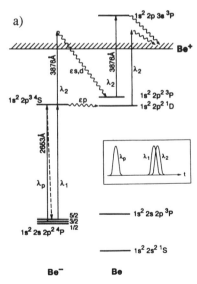

Fig. 3 a) Energy level diagram of beryllium, showing the relevant states in the negative ion and the neutral atom. The probe lasers λ_1 and λ_2 perform resonant photodetachment and ionization of the Be⁻(^4P_J) states, while the pump laser λ_p redistribute ions from the $^4P_{3/2}$ to the complete ^4P_J multiplet.
b) The Be⁺ yield in wavelength scan of the probe laser λ_1 through the ^4P_J - $^4S_{3/2}$ transitions when redistribution is applied 50 ns before the probe.

nonlinear processes in the continuum[6]. Optical pumping was used to transfer the population from the long-lived $^4P_{3/2}$ component via the fast radiatively decaying ^4S state ($\tau_{rad} \sim 1.5$ ns) to the complete ^4P_J multiplet. The population transfer was followed by resonant, stepwise two-photon detachment of the specific ^4P_J fine-structure components with absorption of two laser photons (λ_1 and λ_2). Subsequent resonant ionization of the neutral atoms through the Be($1s^2 2p3s$ ^3P) autoionizing state was applied, yielding a positive-ion signal reflecting the ^4P_J population (see Fig.3b). Lifetimes of the short-lived ^4P_J components were measured by varying the delay between the pump and probe lasers to be $\tau(^4P_{1/2}) = 0.73(8)$ μs and $\tau(^4P_{5/2}) = 0.33(6)$ μs. The fine-structure splittings are determined to be 0.59(7) cm^{-1} (J=5/2-3/2) and 0.74(7) cm^{-1} (J=3/2-1/2). By monitoring of the Be($1s^2 2p^2$ ^1D) population the autoionization rate of the Be$^-$(^4S) state is estimated to be ~$10^8 s^{-1}$.

Recently the RIS technique has been used to investigate resonant excess-photon detachment (EPD) processes via short-lived autoionizing states[7]. The studies have been performed with laser intensities of 10^5- 10^7 W/cm^2, which is about 10^3 times lower than in previous investigations of resonant EPD[8]. Figure 4a shows the photo-ion signal following the population of the Sr($5s5p$ $^3P_1$) level in a two-photon detachment process when the λ_1 dye laser is scanned in the region close to the $5s^2 5p$ $^2P_{1/2}$ - $5s5p^2$ $^4P_{3/2}$ intercombination line. Figure 4b shows the signal proportional to the residual population of the Sr$^-$($5s^2 5p$ $^2P_{1/2}$) level after the λ_1 dye laser pulse was applied. The experimental curves have been fitted using the Fano formula for autoionizing resonances observed as a depletion in the ground-state population: N=$N_o$exp($-\sigma_F W_L$), where N is the ground-state population, W_L the laser-pulse energy and σ_F= (q-$2\Delta/\Gamma)^2/(1+(2\Delta/\Gamma)^2)$ is the Fano expression for photodetachment cross section, with Δ being the laser detuning from the autoionizing-state position. The autoionizing width Γ= 0.5 cm^{-1} and the Fano parameter q=1.7 were obtained from the experimental-data fittings. The arrows in Fig.4a,b indicate the position of the $5s5p^2$ $^4P_{3/2}$ autoionizing state at λ_1=929.90 nm. Figure 4a clearly shows a displacement of the EPD signal in the direction from the autoionizing-state position toward the minimum of the Fano profile and a narrowing of the EPD peak below the natural width of the autoionizing state with increasing laser power. The observed displacement and narrowing are results of the competition between one-photon and two-photon detachment processes during the pulsed-laser excitation. This competition normally leads to a weak dependence of the excess-photon absorption signal on the laser power since the one-photon depletion of the target takes place already at the low intensities present in the beginning of the laser pulse. With λ_1 matching the minimum of the Fano profile, more than 50% of the negative ions was detached by absorption of two photons at a laser power of only 10^7W/cm^2.

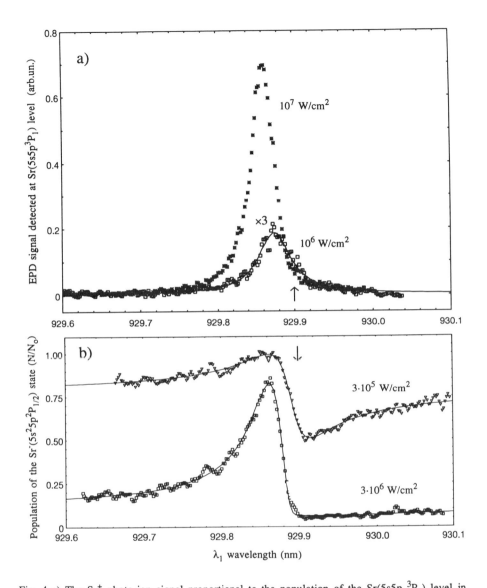

Fig. 4 a) The Sr⁺ photo-ion signal proportional to the population of the Sr(5s5p $^3P_1$) level in the two-photon detachment process when the first dye laser λ_1 is scanned in the region of the $5s^25p$ $^2P_{1/2}$ - $5s5p^2$ $^4P_{3/2}$ intercombination line of Sr⁻. The Sr(5s5p $^3P_1$) atoms are detected using resonant ionization via the Sr(5s14s $^3S_1$) Rydberg state.

b) The Sr⁺ photo-ion signal proportional to the residual population of the Sr⁻(5s²5p $^2P_{1/2}$) state after the first dye laser λ_1 is applied. The Sr⁻(5s²5p $^2P_{1/2}$) population is monitored using photodetachment to the Sr(5s5p $^3P_0$) level with the 532nm Nd:YAG laser light followed by resonant ionization via the Sr(5s14s $^3S_1$) Rydberg state. The maximum field intensities of the first dye laser λ_1 are indicated near the experimental curves.

107

The experimental work has been performed as part of the research program of the Aarhus Center for Advanced Physics funded by Danish National Research Foundation. We would like to gratefully acknowledge discussions and work in different experiments with C.A.Brodie, H.K.Haugen, U.V.Pedersen and J.D.Voldstad.

* Permanent address: Institute of Spectroscopy, Russian Academy of Sciences, 142092 Troitsk, Moscow Region, Russia

1. Yu.A.Kudryavtsev, V.S.Letokhov, and V.V.Petrunin JETP Lett. **42**, 26(1985)
2. S.A.Aseyev, Yu.A.Kudryavtsev, V.S.Letokhov and V.V.Petrunin, Opt.Lett.**16**, 514 (1991); and references therein.
3. P.Kristensen, V.V.Petrunin, H.H.Andersen, and T.Andersen, Phys.Rev. A **52**, R2508 (1995).
4. V.V.Petrunin, H.H.Andersen, P.Balling, and T.Andersen, Phys. Rev. Lett. **76**, 744 (1996).
5. V.V.Petrunin, J.D.Volstad, P.Balling, P.Kristensen, T.Andersen, and H.K.Haugen, Phys.Rev.Lett. **75**, 1911 (1995).
6. H.H.Andersen, P.Balling, V.V.Petrunin, and T.Andersen, J.Phys.B **29**, L415 (1996).
7. V.V.Petrunin, P.Kristensen, C.A.Brodie and T.Andersen, to be published.
8. H.Stapelfeldt and H.K.Haugen, Phys.Rev.Lett.**69**, 2638(1992); H.Stapelfeldt, P.Kristensen, U.Ljungblad, and T.Andersen, Phys.Rev.A**50**, 1618 (1994)

Doppler-free Resonance Ionization Spectroscopy of the He 1s² ¹S - 1s2s ¹S Transition at 120.3 nm

S. D. Bergeson, A. Balakrishnan, K. G. H. Baldwin[1],
T. B. Lucatorto, J. P. Marangos[2], T. J. McIlrath,
T. R. O'Brian, S. L. Rolston, and N. Vansteenkiste[3]

National Institute of Standards and Technology
Gaithersburg, MD 20899

Abstract: An accurate determination of the Lamb shift in the ground state of He is a sensitive check of QED calculations in the 2-electron, 3-body regime. We have begun an experiment that should determine the Lamb shift in the ground state of He with an accuracy better than the uncertainty in the best QED calculations. The measurement will employ a 2-photon Doppler-free resonant excitation of the He 1s² ¹S -> 1s2s ¹S transition at 120.3 nm, followed by ionization of the 1s2s ¹S state.

INTRODUCTION

In 1947, Lamb and Retherford demonstrated that Dirac theory failed to accurately describe the energy levels in atomic hydrogen (1). Contrary to Dirac theory, the $2\,^2S_{1/2}$ and $2\,^2P_{1/2}$ levels are not degenerate, but the $2\,^2S_{1/2}$ lies above the $2\,^2P_{1/2}$ level by about 1 GHz. For high accuracy calculations, the Dirac theory has been replaced by Quantum Electrodynamics (QED), which has been very successful in calculating energy levels in simple atomic systems (2). The difference between observed energy levels and non-QED calculations is called the Lamb shift.

The largest term in the QED calculation of the Lamb shift is the electron self-energy (3). The idea is that the electron can interact with itself. In the language of QED, the electron radiates virtual photons, and absorbs those same photons. This emission and absorption of virtual photons happens on a time scale that is fast compared to the mean orbital time of the electron around the nucleus. The random momentum kicks to the electron from emission and absorption of these virtual photons has the effect of "spreading out" the electron so it no longer behaves as a point particle. Rather, the electron appears to have a finite charge sphere. When

---

[1] Permanent address: Australian National University, Canberra, AUSTRALIA
[2] Permanent address: Imperial Colege London, London, ENGLAND
[3] Permanent address: Institut d'Optique, Orsay, FRANCE

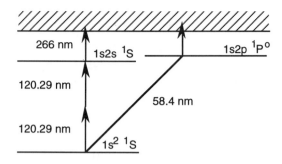

FIGURE 1. A partial energy level diagram of the singlet system in He.

the electron is near or inside the nucleus, the potential is no longer Coulomb, and the finite charge radius of the electron in this case reduces the binding of the electron. This effect is most pronounced in the S states.

The next largest term in the QED calculation of the Lamb shift is the vacuum polarization. Short-lived electron-positron pairs spontaneously appear out of the vacuum. However, they can live long enough to align themselves in the field of the nucleus. This slightly increases the binding of the orbital electron.

The QED calculation of the Lamb shift in He is somewhat more complicated. Most of the Lamb shift can be calculated as the hydrogenic (1 electron) Lamb shift, corrected for the nuclear charge, including the screening of the nucleus by the second electron (4). In addition, there is a qualitatively new contribution from the electron-electron interaction (5).

A partial energy level diagram of the singlet series in He is shown in Figure 1. The Lamb shift in the ground state of He is an order of magnitude larger than the Lamb shift for any other He state. The binding energy of the ground state of Helium was determined with a standard uncertainty of 4.5 GHz by Herzberg in 1958 (6). This uncertainty is about one tenth the calculated size of the Lamb shift. In 1993, Eikema et al. made an accurate determination of the ground state energy with respect to the 1s2s ^{1}P^o level (7). Their measurement also used a resonance ionization technique with a laser at 58 nm as the first step. Subsequent improvements in their experiment have refined their measurement to an uncertainty of 45 MHz (8). This uncertainty is the same size as the uncertainty in the best QED calculations. Among the largest sources of error in these pioneering measurements are the AC Stark shift, the Doppler shift, and the reference line calibration. As discussed below, these sources of uncertainty are markedly reduced in our measurement.

We are measuring the energy of the ground state with respect to the 1s2s ^{1}S level, using a 2-photon transition at 120.3 nm. The realization of the 2-photon transition requires a narrow-band source of 120 nm photons. A schematic diagram of the optical system is shown in figure 2. Our laser system begins as a 20 mW cw laser diode at 842 nm. The cw laser is pulse amplified in a Nd:YAG-pumped dye amplifier to about 50 mJ in a 5 ns pulse. The 4th harmonic is generated in 2 successive BBO crystals. The 4th harmonic (210 nm) and the fundamental (842 nm) are mixed in a phase-matched Kr/Ar 4-wave mixing cell to produce about 10 W

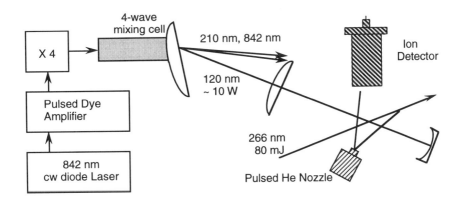

FIGURE 2. Schematic diagram of the laser and optical system.

peak power of 120 nm radiation. The output coupler of the 4-wave mixing cell is a MgF$_2$ lens, which is tilted and used off axis. In this arrangement, light impinges normal to the lens surface and is collimated. The tilted back surface of the lens disperses the beams. The 842 nm and 210 nm beams are deflected out of the optical path, and are blocked. Only the 120 nm beam enters the interaction region.

The 120 nm beam is focused by a second MgF$_2$ lens into the interaction region. In a standard Doppler free arrangement, a spherical mirror is used to re-image the focal spot back onto itself, producing counter-propagating beams in the interaction region. The 120 nm pulse is synchronized with a super-sonic He gas pulse in the interaction region, where the metastable He atoms will be prepared. A delayed laser pulse at 266 nm ionizes the metastable He, and the ions are collected by a time of flight mass spectrometer.

The expected signal levels can be estimated. The 2-photon transition rate from the 1s2s ^{1}S metastable level to the 1s^2 ^{1}S ground level has been calculated (9). The number of atoms making the transition from the ground state to the metastable level can be written as a product of the transition probability, the square of the photon flux, the number of atoms participating in the excitation, the interaction time, and the reciprocal of the laser linewidth. For 10 W of 120 nm radiation in a "transform limited" 4 ns pulse focused to 40 µm, thousands of metastable atoms will be created with each laser pulse. We can ionize the metastable atoms with near unit efficiency, making the He* signal readily detectable.

The accuracy of the measurement will hinge critically on understanding and controlling the systematic effects. The Doppler effect, the AC Stark shift, and the reference calibration line uncertainty, which limited the measurements of Eikema et al. (7,8) are dramatically reduced in our experiment. The 2-photon transition can be made Doppler free. For 10 W of 120 nm radiation focused to a 40 µm spot, the worst case AC Stark shift is 0.3 MHz. We are using a Ne reference calibration line, which is currently known only to 30 MHz. However, plans are underway to calibrate the line to better than 1 MHz.

The largest expected uncertainty in the measurement will stem from frequency chirps introduced by pulse amplification and frequency up-conversion.

Professor Edward Eyler and co-workers have demonstrated that frequency chirps in pulsed amplification and second harmonic generation can be large in sub-optimal amplifier arrangements. However, by proper selection of dye parameters and pump laser pulse shape, the chirps can be minimized (10). Simple models may be used to predict the chirp, making correction of the 842 nm laser frequency feasible at the few MHz level. This translates into an uncertainty in the range of 10 to 20 MHz in the 7th harmonic at 120.3 nm. Because the energy of the 1s2s ^{1}S level is known with respect to the ionization continuum to better than 1 MHz (11), the uncertainty in our determination of the Lamb shift in the ground state of He is expected to be in the range of 15 to 30 MHz.

Summary

We will be making a Doppler-free 2 + 1 resonance ionization spectroscopy measurement that utilizes the two-photon 1s^2 ^{1}S - 1s2s ^{1}S transition as the first step. This measurement will yield an accurate determination of the Lamb shift in the ground state of He. The laser system is completely operational. The vacuum system and optics are all in place. We are beginning to look for the He$^+$ signal. Careful control of the systematic effects in the measurement will lead to an accuracy better than the current best experimental results and better than the estimated uncertainties in the QED calculations of the Lamb shift in the ground state of He.

REFERENCES

1. Lamb, W. E., and Retherford, R. C., *Phys. Rev.* **72**, 241 (1947); Retherford, R. C., and Lamb, W. E., *Phys. Rev.* **75**, 1325 (1949); Lamb, W. E., and Retherford, R. C., *Phys. Rev.* **79**, 549 (1950).
2. Eides, M. I., and Shelyuto, V. A., *Phys Rev . A* **52**, 954 (1995).
3. Mohr, P. J., "Quantum Electrodynamics Calculations," in *The Spectrum of Hydrogen: Advances*, ed. G. W. Series, New York: World Scientific, 1988, ch. 2, pp. 111-136.
4. Drake, G. W. F., and Swainson, R. A., *Phys. Rev. A* **41**, 1243 (1990).
5. Drake, G. W. F., and Makowski, A. J., *J. Opt. Soc. Am.* **B5**, 2207 (1988).
6. Herzberg, G., *Proc. R. Soc. London A* **248**, 309 (1958).
7. Eikema, K. S. E., Ubachs, W., Vassen, W., and Hogervorst, W., *Phys. Rev. Lett.* **71**, 1690 (1993).
8. Eikema, K. S. E., Ubachs, W., Vassen, W., and Hogervorst, W., *Phys. Rev.* submitted
9. Drake, G. W. F., Victor, G. A., and Dalgarno, A., *Phys. Rev.* **180**, 25 (1969)
10. Gangopadhyay, S., Melikechi, N., and Eyler, E. E., *J. Opt. Soc. Am.* **B11**, 231 (1994); Gangopadhyay, S., Melikechi, N., and Eyler, E. E., *J. Opt. Soc. Am.* **B11**, 2314 (1994); Melikechi, N., Gangopadhyay, S., and Eyler, E. E., *J. Opt. Soc. Am.* **B11**, 2402 (1994); Gangopadhyay, S., *Optical Phase Distortions in Nanosecond Laser Pulses and Their Effects on High Resolution Spectroscopy*, Ph. D. Thesis, Univ. Delaware, 1995.
11. Sansonetti, C. J., and Gillaspy, J. D., *Phys. Rev. A* **45**, R1 (1992); Lichten, W., Shiner, D., and Zhi-Xiang Zhou, *Phys. Rev. A* **43**, 1663 (1991).

SESSION IV:
ENVIRONMENTAL APPLICATIONS

Multiple Resonance RIMS Measurements of Calcium Isotopes Using Diode Lasers

B. A. Bushaw*, F. Juston[†], W. Nörtershäuser[†], N. Trautmann[‡],
P. Voss-de Haan[†], and K. Wendt[†]

*Pacific Northwest National Laboratory, PO Box 999, Richland, WA 99352
[†]Institut für Physik, [‡]Institut für Kernchemie, Universität Mainz, D-55099 Mainz, Germany

Abstract. Single-, double-, and triple resonance excitation schemes, using extended-cavity and frequency-doubled diode lasers, have been investigated for highly selective RIMS detection of calcium isotopes. The single-resonance scheme has been applied to isotope ratio measurements in blood serum and meteorite samples. The double- and triple-resonance schemes have been evaluated for improvements in selectivity and sensitivity, with a goal of providing overall isotopic selectivity $> 10^{15}$ for radiochemical dating measurements with ^{41}Ca.

INTRODUCTION

There are a number of applications for the precise and sensitive measurement of calcium isotopes. These include measurement of the long-lived radionuclide ^{41}Ca for geological and anthropological dating and for evaluation of exposure histories of extraterrestrial materials(1), measurement of ratios of the minor stable isotopes in meteorite inclusions for testing nucleosynthetic models(2), and the use of stable isotope tracers in medical studies(3). The analytical requirements vary widely for each of the applications, but include isotopic selectivity of $>10^{15}$ for dating measurements with ^{41}Ca, suppression of titanium isobars in meteorite isotope ratio measurements, and high precision ratio measurements for medical applications.

Double-resonance RIMS using single-frequency cw dye lasers has previously demonstrated detection limits in the attogram range for $^{90}Sr(4)$, and optical isotope selectivity $>10^9$ in the measurement of Pb isotopes(5). However, the complexity, size, cost, and reliability of cw dye lasers make them less than desirable for routine analytical measurements. Thus, in this work we describe efforts to perform similar measurements using solid-state diode lasers to replace the cw-dye lasers. We have investigated a number of ionization schemes for calcium, including single-, double- and triple-resonance excitation (Figure 1) using exclusively single-mode diode lasers for the resonance excitation steps. All diode lasers were operated with grating-tuned extended cavities and the first-step excitation of the 422.7 nm Ca resonance line required the construction of an external cavity $KNbO_3$ frequency doubler for a diode laser operating at 845.4 nm. Second- and third-step resonances could be excited with

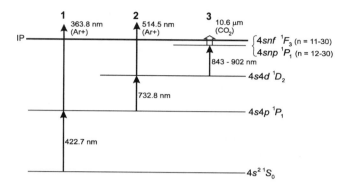

FIGURE 1. Single-, double, and triple-resonance RIS schemes for calcium using diode lasers for the resonance steps. In the double- and triple-resonance schemes, the diode lasers, as well as the indicated fixed-wavelength lasers, can effect ionization.

diode lasers at their fundamental wavelengths. Ionization out of the highest-lying level can then be accomplished with a higher power cw laser (e.g., Ar ion or CO_2) of fixed wavelength, and the ions produced are analyzed with a quadrupole mass spectrometer. Samples were atomized with a previously described(6) graphite tube furnace, which was suitable for both metallic samples (continuous atomization for spectroscopic studies) as well as discrete aqueous samples (analytical studies).

SINGLE-RESONANCE STUDIES

The single-resonance scheme has been evaluated for precision determination of the minor stable calcium isotopes using computer-controlled two-dimensional mass – laser frequency scanning, as shown in Figure 2. Integration over the individual peaks and correcting for counter dead time and mass dependent interaction time has yielded isotope ratios with a precision of ~0.2% for prepared samples. In most cases, the absolute values of the ratios agree with accepted values to within the precision, however, ^{43}Ca values are low (~15%) due to hyperfine optical pumping and some anomalies (~2.5%) for ^{48}Ca have been observed. These anomalies indicate that calibration with reference standards will be necessary to achieve ratio accuracy that approaches the precision. Detection limits of a few picograms have been demonstrated for discrete aqueous samples, limited primarily by the relatively inefficient photoionization step, which used only ~200 mW of 363.8 nm radiation. Demonstration measurements were performed on washings taken from the Orgueil-2 meteorite and blood serum samples. In the former, low quantities of Ca (~2.5 μg) limited the statistical precision and no isotope anomalies greater that the 3% statistical limit were observed. In contrast, 100 μl of blood serum was found to contain sufficient calcium for the determination of $^{44}Ca/^{40}Ca$ with a precision of ~0.1%. In additional measurements, relevant to meteorites, prepared samples containing equal amounts of titanium and calcium were measured. No interference

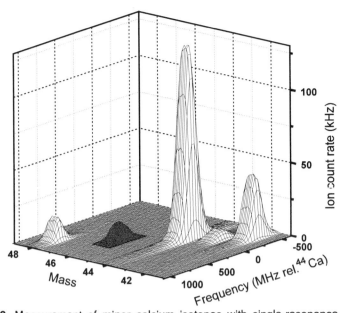

FIGURE 2. Measurement of minor calcium isotopes with single-resonance excitation. The shaded area is where signal integration is increased 50-fold for detection of the 3×10^{-5} abundant ^{46}Ca isotope.

from the Ti isobars was observed. This is in contrast to direct mass spectrometric methods where similar samples have shown errors as large as 25% for ^{46}Ca determinations(7).

MULTIPLE-RESONANCE STUDIES

Double- and triple-resonance schemes were investigated for improving optical isotopic selectivity, as will be needed for measurements of ^{41}Ca. Figure 3 shows a two-dimensional spectrum for minor calcium isotopes obtained for double-resonance excitation. The increase in selectivity, compared to Figure 2, is apparent even though no mass-selective detection has been used. Similar measurements were performed for the triple-resonance scheme, where the first-step laser was locked to ^{40}Ca absorption in a reference atomic beam and the 2nd and 3rd step lasers were then scanned to study the triple-resonance line shape. The observed linewidth was 8.5 MHz (FWHM) and is approximately a factor of 2 greater than that predicted by detailed density matrix calculations. After accounting for laser (jitter) linewidth, agreement between theory and experiment is good, indicating that predicted optical selectivity $>10^{10}$ for $^{41}Ca/^{40}Ca$ should be achievable. However, this is yet to be demonstrated experimentally. In these initial triple-resonance measurements, only the available diode laser power (~10mW) was used for the photoionization step, however, the efficiency was already comparable to the of the single-resonance scheme. Use of a 10W cw CO_2 laser for the ionization step should improve efficiency about 4 orders of magnitude.

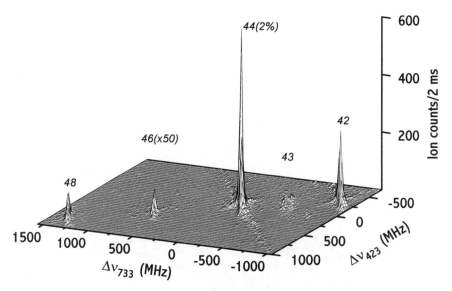

FIGURE 3. Isotope shifts for the minor calcium isotopes observed in double-resonance three-photon ionization. Hyperfine structure for ^{43}Ca is partially resolved.

CONCLUSIONS

Single-, double-, and triple-resonance ionization schemes using diode lasers for the measurement of calcium isotopes have been demonstrated. The single-resonance scheme has been shown capable of precision isotope ratio measurements and has sufficient sensitivity for medical applications, however, measurement of minor isotope ratios in single meteorite inclusions will require further improvements in ionization efficiency. Preliminary studies of the multiple-resonance schemes have shown a dramatic increase in optical isotope selectivity. Continuing work will focus on using the triple-resonance scheme, with CO_2 laser ionization, for ultra low-level measurements of ^{41}Ca.

ACKNOWLEDGEMENT

This work was supported by the Deutsche Forschungsgemeinschaft under grant TR-336 and the U. S. DOE under contract DE-AC06-76RLO 1830.

REFERENCES

1. Fink, D., Klein, J., and Middleton, R., *Nucl. Instr. Meth. Phys. Res. B* **52**, 572-582 (1990).
2. Sorlin, O., et al., *Phys. Rev. C* **47**, 2941-2953 (1993).
3. Yergey, A. L., et al., *J. Nutr.* **124**, 674-682 (1994).
4. Bushaw, B. A., *Inst. Phys. Conf. Ser.* **128**, 31-36 (1992).
5. Bushaw, B. A., and Munley, J. T., *Inst. Phys. Conf. Ser.* **114**, 387-392 (1991).
6. Bushaw, B. A., and Gerke, G. K., *Inst. Phys. Conf. Ser.* **94**, 277-280 (1989).
7. Niederer, F. R., and Papanastassiou, D. A., *Geochim. Cosmochim. Acta* **48**, 1279-1293 (1984).

Coupling of Gas Chromatography with Jet-REMPI Spectroscopy and Mass Spectrometry

Ralf Zimmermann[A,B,C], Hans Jörg Heger[A], Egmont R. Rohwer[A,D],
Edward W. Schlag[A], Antonius Kettrup[B,C], Ulrich Boesl[A]

[A]Institut für Physikalische und Theoretische Chemie, Techn. Univ. München, D-85747 Garching
[B]Lehrstuhl für Ökologische Chemie u. Umweltanalytik, Techn. Univ. München, D-85354 Freising
[C]Institut für Ökologische Chemie, GSF-Forschungszentrum für Umwelt und Gesundheit,
    D-85758 Oberschleißheim, Germany
[D]Department of Chemistry, University of Pretoria, 0002-Pretoria, Republic of South Africa

Abstract:. A novel interface for hyphenation of gas chromatography (GC), resonance enhanced multi-photon ionization (Jet-REMPI) spectroscopy and mass spectrometry (MS) for organic trace analysis is presented. Technical details and results are given in this contribution.

Hyphenation of different separation- and detection-methods is a successful strategy for the development of new analytical techniques. In this paper a combination of resonance ionization spectroscopy, gas chromatography and mass spectroscopy is presented. This GC-Jet-REMPI-MS represents a three dimensional analytical technique based on the compound selective parameters of retention time, resonance ionization wavelength (~ high resolution UV-spectroscopy) and molecular mass [1,2].

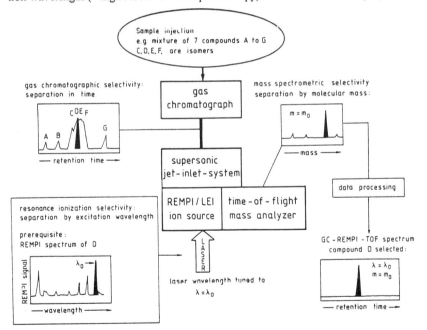

Figure 1
Schematic representation of the GC-Jet-REMPI-MS technique: The sample is injected into the GC and the compounds are separated in time. However, in chromatograms of complex mixtures coelution often occurs. After Jet-cooling, REMPI-ionization is applied, introducing high optical selectivity. The formed ions are mass analyzed in a time-of-flight mass spectrometer, introducing mass selectivity. Jet-spectroscopic and mass spectrometric steps can be seen as a filter-function eliminating all non target compound signals from the chromatogram.

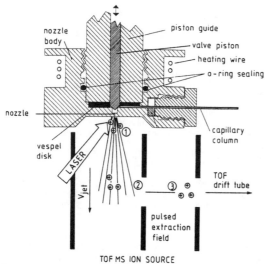

nozzle
body

piston guide

valve piston

heating wire

o-ring sealing

nozzle

vespel
disk

LASER

V_{jet}

capillary
column

TOF
drift tube

pulsed
extraction
field

TOF MS ION SOURCE

Figure 2
Cross-section of the GC-Jet-valve and a schematic representation of the free-jet ion source. With this arrangement ionization occurs directly underneath the orifice in the region of high gas density①. The formed ions drift by their jet-velocity② in the center of the ion source and are extracted by a pulsed acceleration field into the TOFMS ③

Although there have been several previous attempts to combine gas chromatography with jet-spectroscopy (e.g. [3]), it has not been possible to achieve full GC and optical selectivity together with high detection sensitivity. The pulsed valve plays a crucial role in a successful GC-Jet interface [1,2]. In our present design the dead volume of the valve was minimized in order to avoid GC-peak broadening. Make-up gas and the resulting dilution of the GC-eluent is thus avoided. Sufficient cooling could be obtained by jet expansion of the GC carier gas itself (5 K rotational temperature measured for benzene, see below).The chromatographic eluent is injected as short jet gas-pulses (20Hz) into the TOFMS-ion source. These sample packets are intersected with a pulsed laser beam and the resulting ions are mass-analyzed in a TOFMS. Figure 2 shows a cross-section of the developed GC-Jet-valve. A practical analytical application (gasoline) is given in figure 3. It is shown that Jet-REMPI is highly compound selective, even able to distinguish isomeric compounds. This is of particular interest for environmental analysis, because isomeric compounds often exhibit very different toxicological properties (e.g. consider the highly carcinogenic benzo[a]pyrene

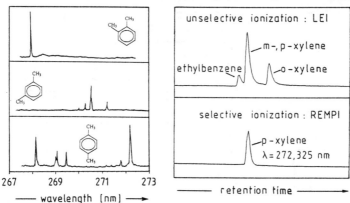

267   269   271   273

⟶ wavelength [nm] ⟶

unselective ionization : LEI

m-, p-xylene

ethylbenzene   o-xylene

selective ionization : REMPI

p-xylene
λ=272,325 nm

retention time

Figure 3
*Jet-REMPI selectivity achieved for coeluting isomers: **a)** One-color two-photon REMPI spectra of the isomeric xylenes (see [4]) **b)** Chromatograms of a gasoline sample under general (EI, top) and selective ionization (REMPI, bottom) conditions. In the latter case only p-xylene is ionized, allowing its detection in the presence of coeluting m-xylene.*

versus the harmless benzo[e]pyrene). We have shown that the GC resolution is not appreciably affected by the heated interface. Measurement of the rotational temperature of benzene (continuously seeded into the GC-Jet carrier gas) as a function of the delay time between the valve trigger and the laser pulse pointed out that under gas flow conditions optimized for GC (30m DB5 collumn, i.d. 0.25 mm), only a relatively low pressure is obtained in the valve because not enough gas can be delivered by the capillary. The cooling is therefore not optimized, especially at the beginning and the end of the jet-gas pulse. However, at the gas pulse center sufficient cooling is achiveable (figure 4). The need of of expansion gas addition and further sample dilution is thus avoided. The limited gas flow complicates the effect of increasing valve opening times as the resulting pressure drop in the valve reduces the rotational cooling. In figure 5 the effect of jet-pulse duration is shown. Longer valve opening times (figure 5a) decrease the cooling properties and shorten retention times. At some stage, increased opening times do not affect the retention time any further. The bottleneck of the gas flow has moved completely into the capillary (i.e. the end part of the capillary is partially evacuated as in the case of a conventional 'direct' GC-MS coupling. The reported GC-Jet-REMPI-MS results (figure 3,4 and 5 as well as [1,2]) have been recorded with a classical jet spectroscopic arrangement, using a skimmed Jet with a differentially pumped two-chamber vacuum system and a distance between the jet-nozzle and the ionization region of about 7 cm. A detection limit for toluene of 200 fg (S/N=2) was achieved, although our REMPI-TOF spectrometer is not optimized for sensitivity. One important point in achieving such high sensitivities is the high jet-velocity selectivity in our instrument. The ions are extracted perpendicular to the jet-expansion axis. The ion drift path in the TOFMS is

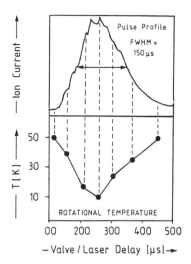

Figure 4
Measurement of the rotational temperature of benzene as a function of the delay time between the valve trigger and the laser pulse. A reduced cooling - especially at the beginning and the end of the jet-gas pulse - is observed due to relatively low pressure in the valve and gas shortage under flow conditions optimized for GC. However, at the gas pulse center sufficient cooling for analytical separation can still be achieved.

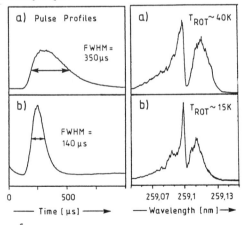

Figure 5
*The gas shortage in the GC valve causes a decrease of the measured rotational cooling if the valve opening time is enlarged: time profile of a jet gas pulse with long **a)** and short **b)** duration and the corresponding rotational benzene $\nu 6_0^1$ band rotational contours.*

121

adjusted by deflection plates so that only ions with the full jet velocity component (i.e. those that originate from the "cold" part of jet) can reach the detector. Ions originating from 'stationary' molecules (e.g. from source contaminations or background gas) are rejected, reducing the noise considerably. So far, maximum selectivity with the hyphenation of GC, Jet-REMPI and TOFMS could be demonstrated. No attention was paid, however, to the improvement of the sensitivity of the instrument. Beside a careful design of the ion optics, avoiding small apertures, meshes and foci, the main emphasis lies on increasing the number of ions per jet pulse. Improvements can be achieved in the time domain (jet pulse as short as possible) and in the space domain. The gas density as a function of the distance, d, from the nozzle is proportional to $1/d^2$, an experimental verification for $d = 25$ mm to 60 mm is shown in [5]. Consequently, the ionization region should be as close to the nozzle as possible. With the free-jet configuration shown in figure 2 an investigation of the properties of the free-jet expansion in the first millimeters was made (figure 6). It points out, that a distance of at least 4 mm (20x *orifice diameter*) should be chosen in order to avoid selectivity losses either due to incomplete cooling or ion-molecule reactions. An instrument optimized for selectivity as well as sensitivity is now under construction.

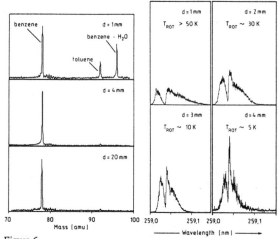

Figure 6
Properties of the free jet in the first millimeters of the expansion, a)mass spectra, showing ion-molecule reactions, b)rotational contour of benzene, used to estimate the rotational temperature

In conclusion, the presented GC-Jet-REMPI-MS technique is well suited for the analysis of target compounds in complex matrices. An interesting application would be a GC-Jet-REMPI-MS for on-line emission control of waste incinerators. This is part of our current field of investigation [6]

ACKNOWLEDGEMENT

E.R.R. wishes to thank the Alexander von Humboldt-Stiftung for a fellowship and his host, Prof E.W. Schlag, for the hospitality shown at the Institut für Physikalische und Theoretische Chemie.
This project is supported by a grant from the Deutsche Bundesstiftung Umwelt.

REFERENCES

[1] R.Zimmermann, C.Lermer, K.-W.Schramm, A.Kettrup, U.Boesl, Europ. Mass Spectrom. 1 (1995) 343.
[2] R.Zimmermann, H.J.Heger, A.Kettrup, K.-W.Schramm, E.R.Rohwer, E.K.Ortner, U.Boesl, Proceedings of the 18th Internat. Symp. on Capillary Chromatography, Rival del Garda May 1996, Hüthig-Verlag (1996) 61.
[3] T.Imasaka, *Spectrochim. Acta. Rev. 14*, (1991) 261
[4] G.Blease, R.J.Donovan, P.R.R.Langridge-Smith, T.Ridley, J.P.T.Wilkinson; Proceedings of the 5th Int. Symp. on Resonance Ionization Spectroscopy, *American Inst.of Physics Conf. Ser.* No.84 (Sect. 6) (1986) 217.
[5] H.Oser, R.Thanner, H.-H.Grotheer; Proceedings of the 8th Int. Symp. on Transport Phenomena in Combustion. SF (1995)
[6] R.Zimmermann, E.R.Rohwer, H.J.Heger, E.W.Schlag, G.Gilch, A.Kettrup, U.Boesl, this Proceedings (1996)

Resonance Ionization Laser Mass Spectrometry: New Possibilities for On-Line Analysis of Waste Incinerator Emissions

Ralf Zimmermann[A,B,C], Egmont R. Rohwer[A,D], Hans Jörg Heger[A], Edward. W. Schlag[A]
Antonius Kettrup[B,C], Gerhard Gilch[A], Dieter Lenoir[B], Ulrich Boesl[A]

[A] Institut für Physikalische und Theoretische Chemie, Techn. Univ. München, D-85747 Garching,
[B] Institut für Ökologische Chemie, GSF-Forschungszentrum für Umwelt und Gesundheit,
    D-85758 Oberschleißheim
[C] Lehrstuhl für Ökologische Chemie und Umweltanalytik, Techn. Univ. München, D-85354 Freising
[D] Department of Chemistry, University of Pretoria, 0002-Pretoria, Republic of South Africa

Abstract: A concept for the use of Resonance Ionization Laser Mass Spectrometry tor on-line emission analysis of chlorinated aromatic compounds in waste incinerator flue gas is presented. New analytical results suggest that low chlorinated benzenes can be used as indicator parameter for dioxin emissions.

The polychlorinated dibenzodioxins and -furans (PCDD/F or dioxins) are highly toxic trace by-products of combustion processes. Waste incinerators (WI) are notorious for their dioxin emissions. This is on the one hand due to the inhomogeneous furnace conditions (i.e. changing heat value of the feed) and on the other hand caused by the high content of organic material, chlorine and catalytically active metals (e.g. copper) in the feedstock. Although the PCDD/F formation mechanisms are still under discussion, there is a general agreement that dioxins most likely are formed on the surface of fly-ash particles in the flue gas at temperatures around 300°C. It is of great interest to get a deeper insight in the chemistry of dioxin formation, particularly in order to minimize the emission of these compounds. An important prerequisite for either PCDD/F formation studies or efficient emission control is a fast measurement technique. Conventional dioxin analysis is an extremely time consuming technique, giving results in a time scale of days. It has been shown previously that the resonance-enhanced multi-photon ionization technique in combination with a mass selective ion detector (time-of-flight mass spectrometer, MS) allows highly time resolved emission monitoring of automotive exhaust gases ([1], see fig. 2). A measurement rate up to 50 Hz with a detection limit in the low ppmv/high ppbv concentration range for aromatics has been achieved. In this contribution the applicability of REMPI-MS for on-line monitoring of chloroaromatics in the flue gas of waste incinerators is discussed. Figure 3B shows that rapid changes in the CO emission level are observed in industrial WI's, probably due to rapid variation in feed/combustion conditions. An aim of our present research is to correlate these events with changing dioxin emissions. A REMPI-MS based real-time monitor should be reliable and should require minimal operator intervention. Two-color (i.e. two-laser) REMPI-ionization, as recommended for selective detection of many higher

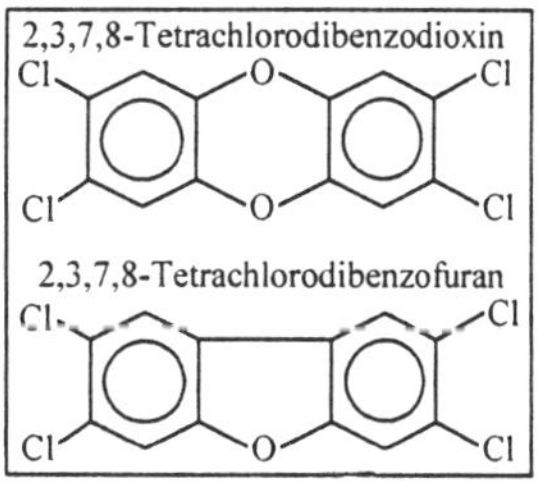

2,3,7,8-Tetrachlorodibenzodioxin

2,3,7,8-Tetrachlorodibenzofuran

Figure 1
Molecular structure of the most toxic dibenzodioxin and -furan congeners.

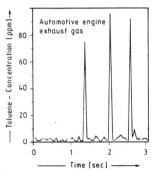

Figure 2
Time-resolved measurement of toluene in the exhaust gas of an internal combustion engine (taken from [1]). The observed spikes are due to emission of unburnt fuel after ignition failures

chlorinated compounds [2], is disadvantageous in this context due to its high adjustment effort. As one-color ionization is feasible for many low chlorinated aromatics, we

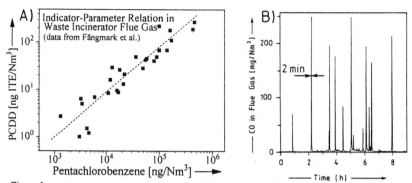

Figure 3
A) Indicator parameter relation between pentachlorobenzene and polychlorinated dibenzodioxins in waste incinerator flue gas (data from [3]). **B)** Concentration of CO, measured in the stack gas of a WI-facility as a function of the time. This measurement proves that combustion conditions vary on a time scale of minutes.

emphasize the use of the indicator parameter approach, based on the constant relation of different chloroaromatics in the flue gas. As e.g. shown in figure 3A, the concentration of dioxins can be estimated by measurement of pentachlorobenzene. However, on the one hand only the mono- and dichloro-benzenes are accessible in a one-color REMPI-step and on the other hand the indicator parameter relation is established for penta- and hexachlorobenzene at the incinerator under investigation [4]. We consequently started a campaign to measure all chlorobenzene congeners in the stack gas of the same incinerator. First results of these measurements are shown in figure 4. The chlorobenzenes were quantified by thermal desorption and conventional GC-MS (Perkin Elmer ATD 400 thermal desorber, Auto System GC and Qmass 910 Mass Spectrometer). All 12 congeners could be separated on the 60m x 0.32mm x 0.15µm DB 1701 column (J&W Scientific). A part of a chromatogram (single ion monitoring) is shown in figure 4A. The stack gas samples were collected for 10 min. at 100 cm³min⁻¹ on Tenax AT traps. Tests were conducted to ensure no breakthrough occurred for monochlorobenzene at the elevated accumulation temperature (84°C) required to prevent water condensation. A liquid sample (1µl) containing known amounts of each homologue was injected on an identical Tenax trap to perform external calibration. The GC-MS measurement (without trapping) takes more than half an hour. In figure 4B the found chlorobenzene concentrations of six successive measurements are shown, exhibiting a constant pattern of the chlorobenzenes. This pattern suggests that the indicator parameter approach may be extended to the mono- and dichlorobenzenes. Additionally the concentration of the latter compounds is about three to ten times higher than that of the penta and hexachlorobenzenes. This would allow a relatively simple setup of the on-line REMPI-MS instrument. The one-color REMPI detection limit for monochlorobenzene (600 pptv [5]) is in the same order of magnitude as the concentrations observed in the stack gas (see fig. 4). Figure 5 shows some one-color REMPI results on chlorinated aromatics obtained by direct sample inlet (fig. 5A) and by the GC-Jet-REMPI-MS device (fig .5B,C), described in detail in a second contribution to this proceedings [6] and in [7].

Two approaches are possible for on-line measurement by REMPI-MS (see fig. 6):
A) A fast sample concentration technique, similar to the case of our conventional chlorobenzene measurement, with subsequent thermodesorption and fast gas chromatography Jet-REMPI-MS [7] detection. This would be our approach for compounds that are present in extremely low concentrations and/or are spectroscopically difficult as for example the higher chlorinated benzenes or dibenzodioxins/-furans (two-color REMPI required). Due to the high selectivity/sensitivity of the REMPI-process [2] a

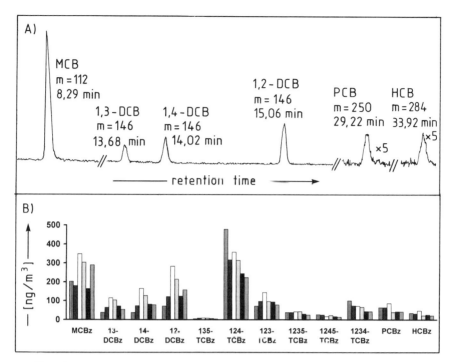

Figure 4
A) Part of the single ion monitoring (SIM) trace of the chlorobenzenes in the stack gas of a waste incinerator. We developed a special sample enrichment technique which allows direct quantification by thermodesorption GC-MS. **B)** Concentration pattern of all chlorobenzene congeners in the flue gas (six successive measurements), suggesting the applicability of the indicator parameter approach for low chlorinated benzenes.

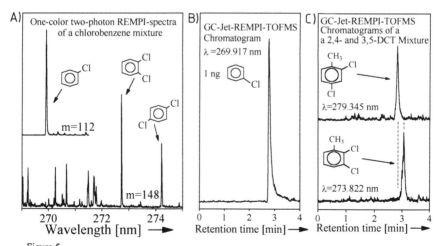

Figure 5
A) Mass resolved one-color REMPI-spectra of a mixture of chlorobenzenes, demonstrating the spectral selectivity. Higher chlorinated benzenes are not accessible in a one color process [2] **B)** GC-Jet-REMPI-MS chromatogram of 1 ng monochlorobenzene. **C)** Isomer selective ionization, demonstrated by GC-Jet-REMPI-MS chromatograms of a dichlorotoluene mixture.

cycle time in the one to five min. range is possible for detection of low chlorinated benzenes in the stack gas. B) A continuous monitoring set up, similar to the automotive exhaust gas measurement technique [1], but with a supersonic-jet sample inlet system, can be applied for detection of low chlorinated benzenes, phenols and some small polycylic aromatics (e.g. naphthalene) in the uncleaned flue gas. Here the higher concentration would allow the direct monitoring of these compounds, useful e.g. for active control of the subsequent flue gas scrubber. Our conventional GC-MS results indicate, that the combined jet- and mass spectroscopic selectivity is sufficient for direct REMPI-MS measurement of the chlorobenzenes in the complex flue gas matrix.

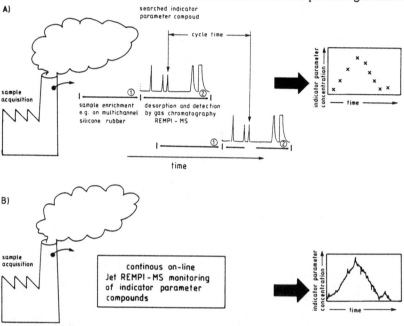

Figure 6
The Jet-REMPI-MS approach can be applied in two modes for on-line emission monitoring of exhaust gases: a) on-line monitoring with fast sample enrichment/thermodesorption and subsequent GC-Jet-REMPI-MS [7] detection b) real-time on-line monitoring directly by Jet-REMPI-MS

ACKNOWLEDGEMENT

E.R.R. wishes to thank the Alexander von Humboldt-Stiftung for a fellowship and his host, Prof E.W. Schlag, for the hospitality shown at the Institut für Physikalische und Theoretische Chemie.This project is supported by the Deutsche Bundesstiftung Umwelt, Osnabrück

REFERENCES

[1] U.Boesl, C.Weickhardt, R.Zimmermann, S.Schmidt, H.Nagel; SAE-Technical Paper Series 930083 (1993) 61. and C.Weickhardt, U.Boesl, E.W.Schlag; *Anal. Chem. 66* (1994) 1062.
[2] R.Zimmermann, C.Weickhardt, U.Boesl, D.Lenoir, K.-W.Schramm, A.Kettrup, E.W.Schlag; *Chemosphere* 29 (1994) 1877.
[3] I.Fängmark, B. van Bavel, S.Marklund, B.Strömberg, N.Berge, C.Rappe; *Environ. Sci. and Tech.* 27 (1993) 1602.
[4] A.Kaune, D.Lenoir, U.Nikolai, A.Kettrup; *Chemosphere 29* (1994) 2083.
[5] B.A.Williams, T.N.Tanada,T.A.Cool;*24th-Symp.(International) on Combustion,*The Combustion Institute (1992) 1587.
[6] R.Zimmermann, H.J.Heger, E.R.Rohwer, E.W.Schlag, A.Kettrup, U.Boesl, this Proceedings (1996)
[7] R.Zimmermann, C.Lermer, K.-W.Schramm, A.Kettrup, U.Boesl, Europ. Mass Spectrom. 1 (1995) 343.

Diode Laser Excited Optogalvanic Spectroscopy of Glow Discharges

C. M. Barshick, R. W. Shaw, L. W. Jennings, A. Post-Zwicker, J. P. Young, and J. M. Ramsey

Chemical & Analytical Sciences Division, Oak Ridge National Laboratory, P.O. Box 2008, Oak Ridge, TN 37831-6142

Abstract.   The development of diode-laser-excited isotopically-selective optogalvanic spectroscopy (OGS) of uranium metal, oxide and fluoride in a glow discharge (GD) is presented.   The technique is useful for determining $^{235}U/(^{235}U+^{238}U)$ isotope ratios in these samples. The precision and accuracy of this determination is evaluated, and a study of experimental parameters pertaining to optimization of the measurement is discussed. Application of GD-OGS to other f-transition elements is also described.

INTRODUCTION

The application of diode lasers to isotopically-selective atomic spectroscopy has been of interest to us for several years. Previously we carried out isotopically-selective resonance ionization mass spectrometry of lanthanum using diode lasers in the first step of a multistep resonant excitation.[1] At the last RIS meeting, our initial studies of the use of diode laser excitation for isotopically-selective optogalvanic spectroscopy (OGS) of uranium in a glow discharge (GD)[2] were presented. The study of uranium GD-OGS continues to be a main area of investigation. The technique of using diode lasers for GD-OGS studies has applications to other elements and for other purposes besides analytical determinations. Our work in these areas, particularly with lanthanides and uranium is the subject of this report.

EXPERIMENTAL AND RESULTS

The details of the experimental set-up for these studies have been described elsewhere.[2,3,4] Based on this work, a prototype, transportable instrument which should lead to a field determination of uranium isotope ratios using GD-OGS has been designed and fabricated.

Previously we reported the observation of GD-OGS for uranium or its oxide at wavelengths of 778.9 and 776.19 nm.[2,3] We have now carried out precision and accuracy studies for the measurement of $^{235}U/(^{235}U+^{238}U)$ isotope ratios in depleted and enriched uranium, uranium oxide, and uranium fluoride using diode laser excitation. An example of these results, at 776.19 nm, is shown in the calibration curve given in Figure 1. As can be seen, the data shows a good linear fit (R value of 0.99989 for a slope of 0.99363) from depleted to 20% enriched (in ^{235}U) uranium oxide. The precision for the smaller ratios is poorer than for the enriched sample. This is to be expected since the ^{235}U OGS signal is proportional to its concentration. In these studies, it is notable that uranium oxide, fluoride or metal can be analyzed equally well. Run-to-run, day-to-day, and sample-to-sample precision was evaluated using a 30 mW diode laser at 776.14 nm; the beam was

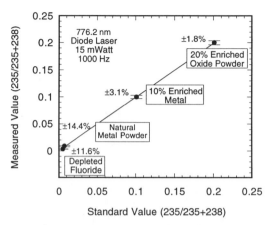

FIGURE 1. Measured $^{235}U/(^{235}U+^{238}U)$ ratio versus the reference value (certified value or thermal ionization mass spectrometric measurement).

focused to give 400 W/cm^2 in the cathode.[4] The sample was a NIST-200 uranium oxide, 20% ^{235}U. Run-to-run relative standard deviation (RSD) of the $^{235}U/(^{235}U+^{238}U)$ ratio at 1σ ranged from ±1.9% to 5.5% over 5 consecutive days. These data yielded a mean precision of ±2.6% RSD. Sample-to-sample precision, based on a number of analyses of three different sample cathodes was ±3.5% RSD. Determination of uranium isotope ratios using GD-OGS is therefore adequate for screening measurements. The sample is also available for a more precise isotopic ratio determination if that becomes necessary.

Although it is apparent from Figure 1 that a reasonable estimation of small $^{235}U/(^{235}U+^{238}U)$ isotope ratios can be made using this technique, it would be advantageous to improve the precision. The precision is ultimately based on the signal-to-noise ratio, and one way to improve this aspect would be to increase the signal. We therefore investigated several diode-laser assessable OGS transitions looking for an enhanced OGS signal; a reasonable isotope shift (I.S.) was also required. A promising uranium transition at 831.84 nm was found. An OGS spectrum, average of 5 scans, for depleted uranium metal powder is shown in Figure 2. The I.S. is 10.1 GHz, compared with -12.6 GHz for 776.19 nm; 150 mW tunable diode lasers in this wavelength region can readily be obtained. Using depleted uranium oxide, the measured $^{235}U/(^{235}U+^{238}U)$ ratio at 831.84 nm was 0.0026; the laser power was 84 mW. The thermal ionization mass analysis value was 0.0027. Run-to-run reproducibility of the depleted sample was ±7.8% (1σ)for 11 measurements.

There appears to be an art connected with generating useful glow discharges for OGS studies. Besides the obvious parameters of pressure, flow rate, and voltage, there are other parameters such as cathode composition, GD cell geometry, ubiquitous presence or generation of H_2O, etc., that influence the quality of the discharge and therefore the intensity of the OGS signal. We have studied some of these effects. They have been reported in detail[5] and are summarized here. Although metal samples generally make a satisfactory discharge, salts (insulators) require the addition of an electrical conductor to maintain a useful discharge. We originally made our uranium oxide cathodes by pressing a 50-50 weight % mixture of oxide and silver metal powder into a pellet. An ideal metal additive to a uranium

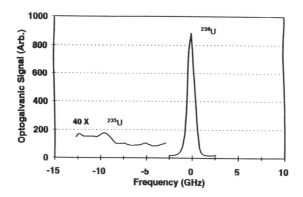

FIGURE 2. Uranium (0.27% ^{235}U) optogalvanic spectrum (average of five scans) recorded for the 831.84 nm line using a diode laser with a power density of ~2.1 x 10^3 W/cm^2.

oxide sample should also exhibit a strong bond with atomic oxygen and have a high sputtering rate in the discharge. We examined the influence of Ag, Ti, Ta, and mixtures thereof on the intensity of the OGS signal of uranium in an oxide sample at 831.84 nm. A 50-50 weight % Ag-Ta mixture was found to provide a useful compromise of sputter rate and stability of the metal-oxygen bond to compete with the stability of the U-O bond.

Many of our laboratory studies of the uranium GD-OGS technique have used a liquid nitrogen (LN) cooled coil placed in the vicinity of the cathode to improve the OGS signal. The cooled coil effectively removes condensable vapors from the region of the cathode. It is not convenient to have LN cooling in a fieldable apparatus, so an evaluation of a non-evaporatable Zr/V/Fe alloy getter (Model ST172/NP/HITS-L/7.5-7/150C, SAES Getters, Colorado Springs, CO) was carried out. Relative U signal improvements for uranium oxide as a function of various parameters are given in Table 1, the cathode in this study contained a mixture of oxide and Ag.

TABLE 1. Enhancement of ^{238}U OGS Signal at 831.84 nm as a Function of GD Operating Condition.

Condition	Relative OGS Signal	Discharge Current, mA
Discharge only	(1.00)	14.83
Zr/V/Fe alloy getter	1.9	14.79
Liquid-nitrogen cooling	5.2	14.65

Although the signal is strongest with liquid-nitrogen cooling, it is improved with the getter that is incorporated in the fieldable instrument. In other studies with a cathode that contained uranium oxide with the 50-50 weight % Ta and Ag, it was found that after discharge initiation the uranium OGS signal reached maximum intensity in about 30 minutes using the Zr/V/Fe getter compared to one hour without the getter.

The GD-OGS technique should be applicable to most elements and compounds provided that a laser excitation source is available. We have limited our studies to OGS atomic transitions that could be excited by diode lasers. For the f-transition elements, a number of transitions accessible to diode lasers are available. We have carried out some survey GD-OGS studies of Ce, Dy, and Gd. We are planning a study with Ce that can give information pertaining to GD characteristics. We have observed several OGS transitions for Ce; they are given in Table 2.

TABLE 2. Observed GD-OGS Transitions of Cerium

Wavelength, nm	Spectrum	Transition, cm^{-1}
791.352	I	6663 - 19296
792.772	I	4455 - 17066
802.556	II	0 - 12457
822.070	I	3210 - 15371
822.429	II	3704 - 15859
825.489	I	0 - 12114

There is a ground state (g.s.) transition for both Ce I and Ce II which we will utilize to probe the relative population density of these two oxidation states. There have been previous studies directed to characterizing glow discharges.[6] They generally involved profiling free atom density in the volume for the discharge; this was done by atomic absorption or emission. The ion population was measured, in general, by mass spectrometry. In the Ce example given above, we plan to study the ratio of atom to ion population as a function of discharge conditions. From this type of information it should be possible to create experimental conditions that favor ion or atom population. This approach samples only the g.s. population of both species.

ACKNOWLEDGEMENTS

Research sponsored by U.S. Department of Energy, Office of Research and Development. Oak Ridge National Laboratory is managed by Lockheed Martin Energy Research Corporation for the U.S. Department of Energy under contract DE-AC05-96OR22464.

REFERENCES

1.   Shaw, R. W.; Young, J. P.; Smith, D. H.; Bonanno, A. S.; and Dale, J. M.; *Phys. Rev. A*, **41**, 2566-2573 (1990).
2.   Young, J. P.; Barshick, C. M.; Shaw, R. W.; and Ramsey, J. M.; Amer. Inst. of Physics Conf. Proceedings 329 (Reson. Ioniz. Spectrosc. 1994), H.-J. Kluge, J. E. Parks, K. Wendt, eds., p. 111-115.
3.   Barshick, C. M.; Shaw, R. W.; Young, J. P.; and Ramsey, J. M.; *Anal. Chem.*, 66, 4154-4158 (1994).
4.   Barshick, C. M.; Shaw, R. W.; Young, J. P.; and Ramsey, J. M.; ibid. 67, 3814-3818 (1995).
5.   Shaw, R. W.; Barshick, C. M.; Jennings, L. W.; Young, J. P.; and Ramsey, J. M.; *Rapid Comm. in Mass Spectrometry*, 10, 316-320 (1996).
6.   Smith, R. L.; Sexner, D.; Hess, K. R.; *Anal. Chem.*, 61, 1103-1108 (1989).

SESSION V: SURFACE ANALYTICAL APPLICATIONS

Challenges for Materials and Device Characterization in the Semiconductor Industry

T. J. Shaffner

Materials Science Laboratory, Texas Instruments, Incorporated
Box 655936, MS-147, Dallas, TX 75265

Abstract. Materials and device characterization serves the essential role of defining how a manufactured integrated circuit differs from its intended design and function. Over the years, a variety of techniques based on probes of electrons, ions and X-rays have evolved to fill this need. Each has a specialized application for resolving specific manufacturing problems related to smaller geometry, material impurities and silicon crystal defects. This paper illustrates how these are addressing strategic needs of the semiconductor industry, and outlines some of the challenges characterization specialists face.

INTRODUCTION

The acclaimed National Technology Roadmap for Semiconductors, known as the NTRS (1), affirms the customary semilogarithmic projection of gate length reduction with each new dynamic random access memory (DRAM) technology node. Downsizing is the most frequently cited driving force for smaller analytical probes, but it is evident from Figure 1 that similar trends with contamination, metal impurities and point defects are expected to be forces of equal, if not greater magnitude in the near future. Others agree, citing that microcontamination and particle control tools will drive the infrastructure of the industry (2,3).

In this review, we illustrate how key characterization techniques are evolving in response to requirements for smaller geometry and reduced levels of contamination and defects. Techniques of resonant ionization spectroscopy covered in RIS-96 are in a timely situation to address such problems, which may involve no more than a small cluster of impurity atoms unfavorably positioned near a gate oxide or critical junction. For more description of these and other characterization issues, the reader is also referred to a sampling of fine books and articles (4-6).

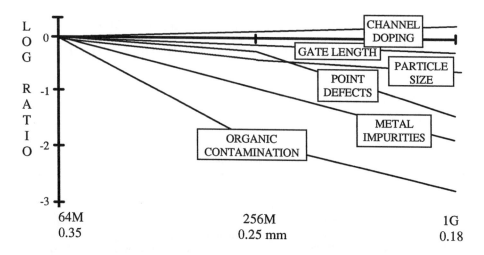

FIGURE 1. The National Technology Roadmap for Semiconductors defines challenges for materials and device characterization (1). The vertical axis is the ratio of each metric (identified by a box) relative to today's performance. Downward slopes denote reduction in the metric as a function of technology node.

SHRINKING GEOMETRY

During the past two decades, the scanning electron microscope (SEM) displaced the optical microscope as the tool of choice for critical dimension measurements and circuit failure analysis. However, we know today that even the SEM is unable to provide quantification of linewidths near 0.5 µm, when 10% or better tolerance is routinely required. As with optical microscopy, the procedure relies on techniques of matching an ideal step function to a fuzzy intensity profile across the edge. The ambiguity arises from the diffraction limit in optical microscopy and electron scattering in the SEM, and only marginal improvements with these appear likely in the near future. A similar difficulty in defining dimensions of ultra-shallow junctions less than 100 nm deep is also receiving increased attention (7).

Scanning probe microscope (SPM) techniques are being extensively tested for metrology applications, but offer different challenges for achieving robust piezo-manipulators with reproducible linearity, and for deconvolving the effects of shape and atomic scale protrusions on each individual tip (8,9). We can anticipate rapid advances in probe methodologies, but today, engineers still rely mostly on transmission electron microscopy (TEM) as an absolute reference for nanometer quantification.

The difficulty in preparing samples for TEM is well known, particularly for dense integrated circuits where a specific site failure has occurred. It was believed for years that preparing a thinned section through a single faulty bit in a megabit DRAM was impractical, if not impossible to achieve. Early progress with this problem was reported in 1989 by Benedict et al. (10), who developed an ingenious tripod tool that permits precise control of repetitive polishing and inspection. More recently, focused ion beam (FIB) columns have been developed capable of carving precise free standing films as thin as 100 nm at any orientation and position within the volume of an integrated circuit.

IMPURITIES AND DEFECTS

Although electron microscopes provide insight into nanostructural problems, they remain unable to identify chemical species at the single atom level. Scanning probe techniques are at the threshold of achieving this, but current demonstrations based on single molecule imaging (11), photoluminescence (12), and Raman spectroscopies need to mature considerably before routine applications are possible. Today, secondary ion mass spectrometry (SIMS) still offers the best compromise between small spot analysis and elemental sensitivity for broad semiconductor applications.

Modern magnetic sector and quadrupole SIMS instruments achieve 0.8 µm spatial resolution with sub-parts-per-million-atomic (*ppma*) sensitivity for most elements. Sputter liberated molecular species which have a similar charge-to-mass ratio, like ^{30}SiH and ^{31}P, or ^{56}Fe and $^{28}Si_2$, can be resolved in a high mass resolution mode, but only with significant sacrifice in sensitivity. New techniques are evolving to address this molecule interference limitation, including many of the laser assisted SIMS approaches covered in this conference.

Thin film compositional analysis without standards is a recognized strength of Rutherford backscattering, but perhaps less appreciated, is the exceptional sensitivity to monolayer impurities and surface defects (13). For elements from *Cl* to *Pb* on *Si*, a monolayer or less results in a peak with excellent signal-to-noise that is isolated from matrix backscattering. Typically, a *He* projectile is applied at 2-3 MeV, but further improvement in cross section and elemental sensitivity is possible using a heavier ion accelerated at lower energy. This is incentive for the heavy ion backscattering spectroscopy technique (HIBS) which is under development primarily at Sandia National Laboratory (14), where 400 keV ^{12}C is used for a 1,000X enhancement. Others are using 200 keV *He* to improve depth resolution (~1 nm) in quantitative profiles of ultra-thin oxynitride capacitor dielectric films. These techniques promise to extend applicability to impurities and stoichiometry of ultra-thin films beyond capabilities presently realized by total reflection X-ray fluorescence (TXRF) instruments, which are in common use in most wafer fabs today.

Microdefects in present-day silicon are sparsely distributed and extremely small, typically being several to hundreds of nanometers across. It is difficult to even locate these by conventional microprobe techniques, and weak signals from such small points add to the analysis dilemma. Methods of X-ray diffraction and topography circumvent these problems, because they sense minute strain fields that extend far beyond the defect center, and also simultaneously detect hundreds or thousands that are distributed throughout a large sampling volume. Techniques in common use in the semiconductor industry include Berg-Barrett, double and triple crystal topographies, and low angle X-ray reflectometry.

SUMMARY

The NTRS stresses the importance of point defects, metal impurities and organic contamination in future manufacturing. This is not to diminish the significance of single particles, but if in fact manufacturing needs evolve as outlined in Figure 1, characterization specialists need to commit equal if not more resources to the development and application of such characterization tools. The Sematech consortium centered in Austin, Texas has responded to this challenge with an analytical equipment road map that defines how resources might be best allocated in addressing the buildup of such an infrastructure (15).

Road maps exemplify our *evolutionary* approach to characterization tools, because it is difficult to forecast with any accuracy which new science and techniques will surprise us in the near future. As much as we would like to deliberately invent new ways to achieve ULSI characterization goals, they are more likely to spring from the combination or revitalization of older technologies, sometimes in unrelated fields. This random appearance of new techniques, as illustrated in Figure 2, is why cutting edge characterization organizations seek to function as a collection center for new ideas, inventions and measurement science and technology. The buildup of such infrastructure, and the merging of disciplines provides fertile ground for this *revolutionary* aspect of technique advancement.

One typically predicts where we will be five or ten years from now, based on extrapolation of known technology and historic trends. We might only guess which evolutionary and revolutionary changes to expect, but it is known that characterization technology provides a technical foundation for the fabrication tools of tomorrow. In our daily activities, we must continue to weigh the strengths and weaknesses of characterization techniques not only for the problem at hand, but also with an eye toward the future engineering and scientific foundations of the semiconductor industry.

REFERENCES

1. The National Technology Roadmap for Semiconductors, The Semiconductor Industry Association, San Jose, CA, 1994.
2. Larrabee, G. and Chatterjee, P., "DRAM Manufacturing in the '90s - Part 2: The Road Map," Semiconductor International, May, 1991, p. 90.
3. McDonald, R., "How will You Examine ICs in the Year 2000?", Semiconductor International, January, 1994, p. 46.
4. Semiconductor Characterization, Present Status and Future Needs (W.M. Bullis, D.G. Seiler, A.C. Diebold, eds.), American Institute of Physics, New York (1996).
5. Encyclopedia of Materials Characterization: Surfaces, Interfaces, Thin Films, Butterworth-Heinemann, Boston, (C.R. Brundle, C.A. Evans, Jr., and S. Wilson, eds.) 1992.
6. Shaffner, T.J., Diebold, A.C., McDonald, R.C., Seiler, D.G., and Bullis, W.M., "Business and Manufacturing Motivations for the Development of Analytical Technology and Metrology for Semiconductors," in Proceedings of the International Workshop on Semiconductor Characterization (D.G. Seiler, ed.), American Institute of Physics, New York, 1996, p. 1.
7. Subrahmanyan, R., and Duane, M., "Issues in Two-Dimensional Dopant Profiling," Diagnostic Techniques for Semiconductor Materials and Devices (D.K. Schroder, J.L. Benton, and P. Rai-Choudhury, eds.) vol. 94-33, The Electrochemical Society, Pennington, NJ, 1994, p. 65.
8. Marchman, H., "Critical Dimensional Measurements: Test Structures, Metrology Instrument Correlations, and Calibration Techniques," NIST Workshop on Industrial Applications of Scanned Probe Microscopy, Gaithersburg, MD, 1994, p. 35.
9. Burggraaf, P., "Thin Film Metrology: Headed for a New Plateau," Semiconductor International, March, 1994, p. 56.
10. Benedict, J.P., Klepeis, S.J., Vandygrift, W.G., and R. Anderson, R., "A Method for Precision Specimen Preparation for Both SEM and TEM Analysis," EMSA Bulletin, 19(2), 1989, p. 74.
11. Betzig, E., and Chichester, R.J., "Single Molecules Observed by Near-Field Scanning Optical Microscopy," Science, 262, 1993, p. 422.
12. Hess, H.F., Betzig, E., Harris, T.D., Pfeiffer, L.N., and West, K.W., "Near-Field Spectroscopy of the Quantum Constituents of a Luminescence System," Science, 264, 1994, p. 1740.
13. Keenan, J.A., "Backscattering Spectroscopy for Semiconductor Materials," Diagnostic Techniques for Semiconductor Materials and Devices (T.J. Shaffner and D.K. Schroder, eds.), vol. 88-20, The Electrochemical Society, Pennington, NJ, 1988, p. 15.
14. Knapp, J.A., and Banks, J.C., "Heavy Ion Backscattering Spectrometry for High Sensitivity," Nuc. Instrum. Meth., vol. B79, 1993, p. 457.
15. Diebold, A.C., "Metrology Roadmap: A Supplement to the National Technology Roadmap for Semiconductors," SEMATECH Technology Transfer Document #94102578 A-TR, 1994.

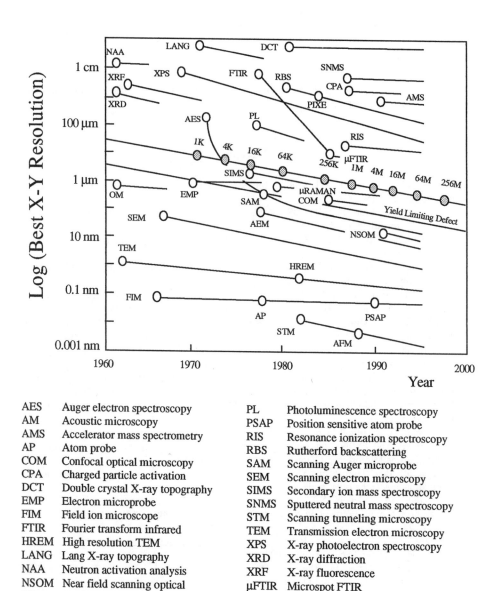

AES	Auger electron spectroscopy	
AM	Acoustic microscopy	
AMS	Accelerator mass spectrometry	
AP	Atom probe	
COM	Confocal optical microscopy	
CPA	Charged particle activation	
DCT	Double crystal X-ray topography	
EMP	Electron microprobe	
FIM	Field ion microscope	
FTIR	Fourier transform infrared	
HREM	High resolution TEM	
LANG	Lang X-ray topography	
NAA	Neutron activation analysis	
NSOM	Near field scanning optical	
OM	Optical microscopy	
PIXE	Particle induced X-ray emission	

PL	Photoluminescence spectroscopy
PSAP	Position sensitive atom probe
RIS	Resonance ionization spectroscopy
RBS	Rutherford backscattering
SAM	Scanning Auger microprobe
SEM	Scanning electron microscopy
SIMS	Secondary ion mass spectroscopy
SNMS	Sputtered neutral mass spectroscopy
STM	Scanning tunneling microscopy
TEM	Transmission electron microscopy
XPS	X-ray photoelectron spectroscopy
XRD	X-ray diffraction
XRF	X-ray fluorescence
μFTIR	Microspot FTIR
μRAMAN	Raman microprobe

FIGURE 2. History of Analytical Techniques for Semiconductor Applications. The revitalization of existing techniques and the development of new methods follow an erratic history relative to the steady decline of minimum DRAM geometry (1K through 256M). Lines illustrate evolutionary changes, while circles show the introduction of revolutionary new techniques for semiconductor applications.

FISH & CHIPS: ANALYTICAL APPLICATIONS OF RESONANCE IONIZATION MASS SPECTROMETRY

H.F. Arlinghaus, X.Q. Guo, T.J. Whitaker, and M.N. Kwoka

Atom Sciences Inc., Oak Ridge, TN 37830, USA

Abstract. Resonance ionization mass spectrometry is becoming recognized as an analytical technique for a wide range of applications. The extremely high element specificity and sensitivity of the resonance ionization (RI) process is especially valuable for ultratrace element analysis in samples where the complexity of the matrix is frequently a serious source of interferences. In this paper, we will describe the implementation of sputter-initiated resonance ionization microprobe (SIRIMP) and laser atomization RIMP (LARIMP) to solve a number of analytical problems and illustrate the technique's salient characteristics with applications ranging from environmental monitoring using fish scales to semiconductor device and DNA diagnostics chips.

INTRODUCTION

Resonance laser postionization mass spectrometry is becoming recognized as an analytical technique with desirable advantages for a wide range of applications. The extremely high element specificity and sensitivity of the resonance ionization (RI) process is especially valuable for ultratrace element analysis in samples where the complexity of the matrix is a serious source of interferences. RI, when combined with a focused ion sputter beam and mass spectrometric detection of ions, is called sputter-initiated resonance ionization microprobe (SIRIMP) and provides an exceptionally efficient analytical technique with the ability to obtain quantitative ultratrace element concentration images with high spatial resolution and virtually no matrix effects. Lower detection limits in the same analysis time can be achieved by replacing the ion beam with a laser beam. This technique is called laser atomization RIMP (LARIMP). We will describe the implementation of SIRIMP/LARIMP and illustrate the techniques' salient characteristics with data of relevance to semiconductor research, geochronology, cancer research and DNA diagnostics.

EXPERIMENT

In SIRIMP/LARIMP experiments, the sample is bombarded with a finely focused pulsed primary ion beam (O_2^+ or Ar^+) or irradiated with an atomization laser beam (248 nm or 193 nm), respectively. The expanding cloud of sputtered material consists of neutral atoms, molecular fragments, and ions; the ions are removed by pulsed electric fields and electrostatic energy analysis. The remaining neutral particles are then probed by the RI laser beams which ionize all the atoms of the selected element within the volume intersected by the laser beams. The resulting ions are analyzed by a time-of-flight (TOF) mass spectrometer. With careful spatial and temporal overlap of the laser beams with the cloud of desorbed material, useful yields (atoms detected/atoms sputtered) between 3% to 8% are obtained, giving sub-parts-per billion detection limits (1-3).

Imaging is achieved by either scanning the ion/laser beam over the sample or by translating the sample under a fixed ion/laser beam. If a liquid ion gun is used for sputtering, nanoanalysis down to 50 nm can be achieved. Charge compensation for insulator analyses is possible using pulsed low energy electrons, which are introduced during the time interval between ion/laser pulses. SIRIMP depth profiles can be obtained by scanning the sample with a continuous ion beam and taking data with a pulsed ion beam in the center of the crater after each raster frame (for more details see (1,4,5)).

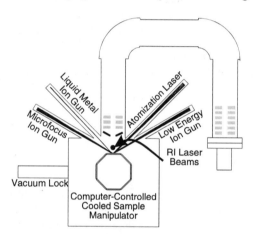

Figure 1. Experimental arrangement.

RESULTS AND DISCUSSION

SIRIMP/LARIMP can be utilized in a broad range of applications that include measuring trace elements in fish scale or whale teeth for environmental monitoring, quantifying ultratrace element concentrations in biological and semiconductor samples, imaging trace element distributions in semiconductor chips, determining isotope ratios for geochronology in geological samples, as well as DNA sequencing and diagnostics using genosensor chips. The following figures depict typical data from some of the above mentioned application fields.

140

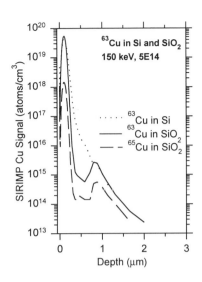

Figure 2. Depth profile of ^{63}Cu in Si and both ^{63}Cu and ^{65}Cu in SiO$_2$.

Figure 2 shows Cu depth profiles obtained from silicon and 0.8 μm silicon oxide on silicon samples. For this experiment, half a silicon wafer and half a silicon oxide wafer were simultaneously implanted with 5x10^{14} ^{63}Cu atoms/cm^2 at 150 keV. This sample was made to: (a) determine the SIRIMP Cu sensitivity factor and (b) investigate the degree of quantitation and matrix independency of the SIRIMP technique. For analysis, both samples were cooled to 107K to reduce migration effects, and depth profiles were taken under the same experimental conditions. The integrated Cu signal for both samples was almost the same (less than a 3% difference which is probably due to the differences in sputter yield), demonstrating the matrix independence and the quantitation accuracy possible with SIRIMP. At the SiO$_2$/Si interface, a higher Cu concentration was observed. The isotopic ratio at this pile-up, near the interface and below, is closer to that of natural copper, which indicates that the sample was contaminated with Cu when the oxide was grown. The Cu data clearly show that laser resonance postionization is a very quantitative and matrix independent technique.

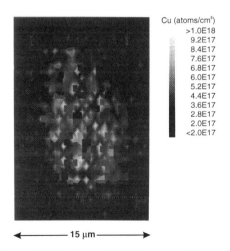

Figure 3. Submicron SIRIMP image of Cu concentration around a Cd inclusion in an CdZnTe film. A concentration of 4x10^{17} atoms/cm^3 correspond to about 10 ppm.

Figure 3 shows a submicron SIRIMP image of copper concentration around a Cd inclusion in an CdZnTe (CZT) film. This image was acquired using a high-resolution liquid metal ion gun. It is clear from the image that the Cu concentration is higher in the Cd inclusion than in the surrounding CZT matrix. These data support the theory that the Cu preferentially migrates to Cd second phase regions inside the CZT matrix. The uneven distribution of the Cu concentration in the Cd precipitate may be due to the amorphous structure of the Cd which could cause accumulation of Cu between the grain boundary.

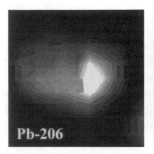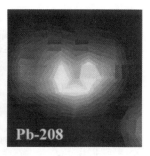

Figure 4. ^{206}Pb, ^{207}Pb, and ^{208}Pb images of a zircon grain. The data were averaged over 100 shots per point and were taken with 5 μm step size over a 40x40 μm area.

The SIRIMP technique can also be applied to measure isotopic ratios in geological samples with high precision and accuracy. Figure 4 shows ^{206}Pb, ^{207}Pb, and ^{208}Pb images of an extraterrestrial zircon grain. The resonant ionization process allowed us to measure isotopic lead concentrations with high useful yields (approximately 3% for Pb) and without molecular interferences. The ^{207}Pb/^{206}Pb ratio averaged 0.098 which was within two standard deviations of the expected value of 0.0878. Note that there appear to be "hot spots" for ^{208}Pb, one coinciding with the peak of ^{206}Pb concentration and the other just to the left of that peak. Both ^{206}Pb and ^{207}Pb were formed in these zircon samples via decay of uranium (^{238}U and ^{235}U respectively) but there is essentially no path for forming ^{208}Pb via any decay pathway from U isotopes. However, ^{208}Pb originating from thorium can exist in some zircons. We used NIST reference standards to demonstrate the precision and accuracy of SIRIMP to be 0.17% and 0.24% respectively, limited mainly by counting statistics. Odd/even isotope effects were not observed for Pb.

Cancer research can also benefit from RIMP analysis. Boron neutron capture therapy (BNCT) is a promising treatment for Glioblastoma Multiforme and other tumors involving the brain. In this therapy, boron, which has been prepositioned by biochemical methods in the tumor, captures a thermal neutron from a beam of thermal or epithermal neutrons and undergoes the ^{10}B(n,α)^{7}Li reaction, producing energetic alphas and Li ions. Since the range of the fission fragments is relatively small (a few μm), the clinical efficacy of the therapy is only guaranteed if the boron is positioned in a sensitive area of the tumor cell, such as the cell nucleus. We have used SIRIMP and LARIMP to determine boron concentrations in various thin tissue sections obtained from $Na_4B_{24}H_{22}S_2$ (BSSB) infused rats (6). Figure 5 shows two orthogonal position-concentration LARIMP boron measurements through a section of a rat brain. Higher boron concentrations are observed at the intersection of the two position scans. This is the choroid plexus region, which produces cerebral spinal fluid (CSF) by an active "ultra-filtration" process from blood. The CSF contains very little protein and no cells. In some sense, this is similar to the kidney (but at a much lower level), where the boron is concentrated in the region in which an "ultra-

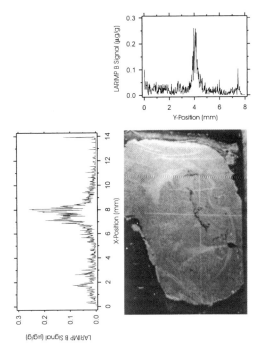

Figure 5. LARIMP boron scans in rat brain.

filtration" process is occurring. Note that the boron peak is much broader in the x-scan than in the y-scan, which correlates with the fact that the beam traverses a much longer region of the choroid plexus in the x-scan than in the y-scan. The minor peaks at x = ~2 correspond to the location of the external capsule (a band of axons) that separates the cortex from the striatum. This is also a watershed zone for two sources of the vascular supply for this area. The somewhat higher boron concentration may be explained by the fact that this is a fine capillary bed, similar to that found in the kidney filtration zone and in the choroid plexus.

SIRIMP can be also applied to detect Sn-labeled DNA at positively hybridized and unhybridized sites on a Form I and Form II DNA genosensor chip which can be used for DNA sequencing and genetic diagnostics. Form I sequencing by hybridization (SBH) is a method where the large genomic DNA is attached to the solid surface, such as a nylon membrane, and the enriched isotope-labeled oligonucleotides of known sequences are allowed to hybridize (7). Form II SBH involves binding a small (typically 4-20-mer) oligonucleotide to a chip (7). By the process of polymerase chain reaction, fragments of the genomic DNA are produced with an attached label. The size of the fragment can vary from a few dozen to several hundred or several thousand nucleotides.

Figure 6 shows a SIRIMP image of ^{118}Sn obtained from a SBH Form II matrix. For this experiment two different 17-mer oligonucleotides were bound to platinum circles on a silicon wafer at a total of six locations (three locations each). Enriched ^{118}Sn-labeled DNA, which was completely complimentary to one of these oligonucleotides, and noncomplimentary to the other, was then hybridized to the chip. After the hybridization, the sample was washed to remove unhybridized probes. The image shows three peaks at the complementary DNA sites. The discrimination between hybridized and unhybridized sites is better than 100. A single point on the top center spot was analyzed twice before the full image was taken, leading in a slight reduction of the signal. Similar data have been obtained for a Form I experiment on nylon.

The data demonstrate that analytical requirements in a number of fields are beginning to exceed conventional capabilities. As this trend continues, the use of SIRIMP/LARIMP will become a necessary tool for a number of applications.

ACKNOWLEDGMENTS

This work was partly supported by NIH, ARPA, and NSF. We thank D. Simons; NIST, S. Sen; Santa Barbara Research Center, N. Shimizu; Woods Hole Oceanographic Institution; G.W. Kabalka, University of Tennessee; R.C. Switzer; Neuroscience, Inc., and K. Beattie; Houston Advanced Research Center, for supplying samples.

REFERENCES

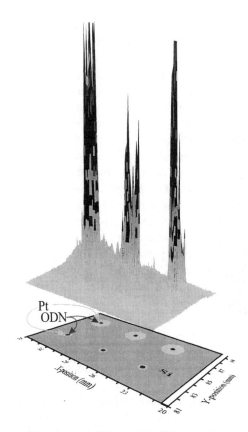

Figure 6. SIRIMP ^{118}Sn image of Sn-labeled DNA hybridized to 6 oligonucleotide spots; 3 complimentary and 3 noncomplimentary.

1. Arlinghaus, H. F., and Joyner, C. F., *J. Vac. Sci. Technol.* **B14**, 294 (1996).
2. Pappas, D. L., Hrubowchak, D. M., Ervin, M. H., and Winograd, N., *Science* **243**, 64 (1989).
3. Pellin, M. J., Young, C. E., Calaway, W. F., Whitten, J. E., Gruen, D. M., Blum, J. B., Hutcheon, I. D., and Wasserburg, G. J., *Phil. Trans. R. Soc. Lond.* **A 333**, 133 (1990).
4. Arlinghaus, H. F., Whitaker, T. J., Joyner, C. F., Kwoka, P., Jacobson, K. B., and Tower, J., in *Proceedings of SIMS X*, in press.
5. Arlinghaus, H. F. and Thonnard, N., in *Laser Ablation Mechanisms and Applications*, eds. J.C. Miller and R.F. Haglund, Jr., Springer, 165 (1991).
6. Arlinghaus, H. F., in *INEL BNCT Research Program Annual Report 1992*, ed. J.R. Venhuizen, EGG-BNCT-2700, p.125-141.
7. Arlinghaus, H. F., Kwoka, M. N., Jacobson, K. B., and Guo, X. Q., to be submitted.

Metastable Excitation of Sputtered Silver Atoms

A. Wucher and W. Berthold

Fachbereich Physik, Universität Kaiserslautern, 67653 Kaiserslautern, FRG

Neutral atoms sputtered from a polycrystalline silver surface under bombardment with 15 keV Ar$^+$ ions were investigated by resonant multiphoton ionization in combination with time-of-flight mass spectrometry. Electronically excited Ag atoms ejected in the metastable $4d^9 5s^2 \, ^2D_{5/2}$ state (excitation energy 3.75 eV) were ionized by a frequency doubled tunable dye laser using a resonant single photon transition into the autoionizing $4d^9 5s(^3D)5p \, ^2D_{5/2}$ state. Ground state silver atoms were detected by means of a resonant two photon ionization scheme involving the intermediate $4d^{10}(^1S)5p \, ^2P^o$ state. Both two and one color schemes were employed, the latter working via resonant excitation of Rydberg states and subsequent field ionization. The *total population* as well as the *velocity distribution* of sputtered metastable and ground state atoms were determined from the saturation behavior of the photoionization process and the flight time of the ejected neutral species between the ion bombarded surface and the ionization volume.

INTRODUCTION

If an energetic ion hits a solid surface, particles may be released into the gas phase by atomic collisions, a process which is usually called "*sputtering*". It has been long known that the flux of sputtered particles contains not only ground state atoms but also a (usually small) fraction of particles (atoms or molecules) which are released in electronically excited states (1). With respect to experimental detection schemes, these can roughly be classified into two categories. *Short-lived* states, on one hand, can easily be detected by the light emission due to radiative deexcitation closely above the surface. The interpretation of the data, however, is always complicated by the complex interplay between radiative deexcitation lifetimes and emission velocities as well as by the important role of cascading transitions from higher lying states. *Metastable* states, on the other hand, are less influenced by transient effects, and are therefore well suited to study the physical mechanisms leading to the excitation of atoms during the atomic collision cascade leading to their sputter ejection. They must, however, in general be detected by laser spectroscopic methods. Corresponding studies have been performed using either laser induced fluorescence (LIF) or resonance ionization mass spectroscopy (RIMS) to investigate the population of metastable excited Fe (2 ,3), Zr (4), Ti (5), U (6) or Rh (7) and Ni (8) atoms sputtered from the respective clean or oxidized metal surfaces. Typical quantities which are adressable by such experiments are the *population* of different excited states and the dependence of the excitation probability on the *kinetic emission energy* of the sputtered particles.

We have recently detected sputtered Ag atoms which are emitted in the first metastable $4d^9 5s^2$ $^2D_{5/2}$ state of silver with an excitation energy of 3.75 eV (9 ,10). To our knowledge, this is the highest lying metastable state of sputtered particles detected so far. In this paper, we will focus on the methods which were employed to identify Ag atoms in this state as well as in the electronic ground state among the total flux of sputtered particles. Results will be shown concerning the total fraction as well as the kinetic energy distributions of metastable and ground state atoms, respectively.

EXPERIMENTAL

The experimental setup used for mass and state selective detection of sputtered neutral atoms has been described in detail elsewhere (11). In short, particles which are sputtered from a polycrystalline silver sample under bombardment with 15 keV Ar^+ ions from a pulsed plasmatron ion gun are postionized by either one or two laser beams directed closely above and parallel to the sample surface. State selective postionization was achieved using either one or two tunable dye lasers employing the resonant photoionization schemes depicted in fig. 1. The ground state $[4d^{10}5s\ (^2S_{1/2})]$ atoms were detected by a resonance enhanced two photon ionization (R2PI) process involving a resonant transition to the intermediate $4d^{10}5p$ $(^2P_{3/2})$ state by an excitation laser (laser 1) tuned to a wavelength of 328.16 nm. Ionization from this state was achieved by two possible ways: First, a second, frequency doubled ionization laser (laser 2) operated around $\lambda \sim 272.2$ nm was employed to excite a non resonant transition to the ionization continuum. This two color scheme was used to determine the total population of the ground state within the sputtered atoms. Second, another resonant transition to a Rydberg state $[4d^{10}28d\ (^2D_J)]$ was induced by absorption of a second photon from the excitation laser. In this one color scheme, ionization was then achieved by the

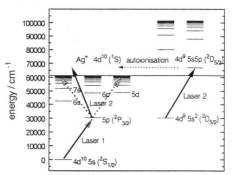

electric field ($\sim$ 215 V/cm) used to **Figure 1.** Relevant energy levels of the extract the photoions into the reflectron silver atom and ionization schemes type time-of-flight (ToF) mass spectrometer. Electronically excited atoms ejected in the first metastable $4d^9 5s^2$ $(^2D_{5/2})$ state of silver were detected by means of a resonant single photon transition to the autoionizing $4d^9 5s5p$ $(^2D_{5/2})$ state (9,12). For this purpose, the excitation laser was blocked and the ionization laser was tuned to a wavelength of 272.27 nm.

For total population measurements, the two laser beams were crossed at an angle of 90° and positioned at a distance of 0.5 mm above the surface. A relatively long primary ion pulse duration of $\tau = 5$ μs was chosen which ensured that the measured signals did not increase with increasing τ, thus indicating that particles of all emission velocities are present in the ionization volume. The ionization laser was focused to a diameter of about 50 μm, while the excitation laser was strongly attenuated and defocused to a diameter of about 1 mm. Therefore, the ionization volume is determined by the spatial overlap of the ionization laser with the sensitive volume of the mass spectrometer (~ 1 mm^3) and, in particular, does not depend on the excitation laser. This is important in order to ensure equal sizes of the effective ionization volume for both metastable and ground state atoms, since the detection of metastables involves the ionization laser alone. For the determination of the kinetic energy distributions, the primary ion pulse width was reduced to 100 ns and the lasers were backed off to a distance of 1.5 mm from the surface. In addition, both laser beams were now collinearly coupled into the chamber and shaped to identical spot sizes (50 μm). This was done in order to ensure that the same procedures could be followed during the somewhat critical alignment of the laser beam with respect to the ToF spectrometer and the ion bombarded spot on the sample surface. The emission velocity of the detected particles was selected via their flight time between the surface and the ionization volume. This was accomplished by introducing a variable time delay t between the primary ion and ionizing laser pulses. The velocity spectrum of sputtered neutral atoms ejected in a specific electronic state was then determined by unblocking the respective laser beam and following the photoion signal (i. e. the integrated ^{107}Ag$^+$ ion peak in the mass spectrum) as a function of t.

RESULTS AND DISCUSSION

Wavelength Spectra

Fig. 2 shows the spectrum of the photoion signal as a function of the wavelength of the ionization laser. During acquisition of this spectrum the excitation laser was blocked. Three distinct resonances are visible, each of which exhibits a different spectral width. Both the energetic positions and the widths of the lines correspond to three resonant single photon transitions from the metastable $4d^95s^2$ $^2D_{5/2}$ state into the autoionizing $4d^95s(^3D)5p$ $^2D_{5/2}$, $^2P_{3/2}$ and $^2F_{7/2}$ states which have been observed previously

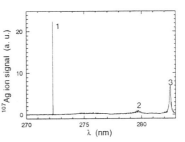

Figure 2. Ag$^+$ signal vs. wavelength of ionization laser

(12). Since no electronic states of Ag exist at or even near the excitation energies corresponding to the observed lines, it is evident that the peaks in fig. 2 cannot be due to the ionization of ground state silver atoms, but must originate from the resonant ionization of the metastable excited state. In the following, we take the transition into the $^2D_{5/2}$ state (line 1 in fig. 2) at a wavelength of 272.27 nm as representative for the metastable state.

Fig. 3 shows the spectrum which is generated if the ionization laser is detuned from the autoionizing resonance and the excitation laser is unblocked and scanned. The resonance at 328.16 nm corresponding to the transition from the ground state to the $4d^{10}(1S)5p$ $^2P^o$ state is clearly visible. Due to the relatively high power density of the excitation laser, this line is significantly broadened due to saturation. Since at the corresponding energy of 30472.71 cm^{-1} no autoionizing states are reachable from the

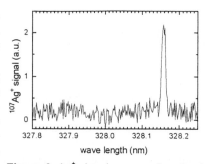

Figure 3. Ag$^+$ signal vs. wavelength of excitation laser

metastable state, this configuration detects the silver atoms ejected in the electronic ground state by two color R2PI. As for the dependence on the wavelength of the ionization laser, unblocking the excitation laser simply adds a constant background to the spectrum of fig. 2 due to the non resonant character of the ionizing transition.

Of particular interest is the spectrum which is obtained when only the excitation laser is used. Fig. 4 shows the result if the electric field extracting the photoions into the mass spectrum is switched off during the ionization process. This was necessary since otherwise the extraction field largely complicated the spectrum by Stark-shift and -splitting. A variety of lines are seen originating from resonance enhanced two photon transitions into a number of Rydberg states which are close enough to the ionization limit to be field-ionized by the extraction field. The envelope of these lines follows the saturation broadened resonance into the intermediate $^2P^o$ state. By extrapolation of known states of the ns and nd Rydberg series (13) (n = 8...12), one can clearly identify each line in fig. 4 as indicated in the figure. Details of this identification

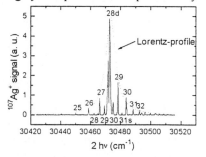

Figure 4. Ag$^+$ signal vs. energy of ionization laser

procedure will be published elsewhere (14). Obviously the most efficient one color R2PI detection of ground state atoms is by resonantly exciting the $4d^{10}28d$ ^2D_J multiplet. The binding energy W_B of these states of about 160 cm-1 leads to a critical field strength for field ionization of about 68 V/mm which is well below the applied extraction field of 215 V/mm. We therefore expect an ionization efficiency of nearly 100 % once a Rydberg state is excited.

Total population

The quantitative determination of the population of a given electronic state from the measured photoion signal is hampered by the *a priori* unknown photoionization cross sections σ_i. One possible way to circumvent this problem is to eliminate the influence of σ_i by driving the photoionization process into saturation. For the silver ground state atoms, fig. 5a shows the measured photoion signal as a function of the excitation laser intensity, while the intensity of the ionization laser was kept at a fixed value. In order reveal the true two color R2PI signal, the background induced by the excitation laser alone - which is due to the one color R2PI scheme (via Rydberg states) described above - was measured and subtracted. It is seen that already at very low power densities the resonant transition into the $^2P^o$ state is completely saturated. The dependence of both the ground state and the metastable

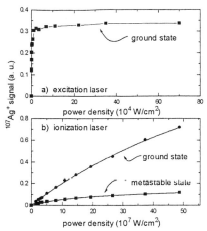

Figure 5. Laser power density dependence of measured signal

state signal on the intensity of the ionization laser is displayed in fig. 5b. It is obvious that in both cases no clear saturation plateaus are observed. This finding is due to signal contributions from the wings of the spatial laser beam profile, which lead to an increase of the effective ionization volume with increasing laser intensity. For the Gaussian beam profiles employed here, the influence of this effect can be included into the theoretically expected laser intensity dependence of the photoion signals. The solid lines depicted in fig. 5b represent least square fits of this dependence to the measured data which include the saturation signal S_{sat} as a fitting parameter. A detailed discussion of the data displayed in fig. 5b as well as the theoretical background behind the fitting procedure is given elsewhere (15). As a first approximation, the resulting ratio between the saturation signals evaluated for metastable and ground state atoms of approximately 7 % can be taken as the relative population of these states among the total flux of sputtered silver atoms.

Kinetic Energy Distributions

The velocity distribution of sputtered atoms was measured by varying the time delay between the primary ion pulse initiating the sputtering process and the firing time of the ionization laser. Fig. 6 shows such a flight time distribution for both the silver ground state and metastable atoms. The geometric alignment was made such that only atoms ejected in a narrow solid angle around the direction along the surface normal were detected. It is seen that the velocity distributions largely differ. More specifically, the metastable atoms possess a lower average velocity than the ground state atoms. This finding is of particular interest since it appears to be at variance with all published data. Previous experiments on a number of different sputtered excited metastable atoms have revealed kinetic energy distributions which were either *identical* with or *broader* than that of the respective ground state atoms (16). These results have been interpreted in terms of a simple deexcitation model: The particles which are excited during the collision cascade leading to their ejection are assumed to be more or less efficientlly deexcited on their flight away from the surface. Due to the limited range of the electronic interaction with the surface, this model implies that faster particles are less efficiently deexcited and therefore predicts an apparent broadening of the velocity distribution. Here, the first case is observed where the velocity distribution of

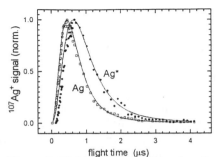

Figure 6. Flight time distribution of meta-- stable and ground state atoms

the excited state is *narrower* than that of the ground state and this model is therefore not applicable. Hence, the observed velocity distribution must be induced by a velocity dependent *excitation* rather than deexcitation mechanism. A possible way to explain the data is to assume that the metastable neutral atoms are formed by resonant neutralization of presursor ions containing a 4d-hole. As discussed in more detail in a forthcoming publication (17), the generation of such ions in the course of the collision cascade seems physically feasible. If neutralization occurs outside the solid, i. e. closely above the surface, a faster particle would be less efficiently neutralized, thus leading to an apparently narrower velocity distribution of the excited atoms.

Literature

1 . Yu, M.L. in *Sputtering by Particle Bombardment III*, eds. R. Behrisch, K. Wittmaack (Springer Berlin 1991), pp. 91
2 . Schweer, B. and Bay, H.L., *Appl. Phys.* **A 29**, 53 (1982)
3 . Young, C.E., Calaway, W.F., Pellin, M.J. and Gruen, D.M., *J. Vac. Sci. Technol.* **A2**, 693 (1984)
4 . Pellin, M.J., Wright, R.B. and Gruen, D.M., *J. Chem. Phys.* **74**, 6448 (1981)
5 . Dullni, E., *Appl. Phys.* **A 38**, 131 (1985)
6 . Wright, R.B., Pellin, M.J., Gruen, D.M. and Young, C.E., *Nucl. Instr. Meth.* **170**, 295 (1980)
7 . Postawa, Z., El-Maazawi, M., Maboudian, R. and Winograd, N., *Nucl. Instr. Meth.* **B 67**, 565 (1992)
8 . He, C., Postawa, Z., Rosencrance, S.W., Chatterjee, R., Garrison, B.J. and Winograd, N., *Phys. Rev. Lett.* **75**, 3950 (1995)
9 . Wucher, A., Berthold, W. and Franzreb, K., *Phys. Rev.* **A 49**, 2188 (1994)
10 . Berthold, W. and Wucher, A., *Phys. Rev. Lett.* **76**, 2181 (1996)
11 . Wucher, A., *Phys. Rev.* **B 49**, 2012 (1994)
12 . Baier, S., Martins, M., Müller, B.R., Schulze, M. and Zimmermann, P., *J. Phys. B: At. Mol. Phys.* **23**, 3095 (1990)
13 . Moore, C.E., *Atomic Energy Levels* Vol. III
14 . Berthold, W. and Wucher, A., to be published
15 . Berthold, W. and Wucher, A., *Nucl. Instr. Meth.* B (1996), in press
16 . Betz, G., *Nucl. Instr. Meth.* **B 27**, 104 (1987)
17 . Wucher, A., Berthold, W. and Sroubek, Z., to be published

Metastable state population and kinetic energy distributions of sputtered Ni and Co atoms studied by resonance ionization mass spectrometry

E. Vandeweert, P. Lievens, V. Philipsen,
W. Bouwen, P. Thoen, H. Weidele and R. E. Silverans

Laboratorium voor Vaste-Stoffysika en Magnetisme, K.U.Leuven
Celestijnenlaan 200 D, B–3001 Leuven, Belgium.

Abstract. We present the development of a sensitive method for the study of atomization processes via the metastable state population and kinetic energy distributions of sputtered atoms, including atoms on high-lying metastable electronic states. The experimental procedure is based on two-colour two-step resonant photoionization. We report on the population distribution of Ar ion sputtered Ni and Co atoms. The measurements show unexpectedly high populations on metastable states with energies above 1.5 eV. The kinetic energy distributions of Ni atoms on such states support the importance of bandstructure effects in the excitation mechanism of atoms after ion beam sputtering.

INTRODUCTION

Although studied both theoretically and experimentally for more than 15 years by now, the fundamental physical processes governing excited atom sputtering are not fully understood yet (1). More experiments under well defined conditions are needed in order to be able to compare different theoretical predictions. Resonance ionization spectroscopic measurements provide excellent experimental tools to determine population and kinetic energy distributions of atoms on *metastable* states (2,3). This yields valuable information on the basic mechanisms responsible for the excitation during or after the atomization process. However, up to now most experimental results are limited to metastable states with excitation energies below 0.5 eV.

We present a highly sensitive procedure for the measurement of population and kinetic energy distributions on metastable states up to a few eV. The procedure is based on two-colour two-step resonant ionization spectroscopy combined with time-of-flight mass spectrometry. It fully exploits the capabilities of the recently built Leuven resonant ionization mass spectrometer (4). Results on population distributions of sputtered Ni and Co atoms and on the kinetic energy distribution of Ni on high-lying metastable states are reported.

EXPERIMENTAL METHOD

For the accurate determination of population distributions on metastable states, we developed a new experimental procedure based on two-step two-colour resonant ionization spectroscopy. In the first step the atoms are resonantly excited from the selected metastable state to an intermediate state. This intermediate state is the same for a selected subset of metastable states, so that a common second step can be used to ionize the atoms. When the excitation steps for the different ionization schemes are saturated, the relative photo-ion signals reflect the relative populations on the corresponding metastable states. This condition is in most cases already fulfilled at moderate photon flux densities (5,6). The population on metastable states decreases drastically with increasing excitation energy. Therefore the efficiency of the ionization process is optimized by tuning the ionizing laser to a transition to an autoionizing state near the first ionization potential (7).

In fig. 1, partial energy level schemes of Ni and Co are shown, including all metastable states and the selected $\mathrm{Ni\,I}\ ^3F^o_{J=2,3,4}$ and $\mathrm{Co\,I}\ ^4D^o_{J=3/2,5/2,7/2}$ intermediate multiplets from which the excited atoms are ionized into autoionizing states. To excite all states to the selected intermediate states, a large tuning range of the excitation laser (from 290 to 1200 nm) is needed.

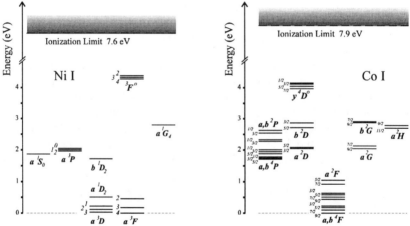

FIGURE 1. Partial energy level schemes of Ni and Co , including all metastable states and the selected $\mathrm{Ni\,I}\ ^3F^o_{J=2,3,4}$ and $\mathrm{Co\,I}\ ^4D^o_{J=3/2,5/2,7/2}$ intermediate multiplets from which the excited atoms are ionized into autoionizing states.

EXPERIMENTAL SETUP

The spectrometer consists of an ultra-high vacuum chamber with a base pressure of about 5.10^{-10} hPa. Thin polycrystalline foils are bombarded by argon ions with energies between 2 and 15 keV (ion current 1.6 µA) at 45° incidence. The kinetic energy distributions were measured using the ion gun in pulsed mode (pulse width of 200 ns). Alternatively, the spectrometer allows to study atoms sublimated from a hot filament under identical conditions. Ejected atoms are resonantly ionized by tunable laser light from Nd:YAG pumped dye laser and optical parametric oscillator systems with frequency doubling options, covering a wavelength range from 225 to 1600 nm. Finally, the photoions are electrostatically extracted and focused into a reflectron time-of-flight mass spectrometer.

RESULTS AND DISCUSSION

A series of experiments with thermally evaporated atoms was performed to validate the reliability of the proposed procedure. Pure Ni filaments were resistively heated to 1400K. Temperatures obtained by comparing the population distributions determined using the two step two-colour schemes with a classical Maxwell-Boltzmann distribution, showed an excellent agreement with the surface temperatures of the filament as measured by optical pyrometry.

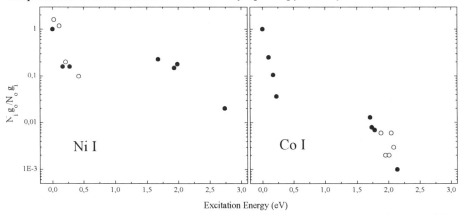

FIGURE 2. Relative population distributions on metastable Ni and Co states as function of the excitation energy after sputtering from polycrystalline targets. Open symbols refer to electronic states with an open shell configuration, closed symbols to states with closed shell configurations.

The procedure was then applied to study the population distribution of sputtered neutral atoms. In fig. 2, we show the relative population on metastable Ni and Co states as function of the excitation energy after sputtering from pure polycrystalline foils with 13 keV Ar ions. These distributions show some remarkable features. First, the population on the metastable Ni states belonging to the first excited multiplet are higher than the ground state population. A similar population inversion was also noticed by C. He et al. after sputtering from a Ni single crystal (2). This was explained by an excitation model where the excitation probabilities are dominated by the nature of the band structure of the metal and by the electronic state of the sputtered atom. A second interesting feature is that the population on excited states above 1.5 eV is remarkably high. The discrepancy between the "temperature", as obtained by comparing the population on these states with a Maxwell-Boltzmann distribution, and the one obtained from the ground-state multiplet, shows that non-thermal mechanisms play a significant role during the sputter process. The sensitivity of the method also allowed the measurement of the kinetic energy distributions of Ni atoms on the metastable states above 1.5 eV. Fig. 3 shows that the time-of-flight distributions of atoms on highly-excited metastable states are identical within the experimental error to the one of the ground-state atoms. Together with the unexpected high population on highly excited metastable states, these kinetic energy distributions give new experimental evidence for the close relation between the electronic structure of the solid and the free atom for the excitation of atoms during the sputtering process. It is indeed expected that the population on atomic states with binding energies close to the binding energies of the valence electrons in the solid is favored (8).

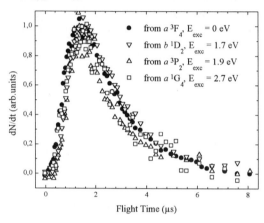

FIGURE 3. Time-of-Flight distributions of Ni atoms sputtered in different highly excited metastable states (open symbols) and in the ground state (filled symbols).

SUMMARY AND OUTLOOK

We reported on a new experimental procedure which was succesfully applied to perform a systematic study of the population and kinetic energy distribution of Ar ion sputtered Ni and Co atoms on metastable electronic states. Future experiments will focus on the influence on these distributions of the environment in which the sputtering process takes place and the matrix from which the atoms are sputtered.

This work is supported by the Belgian National Fund for Scientific Research (N.F.W.O.) and by Concerted Action (G.O.A.) and Inter-University Attraction Poles (I.U.A.P.) Research Programs. P.L. is a Senior Research Assistant and H.W. a Visiting Postdoctoral Fellow of the N.F.W.O. W.B. acknowledges the financial support of the Institute for Science and Technology (I.W.T.).

1. M.L. Yu, in *Sputtering by Particle Bombardment III*, edited by R. Behrisch and K. Wittmaack (Springer-Verlag, Berlin, 1991), Chap. 3.
2. C. He, Z. Postawa, S.W. Rosencrance, R. Chatterjee, B.J. Garrison, and N. Winograd, Phys. Rev. Lett. **75**, 3950 (1995).
3. G. Nicolussi, W. Husinsky, D. Gruber, and G. Betz, Phys. Rev. B **51**, 8779 (1995).
4. E. Vandeweert et al., to be published.
5. G.S. Hurst, M.G. Payne, *Principles and applications of resonance ionization spectroscopy* (Adam Hilger, Bristol, 1988).
6. V.S. Lethokov, *Laser photoionization spectroscopy* (Academic Press, Orlando, 1987).
7. P. Lievens, E. Vandeweert, P. Thoen and R.E. Silverans, Phys. Rev. A., in press.
8. E. Veje, Phys. Rev. B **28**, 5029 (1983).

Resonance Ionization Spectromicroscopy (RISM) with Subwavelength Spatial Resolution: The First Observation of Single Color Centers

V. N. Konopsky, S. K. Sekatskii and V. S. Letokhov

Institute of Spectroscopy Russian Academy of Sciences, Troitsk, Moscow region 142092 Russia

Abstract. We present the first experimental realization of RISM technique based on laser photoelectron projection microscope. Single F_2 color centers on LiF needle tip surface were for the first time observed. Future prospects of the method are discussed.

THE CHOICE OF THE FIRST OBJECTS FOR RISM

Four years ago one of us (V.S.L.) presented at Sixth International Symposium on Resonance Ionization Spectroscopy (Santa Fe, NM, USA, 24-29 May 1992) a talk entitled "Towards laser photoelectron resonance ionization spectromicroscopy (RISM)" (1). At those time it was understood that to realize an idea of RISM (2) (laser resonance photoionization of some specific parts of large molecules or other systems with the subsequent detection of the electrons/ions emitted with subwavelength spatial resolution) it is necessary to work with more "rigid" objects than molecules placed on field emission microscope cathode, because in this case difficulties due to the diffusion, desorption and decomposition of molecules in intense laser fields are too serious to overcome. By this reason we decided to try to realize RISM idea for impurity ions and color centers in transparent broad-band-gap crystals (see Fig.1) which "can, in principle, repeatedly emit photoelectrons under excitation, while remaining rigidly localized in the host matrix" (1). To do this it was necessary first of all to investigate the possibilities of laser selective photoionization of these impurity centers, i. e. to find such conditions of laser irradiation of the corresponding samples when the electron emission observed would be due to the laser photoionization of the impurities but not to the photoionization of the crystalline matrix itself.

Such selective photoionization of impurity centers in crystals has been clearly observed soon for a number of samples. The first example is CaF_2 crystals doped with bivalent rare-earth ions such as Sm^{2+}, Tm^{2+}, Eu^{2+} (3, 4). The threshold of external photoelectric effect for pure CaF_2 crystals lies in the far UV spectral

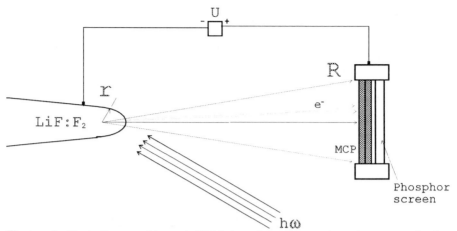

Figure 1. Illustrating an idea of RISM based on laser photoelectron projection microscope. The emitting centers on the needle tip (impurity centers) are indicated by circles.

region, but even visible radiation can cause the direct photoionization of bivalent dopant RE^{2+} ions due to the very large Madelung energy in CaF_2 crystal lattice which is quite comparable with the energy of ionization of free RE^{2+} ions and shifts photoelectric effect threshold to the visible region (3,4).

Another example is transparent broad-band-gap crystals like LiF where stable high-energy defects of the crystalline lattice can be created in large concentration as a result of such crystals irradiation by X-rays or electron beams. The main types of absorbing visible radiation stable defects in LiF crystals are F - aggregated color centers, which are two adjacent anion vacancies in its crystalline lattice that have captured one (F_2^+), two (F_2) or three (F_2^-) electrons (5). Energies of these F-aggregated color centers are rather close to the vacuum level and one can hope to photoionize them when applying visible or near UV laser radiation. This process soon was observed by us during the investigations of laser selective external photoelectric effect for such crystals (3, 6).

Both these crystals ($CaF_2:Sm^{2+}$ and $LiF:F_2$) were chosen by us as first objects to implement RISM idea. In the case of $CaF_2:Sm^{2+}$ we used thick (50 - 100 nm) epitaxial layers of CaF_2 containing dopant Sm^{2+} ions with different concentration which were grown on the tips of sharp Pt/Ir or Si needles. These experiments revealed that CaF_2 looks like very promising coating for the different emission devices (probably not worth or even better than diamond coating) but high photoelectron escape depth in calcium fluoride and relatively small photoionization cross-section prevent an observation of single Sm^{2+} dopant ions in this case. The results of these experiments are more interesting for the field emission specialists and will be published elsewhere.

In the case of LiF crystals with different F_2 color centers concentration we elaborated a method of sharp needles preparation based on the etching of the crystal fragments in strong HCl solutions. The crystals used were synthesized in Institute of General Physics, Moscow. Different possible ways of F_2 color centers photoionization in strong electric field of a projection photoelectron microscope by visible laser light are shown on Fig. 2.

FIRST OBSERVATION OF SINGLE COLOR CENTERS

The schematic diagram of laser photoelectron projection microscope is illustrated on Fig.1. Needles a few millimeters long were fastened with Wood's alloy to a metal electrode to which a voltage in the range 0 to –2.5 kV was applied when studying the photoelectron images of the needle tips and in the range 0 to+20 kV when studying images formed by the positive ions desorbed from the needle tips. The needle under study was irradiated with a CW argon laser, the radiation being focused onto the tip of the needle. The laser radiation intensity at the tip was typically in the range 10^3-10^4 W/cm^2. The photoelectrons emitted from the surface of the needle were directed by means of an electric field to the

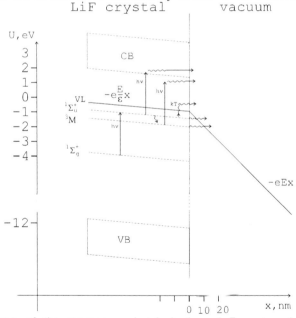

Fig. 2. Pathways of the resonance photoionization of F_2-centers in LiF crystals. Photoionization can take place as a result of either absorption of one more laser quantum by the F_2-center at the excited levels or the tunnel and/or thermostimulated ionization from these levels.

157

grounded entrance of a microchannel-plate-fluorescence-screen assembly at a distance of $R \cong 10$ cm from the tip. The optical image formed at the exit from the assembly was picked up by a TV camera operated in conjunction with a Model Argus-50 image processing computer system (Hamamatsu Photonics K. K., Japan). To avoid difficulties associated with the space charge, we operated with electron currents no greater than 1 pA. An oilless vacuum of some 3×10^{-7} Torr was maintained in the experimental chamber. A detailed description of the design of the laser photoelectron projection microscope and other experimental details can be found in (7).

A typical photoelectron image of the tip of a LiF:F_2 needle (F_2 centers concentration $n \cong 10^{16}$ cm^{-3}) is presented in Fig. 3. As can be seen, the image is formed by a multitude of individual, clearly defined bright and dark spots. No single bright spots were observed in case of samples with high ($\sim 10^{18}$ cm^{-3}) F_2 color centers concentration and nominally pure LiF crystals. At the same time the distances between the individual bright spots on Fig. 3 correspond well to the concentration of F_2 - centers in the sample.

As in the case of classical field electron (ion) microscopy, the magnification M of the microscope is equal to the ratio between the needle-detector distance R and the tip radius of curvature r (8): $M = R/\chi r$, where χ is a

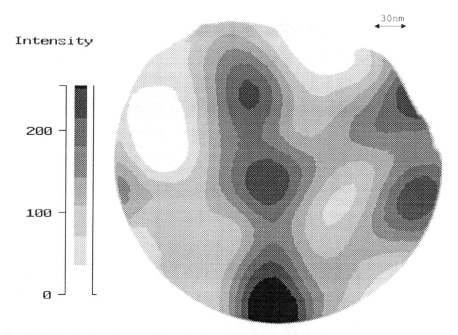

Fig. 3. Photoelectron image of the tip of a LiF:F_2 needle. The radius of curvature of the tip is $r = 600$ nm, the concentration of F_2-centers is $n = 10^{16}$ cm^{-3}.

numerical coefficient ranging between 1.5 and 2. For our experimental conditions, this formula yields $M \cong 10^5$. The spatial resolution of the microscope, i.e., the diameter of the spot on the detector that images a single bright emission point on the surface of the needle tip, is governed by the energy E_0 of the transverse motion of the emitted electron and is given by (8):

$$d = 2\,\chi r \sqrt{\frac{E_0}{eU}} \qquad (1)$$

In our experimental conditions, $U = 2.5$ kV and $E_0 \sim 1$ eV, and so we have $d \approx 40$ nm, which agrees with the size of a single bright spot in Fig. 3.

Using the experimentally measured length of escape of photoelectrons from LiF crystals, $l_{esc} \cong 3$ nm (see [9] and references cited therein) and the concentration of F_2 centers in the samples under study we obtain for the average distance between the images of individual bright spots $l_0 \cong (l_{esc}n)^{-1/2} \sim 100$ nm, i.e. a value of the same order of magnitude as that observed experimentally: one can see from Fig. 3 that the average distance between individual bright spots is of the order of 100-150 nm.

FUTURE PROSPECTS OF RISM TECHNIQUE

Thus the experimental data presented demonstrate the first practical realization of RISM idea: we have not only image LiF:F_2 needle tip surface with high spatial resolution, but clearly identify bright spots on the image as single F_2 color centers. What prospects of RISM technique can be anticipated today based on the existing experimental experience?

Earlier it was decided that one of the most serious objections to implement RISM technique is the problem of samples conductivity: one needs some conductivity to obtain a radial electric field in the vicinity of the tip and thus to attain magnification and spatial resolution. By this reason we always thought about objects under study placed on the metal tip - and always met with the serious difficulties (metal quenching of molecule excitation leads to the necessity of using intense laser fields which leads to the decomposition or desorption of molecules, etc.) But at present we have realized that very poor conductivity is really necessary. One needs the voltage of the order of only 100-200 V on the tip which is sufficient to efficiently detect the photoelectrons emitted on MCP. This means that one can apply the voltage of some kV on the thick part of the sample cone (see Fig.1) and sacrifice all but 100 V to the voltage drop along the sample. Because very small current (of the order of 0.1 pA) is necessary to observe photoelectron images, simple estimations show that needles with a fairly high resistivity, up to around 10^{12} - 10^{13} Ω cm can be investigated using this technique

(7). (The best evidence of this possibility are the photoelecton images of the tips of LiF:F$_2$ needles obtained in the present experiment.) This means that it is possible to investigate such samples as, for instance, complex molecules dissolved in paraffin's or other suitable matrices. One should prepare a small sharp needle from such "molecules-in-the-matrix" sample and study it in the same way as reported here; at present we are preparing such experiment. This approach should materially enlarge the variety of samples studied and help to avoid the difficulties due to the metal quenching.

The use of the classical photoelectron microscope technique instead of projection electron microscopy can also materially enlarge the variety of samples studied because in this case it will be possible to work not only with the specially prepared ultrasharp pointed needles but also with objects of arbitrary shape (and, what is most important, with flat surfaces). Another possibility to extend the class of objects that can be studied and improve the selectivity with which specific centers can be observed is to use tunable femtosecond laser pulses for the purpose of laser resonance photoionization of light-absorbing centers.

ACKNOWLEDGMENTS

We thank Hamamatsu Photonics K.K. for lending us the necessary instrumentation. We are also grateful to the Russian Foundation for Fundamental Research and Department of Defense (USA) for their financial support.

REFERENCES

1. Letokhov, V. S., and Sekatskii, S. K., In: *Resonance Ionization Spectroscopy 1992*, ed. by C. M. Miller and J.E. Parks (IOP Publ., Bristol, 1992); *Inst. Phys. Conf. Ser.* **128**, 241-248 (1992).
2. Letokhov, V. S., *Kvant. Electr.* **2**, 930-935 (1975) (in Russian).
3. Letokhov, V. S., and Sekatskii, S. K., *Opt. Spectrosc.* **76**, 303-310 (1994).
4. Sekatskii, S. K., Letokhov, V. S., and Mirov, S. B., *Opt. Communicat.* **95**, 260-264 (1993).
5. Basiev, T. T., Mirov, S. B., and Osiko, V. V., *IEEE J. QE* **24**, 1052-1064 (1988).
6. Sekatskii, S. K., Letokhov, V. S., Basiev, T. T., and Ter-Mikirtychev, V. V., *Appl. Phys.* *A* **58**, 1026-1029 (1994).
7. Konopsky, V. N., Sekatskii, S. K., and Letokhov, V. S., *Appl. Surf. Sci.* **94/95**, 148-155 (1996).
8. Tsong, T. T., *Atom-Probe Field Ion Microscopy*, Cambridge: Cambridge Univ. Press, 1990.
9. Aleksandrov, A. B., Aluker, E. O., Vasilyev, I. A. et al, *Introduction to Radiational Physics and Chemistry of Alcali-Haloid Crystal Surface*, Riga: Zinatne, 1989.

160

Studies of OH Radicals and H_2O in Catalytic Reactions on Pt using REMPI and Mass Spectrometry

A. P. Elg, M. Andersson and A. Rosén

Department of Physics, Göteborg University and Chalmers University of Technology, S-412 96 Göteborg, Sweden

Abstract. OH radicals and H_2O desorbing from a polycrystalline Pt foil, during the $H_2 + 1/2O_2 \rightarrow H_2O$ catalytic reaction, have been studied using Resonance Enhanced Multiphoton Ionization REMPI. The OH and H_2O desorption rates were measured as a function of gas mixture at catalyst temperatures of 1000-1400K and a total pressure of $5 \cdot 10^{-5}$ mbar. An enhancement of the OH signal by a resonant (2+1) photon ionization was achieved by using the $D^2\Sigma^-(\nu = 1)$ state at 236-239 nm. The produced ions were detected in a time-of-flight mass spectrometer (TOF-MS), in order to avoid interference from non-resonantly ionized species. By using a TOF-MS a simultaneous non-resonant ionization and detection of H_2O, H_2 and O_2 was achieved.

The reaction kinetics probed with REMPI were evaluated, using a theoretic model, and found to be in agreement with earlier Laser-Induced Fluorescence (LIF) measurements at higher pressures. The desorption of H_2O was also studied in the 500-1000K range on polycrystalline Pt, in order to obtain additional information about the catalytic reaction. Further studies were performed in a UHV chamber on a Pt(111) single crystal at 250-500K, by resonant detection of H_2O with (2+1) REMPI around 248 nm via the $\tilde{C}^1B_1$ state.

INTRODUCTION

Heterogeneous catalytic reactions are important in many industrial processes, and are therefore extensively investigated with advanced surface science techniques, often at vacuum or UHV conditions. Single steps of a reaction are commonly studied by means of probing the states by surface analytical techniques. By using visible or UV laser light, studies of desorbing intermediates and products give an indirect way of probing the reaction, where laser-induced fluorescence (LIF) is one of the most commonly used techniques. The most important features of laser based methods, tuned in on resonance with the species, are the access to the state distribution, a high selectivity and sensitivity.

LIF studies of the kinetics of the catalytic formation and decomposition of H_2O on Pt, $H_2 + 1/2O_2 \rightleftharpoons H_2O$, have been used to develop a scheme for probing catalytic reactions (1). A key for an understanding of this reaction is to probe the intermediate OH, which at high surface temperatures, i.e. $T_s > 900K$, desorbs in sufficient amounts for laser spectroscopic detection.

By probing the desorption rate of OH as a function of surface temperature, total reactant pressure or gas mixture, a measurement of the OH desorption gives the relative OH coverage on the surface and conclusions can be drawn regarding the reaction kinetics on the surface. In addition to measurements of OH the water desorption rate is also important for the interpretation of the reaction. It is therefore desirable to be able to probe both desorption rates simultaneously.

The experimental results have been compared with a kinetic model (2), based on the mean-field approximation, with which the reaction is modeled at steady-state conditions. The model includes the kinetics of the different steps in the reaction, such as adsorption, reactions on the surface and desorption.

LIF detection of OH can effectively be used to study this reaction at $p \geq 10^{-3}$ mbar total reactant pressure. However, due to gas phase collisions the LIF signal has to be corrected for rotational and electronic quenching. Since the water formation reaction is strongly exothermic in nature, i.e. 2.5 eV/molecule, the water reaction rate has earlier been probed by microcalorimetry.

In this work resonance enhanced multiphoton ionization (REMPI), which is known for its higher sensitivity, has been used instead of LIF at collision free pressures, i.e. $p \leq 5 \times 10^{-4}$ mbar. The intermediate OH is probed by resonant ionization, and the much more abundant water molecules are simultaneously ionized non-resonantly.

EXPERIMENTAL SET-UP AND RESULTS

The experimental set-up is based on a TOF-MS of the Wiley-McLaren type, in which the Pt sample is a part of the ion-optics (3). A focused laser beam is directed parallel to the catalytic surface where desorbing molecules (OH and H_2O) are ionized. The OH radicals are ionized via a two-photon transition $(X^2\Pi(\nu=0) \rightarrow D^2\Sigma^-(\nu=1))$ followed by one photon ionization (4), whereas water is ionized non-resonantly with three photons. The ions are detected by a micro-channel plate (MCP) in the TOF-MS.

The laser wavelength used in the resonant ionization of the OH molecules has been tuned in on a resonance which probes a single rotational state in the ground state. This is important for an easy interpretation of the kinetics of the reaction. Modeling of the two-photon excitation of OH has

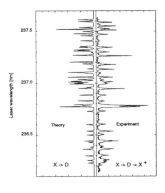

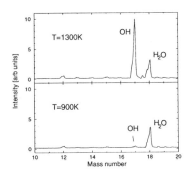

Figure 1: Wavelength spectrum of REMPI of OH at 236-238 nm. Left part shows a simulation of a two-photon excitation and the right part the REMPI signal of the experiment.

Figure 2: Partial mass spectra obtained at sample temperatures of 1300K (top) and 900K (bottom).

been performed to find a proper resonance, and is given in comparison to an experimental spectrum in Fig. 1.

The choice of excitation wavelength and laser intensity is important to avoid dissociation of water, which could create a huge OH background. As a control of that the dissociation of water did not interfere with the signal from desorbed OH, mass spectra were recorded at 1300K, where OH desorption is significant, and at 900K, where OH desorption is negligible. A demonstration of this is given by the two mass spectra in Fig. 2.

The total reactant pressure, the reactant gas mixture and the surface temperature (T_s) are varied during the experiments and the OH and H_2O signals are recorded simultaneously. In Fig. 3 measurements of the OH and H_2O yield are given as a function of the reactant mixture ($\alpha = p_{H_2}/[p_{H_2} + p_{O_2}]$) for two different pressures together with theoretical modeling of the reaction. In the lower part of the figure the results from the REMPI measurements at 5×10^{-5} mbar are plotted together with results from modeling of the reaction (solid lines). For comparison a measurement with LIF and microcalorimetry at 0.13 mbar (5) is given in the upper part.

The maximum of the OH and H_2O yield represents the balance of adsorption and desorption of the reactants. A shift of the maximum OH and H_2O desorption as a function of α is observed as the total reactant pressure is changed. The adsorption is a function of the total pressure and the sticking probability, whereas the desorption rate of the reactants is proportional to the square of the relative coverage and exponentially dependent on the surface temperature and desorption energy. Since the two reactants have different desorption energies, a change in total reactant pressure will affect the species differently, thereby shifting the stoichiometric reactant conditions. These facts are not easily recognized without detailed comparison with the kinetic model. This will together with the temperature dependence (500-1400K) be elaborated elsewhere (6).

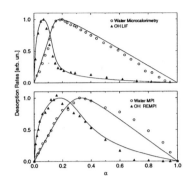

Figure 3: Measurements of the OH and H_2O yield at two different pressures. Normalized LIF and microcalorimetry data (top) at 0.13 mbar and REMPI data (bottom) at 5×10^{-5} mbar.

The studies of the reaction kinetics have also been extended in temperature down to 250K, thereby totally covering 250-1400K. At room temperature or below, the reaction is much slower and therefore a resonant (2+1) detection scheme of H_2O via the $\tilde{C}^1B_1$ state was used.

ACKNOWLEDGEMENTS

We would like to thank Dr Erik Fridell for many fruitful discussions in connection to this work. We gratefully acknowledge financial support from NFR, contract no. 2560-129,-321,-336 and from TFR, contract no. 92-538,93-341.

REFERENCES

1. Fridell, E., Elg, A. P., Rosén, A., and Kasemo, B., *J. Chem. Phys.* **102**, 5827-5835 (1995).

2. Hellsing, B., Kasemo, B., and Zhdanov, V., *J. Catal.* **132**, 210-228, (1991).

3. Elg, A. P., Andersson, A., and Rosén, A., *Submitted to Appl. Phys. B*, (1996).

4. Collard, M., Kerwin, P., and Hodgson, A., *Chem. Phys. Lett.* **179**, 422-428, (1991).

5. Fridell, E., Hellsing, B., Kasemo, B., Ljungström, S., Rosén, A., and Wahnström, T., *J. Vac. Sci. Technol. A* **9**, 2322-2325, (1991).

6. Elg, A. P., Andersson, A., and Rosén, A., *To be published.*

SESSION VI: NEW LASER
SOURCES AND APPLICATIONS

A Paul Trap System for Laser Spectroscopic Measurements

W.Z. Zhao[a], S. Gulick[a], S. Fedrigo[a], F. Buchinger[a], J.E. Crawford[a], J.K.P. Lee[a], and J. Pinard[b]

[a] Foster Radiation Laboratory, McGill University, 3559 University St. Montreal, Canada H3A 2B1
[b] Laboratoire Aimé Cotton, 91405 Orsay, France

Abstract. A Paul trap system, suitable for spectroscopic investigations of very minute samples is described. This trap is loaded by a laser-desorption source, and RIS beams are used to select a single element for study. A phase-locked counting technique is used to minimize linewidth. A new technique based on selective optical quenching has been used to eliminate interfering hyperfine lines.

Introduction

Radiofrequency quadrupole traps have found extensive use in mass spectrometry: commercial instruments are widely used in chemical studies, and specialized Paul traps have been developed for high precision mass measurements made on-line at nuclear accelerators (1). They were first used in optical spectroscopy with lasers in 1977 (2). Since the same ion cloud in the trap is continuously excited by the laser beam, it is possible to study very small samples - even in principle a single trapped ion. However, in studies of isotope shift and hyperfine structure along long isotopic chains, they have not been strong competitors of many more traditional methods (3). One problem has been the difficulty of confining a sample for a long time in the trap, since chemical processes eventually result in ion loss. A second problem has been the Doppler broadening of spectral lines. Ions in the trap undergo a complex oscillation - a superposition of the relatively slow simple harmonic motion in the trap's pseudopotential well (the macromotion) and a

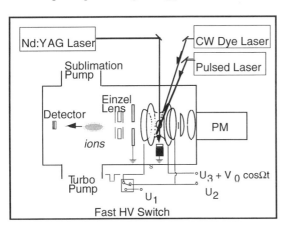

FIGURE 1. Schematic of the trap system

more rapid oscillation at the frequency of the driving RF field (the micromotion). The system we have developed has provided useful information on ways to optimize trap operating parameters and to reduce the Doppler linewidth. We have also found a photochemical technique that can be used in some cases to remove hyperfine lines arising from contaminant isotopes. Our system uses an internal laser-desorption source, combined with pulsed RIS beams to load the trap selectively. We have previously shown that it is possible to use a similar system to inject ions from an external source (4), so that the methods described here could be used for on-line laser spectroscopy for nuclear studies.

The Trap System

The trap's hyperbolic endcaps are separated by $2z_0 = 28.7$ mm, with a ring electrode of inner radius $r_0 = \sqrt{2}z_0$, with apertures to admit laser beams. It is located near a photomultiplier mounted at one end of the chamber. Samples to be studied are placed in a cup below the ring electrode, and heated by the pulses from a Nd:YAG laser. The atoms desorbed enter the trap; these are selectively ionized at the trap center by RIS beams from pulsed dye lasers, so that the trap can be loaded with ions of essentially a single element. The present work was done with various isotopes of hafnium, with targets mostly in the form of $HfCl_4$, or HfO_2; the RIS scheme is shown in Fig. 2.

The trap was operated with an RF voltage of amplitude $V_0 = 440$ V, with frequency $f_0 = 400$ kHz and with a DC offset voltage between ring and endcap of 15 V. With these parameters, the pseudopotential well depth for Hf is about 13 eV for the axial motion and 18 eV for the radial motion. Both a turbomolecular and a Ti sublimation pump are used to evacuate the system to a base pressure of 5×10^{-8} Pa, and the system is baked out during pumpdown at about 200° C. Hydrogen buffer gas is introduced during bakeout, and the spectroscopic measurements are done at a pressure $\sim 10^{-3}$ Pa. Even with careful preparation, chemical processes still limit the lifetime of Hf ions within the trap to ~1h. With repeated YAG laser shots, it is possible to load the trap to a saturated level of about 10^5 ions. The ion number is estimated by pulsing the cloud out of the trap to

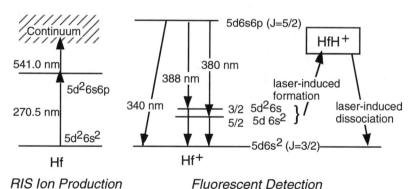

FIGURE 2. Transitions used in the RIS ion production, and the detection scheme

168

the channelplate shown in Fig. 1. This simple time-of-flight (TOF) system also confirms the ion mass.

Studies of Ion Motion and Spectroscopic Linewidth

A cw frequency-doubled laser with $\lambda = 340$ nm excites the Hf ions to a level at 29405 cm^{-1}, and two transitions (380, 388 nm) leading to metastable levels are observed by a photomultiplier (Fig. 2). Hydrogen buffer gas reduces the kinetic energy of the trapped ions by collisions, and also quenches those in the metastable states back to the ground state. We have shown (5) that, as might be expected, the cooling rate is proportional to the buffer gas pressure; the characteristic time to cool Hf$^+$ at a pressure 10^{-3} Pa ~ 100 ms. However, after cooling, space charge repulsion within the trap forces the ions to occupy a region with nonzero radius, and average ion energy ~0.3 eV (for 10^5 ions). The corresponding spectral linewidth is about 3 GHz. It is possible to reduce this linewidth by gating the photomultiplier to accept photons only during the part of the RF period at which the ions have minimum velocity. Gating the 340 nm excitation with an optical shutter is also used to improve the signal:background ratio. The linewidth reduction by this RF gating technique is most easily understood by considering the ion motion in phase-space (v_z vs. z). The solution of the trap field equations shows that an ion lies at a point on a phase-space ellipse; in general, in the absence of collisions, and at the same RF phase, it will be found at a point on the same ellipse at subsequent times. Although the ellipse area remains constant, its orientation and eccentricity change with RF phase, undergoing one complete cycle in one period (Fig. 3). A cloud of many ions will occupy a set of concentric ellipses, all simultaneously undergoing the same change of orientation. If there are collisions with a buffer gas, the ellipses shrink until an equilibrium is reached: the ions' space charge repulsion prevents them from occupying a single point at the trap center. Minimum linewidth (~1 GHz) is achieved by gating the photomultiplier close to the phase at which the velocity spread is a minimum.

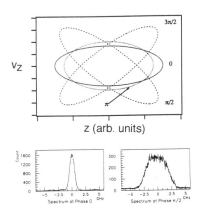

FIGURE 3. Linewidth vs RF phase

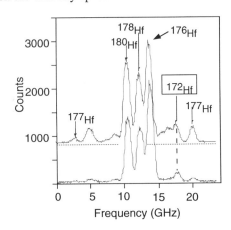

FIGURE 4. Suppressing contaminants

169

Photochemical Suppression of Contaminant Signals

The 380,388 nm fluorescence observed from the ion cloud is not constant in intensity, but exhibits a decay with two lifetimes — a fast component, with $\tau \sim 2m$, and a slow component with $\tau \sim 1h$. The TOF spectrum shows that the longer decay simply represents a chemical conversion of Hf^+ to compounds, predominantly HfO^+, and $HfOH^+$. However, the faster decay occurs only when the 340 nm excitation illuminates the cloud, and stops when the excitation is switched off. The TOF spectrum collected after $\sim 1m$ irradiation shows that Hf^+ in this case is being converted to HfH^+, so the decrease in the fluorescence simply represents the loss of Hf^+ ions. If the excitation is temporarily detuned from resonance, and later retuned, the fluorescent intensity is observed to increase, indicating that the HfH^+ is being photochemically dissociated. We interpret these results by the process shown in Fig. 2. The metastable levels de-excite to the ground state mainly by collisions with H_2; however, there is a branch for the formation of the hydride through the reaction

$$Hf^{+*} (5d^26s, 5d6s^2(J=5/2)) + H_2 \Rightarrow HfH^+ + H.$$

Estimates of the fluorescent photon emission rate, and fits to the fast decay suggest that the branch for the formation of the hydride in our experiment is about 10^{-5} that of the collisional quenching to the ground state.

This effect can be useful spectroscopically. In this experiment, the hyperfine splitting is large enough to permit tuning the 340 nm excitation to the frequency of one hyperfine line of a single isotope (e.g., ^{177}Hf). When this is done, that isotope alone is converted to the hydride, so all its hyperfine lines will be reduced in intensity. In the present experiment, where the aim was to make a precise measurement of the isotope shift of ^{172}Hf, the interfering ^{177}Hf and ^{179}Hf components were reduced in intensity by pre-irradiating the ion cloud for several minutes with the cw laser tuned at a fixed frequency to selected hyperfine lines. The improvement of the signal:background for the ^{172}Hf line is shown in Fig. 5. It should be noted that it is not possible to completely remove interfering lines by this technique. While a second isotope (e.g., ^{179}Hf is being converted to HfH^+, lines from a previously removed isotope (like ^{177}Hf) will increase somewhat in intensity, since its hydride is being photodissociated.

REFERENCES

1. Sharma, K.S. *et al*, "The Canadian Penning Trap Mass Spectrometer", presented at the International Conference on Exotic Nuclei and Atomic Masses, Arles, France. June 19-23, 1995 [in press].
2. Iffländer, R., and Werth, G., Metrologia 13, 167-170 (1977).
3. Otten, E.W., *Treatise on Heavy-Ion Science vol. 8*, ed D.A. Bromley, Plenum Press, New York, 1989, ch. 7.
4. Crawford, J.E. *et al*, Hyperfine Interactions 81, 143-149 (1993).
5. Zhao, W.Z. *et al*, Nucl. Inst. and Meth. B 108, 354-358 (1996).

Plasma Emission in a Pulsed Electric Field after Resonance Ionization of Atoms

O.I. Matveev, W.L. Clevenger, L.S. Mordoh, B.W. Smith, and J.D. Winefordner

Department of Chemistry, University of Florida, Gainesville, FL 32611-7200

Abstract. A technique is presented which is based on the measurement of the emission of excited buffer gas atoms which are created by interactions with electrons in a strong pulsed electric field after laser assisted ionization of analyte atoms. In principle, this emission method can be applied for the detection of single atoms, molecules, or photons.

The detection of a resonance ionization (RI) or laser enhanced ionization (LEI) signal in a flame or buffer gas has been achieved using several different techniques, most of which are dependent on electrodes for signal collection(1-4). Limitations include a decrease in accuracy due to contamination from the electrodes and a decreased atomization efficiency due to the cooling of the media caused by the insertion of electrodes. In addition, the limit of detection (LOD) in a flame with electrical detection is limited by flicker noise and usually is not better than 1000-3000 electrons/pulse; under optimal conditions, flicker noise can be removed(5), improving the LOD to 100-500 electrons/pulse. In order to decrease the LOD in a flame or plasma, a higher voltage can be applied to create an avalanche mode of electrical signal measurement, which requires the highly sensitive preamplifier to be protected against breakdown which in turn deteriorates the LOD. A design which is not limited by the necessity of electrodes can overcome these drawbacks, in addition to being an attractive option in analytical applications which require a sensitive method of detection of a small number of atoms in a sealed cell to avoid exposure to the open air.

For a signal detection method to take advantage of the sensitivity of which RI or LEI is capable, it must involve avalanche amplification of the signal. A new method of RI signal detection which does not depend on electrodes for signal collection is reported here. Electrons created as a result of the resonance photoionization of the analyte in a gas medium are accelerated in a strong pulsed electric field, resulting in collisional excitation and emission of the buffer gas atoms and/or molecules. The intensity of this emission (under certain conditions) yields information about the initial number of electrons created by the laser assisted ionization.

EXPERIMENTAL

An excimer laser (Model 2110, Questek, Billerica, MA) was used to simultaneously pump two dye lasers (DLII, Molectron Corporation, Santa Clara, CA). The first dye laser was tuned to the Hg transition $6^1S_0 \rightarrow 6^3P_1^0$ (λ_1=253.7 nm) and the second dye laser to the transition $6^3P_1^0 \rightarrow 7^3S_1$ (λ_2=435.8 nm). The measured spot size for the first and second laser beams was 3×4 mm and 4×4 mm, respectively. The pulse durations were 10 ns for the first laser and 8 ns for the second laser. The pulse energy for the first laser was 0.15 mJ and 0.2 mJ for the second laser. The laser beams were directed into a quartz cell, which was saturated (at room temperature) with mercury vapor mixed with varying pressures of buffer gas. Neon was chosen as the buffer gas because of its well characterized emission lines in the red and orange region of the spectrum, which can be easily selected using simple glass filters. The cell used is shown in Figure 1.

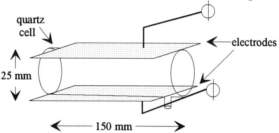

Figure 1. Quartz cell used in experiment.

The cell contained no electrodes but instead used two external Al electrodes to create a pulsed electric field inside the cell. A high voltage pulse generator, constructed in house, was used to produce two bipolar pulses of variable duration from 20 to 170 ns with a voltage up to 20 kV; the time delay of the pulses with respect to the trigger could be varied from 0 to 2 μs and the separation between the pulses could range from 0 to 11 μs. To detect the emission signal, a Hamamatsu 1P28 photomultiplier tube masked by an orange OS-14 filter (LOMO, St. Petersburg, Russia) was used; the OS-14 filter is transparent for all wavelengths above 570-580 nm. The PM tube was connected to a Tektronix TDS 620A digital oscilloscope.

RESULTS AND DISCUSSION

When a pulsed electric field was applied (applied voltage = 13 kV) to the cell, a spatially homogeneous, strong emission signal, dependent on both laser intensity and wavelength, was observed visually. The time-resolved emission characteristics of this signal can be seen in Figure 2; the duration of the signal pulse was dependent on the HV and was shifted 250 ns from the laser pulse. In this figure, the first peak is due to laser-induced luminescence from the cell windows and the second and third, corresponding in time to the positive (first peak) and negative (second peak) polarity electric field pulses, are due to the neon emission signal (see Figure 2a). The signal was proportional to the number of electrons created by RI. The quantitative characteristics of the third peak were poorer than the second one; therefore, all further experiments were carried out with only one pulse (Figure 2b). The maximum emission occurred with 100-150 Torr neon (see Figure 3).

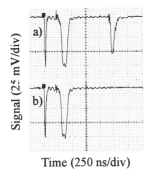

Figure 2. Time-resolved emission characteristics of the emission signal; HV pulse applied is 13 kV.

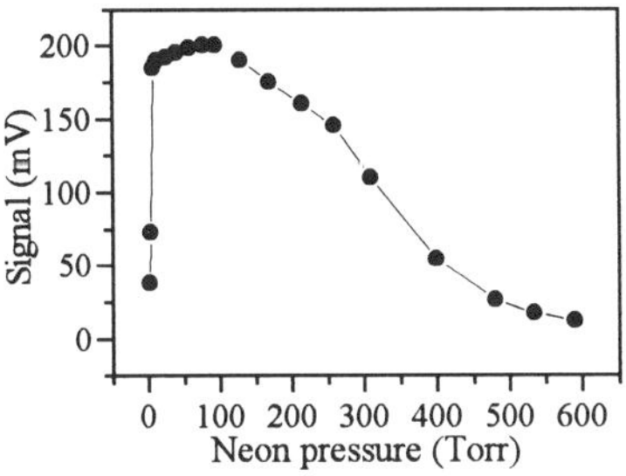

Figure 3. Signal as a function of neon pressure; variable high voltage pulse duration.

A comparison of the optical signal to an electrical signal using a similar quartz cell which contained electrodes(6) is shown in Figure 4. In both cases, a wavelength dependence was observed, although the behavior of the two signals is different. Within 0.2 nm of the Hg line (253.7 nm), the electrical signal dropped four orders of magnitude while the optical signal dropped only two orders of magnitude, while in the wings of the line, the wavelength dependence is similar for the two signals; the difference was assumed to be due to the nonlinear effect of a space charge at higher concentrations (~10^8-10^9 species/cm^3) on the optical signal.

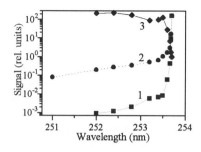

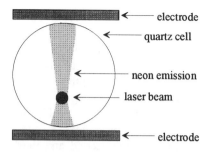

Figure 4. Signal as a function of wavelength. Curve 1 is the electrical signal, curve 2 is the optical signal, and curve 3 is the ratio.

Figure 5. Schematic representation of position-dependent neon emission (yz-plane.)

The neon emission occupied a plane between the two Al plates which was dependent on the position of the laser beams (see Figure 5). For tightly focused beams, a spatial resolution of 1 mm for the neon discharge image in the horizontal plane was achieved. When both lasers were blocked and the maximum signal duration was set, the radiation from the neon plasma was visually observed to become rather inhomogeneous with randomly appearing strata, which can be interpreted as the detection of a very small number of residual electrons randomly distributed along the cell. This indicates the possibility of detecting the presence of possibly even a single electron in addition to its original position. In this way, the possibility of using this technique as an image detector is proven.

ACKNOWLEDGMENTS

This work was supported by a grant from the US Department of Energy (DOE-DE-FGO5-88-ER13881).

REFERENCES

(1) Travis, J., Turk, G., ed. *Laser Enhanced Ionization Spectrometry*, New York: John Wiley & Sons, 1996, pp. 99-160.

(2) Arimondo, E., Di Vito, M., Ernst, K., Ingustio, M., *J. Phys.*, **44**, C7-267-74(1983)

(3) Berglind, T., Casparsson, L., *J. Phys.*, **44**, C7-329-34(1983).

(4) Niemax, K., *Appl. Phys.*, **B32**, 59-62(1983).

(5) Matveev, O.I., Omenetto, N., "Some Considerations on Signal Detection and Signal to Noise Ratio in Laser Enhanced Ionization Spectroscopy," in *AIP Conference Proceedings No. 329 (RIS 94)*, 1994 , pp. 515-18.

(6) Matveev, O.I., Smith, B.W., Omenetto, N., Winefordner, J.D., "Resonance Ionization Detection of 253.7 nm Photons from Mercury Atoms," in *AIP Conference Proceedings (RIS 96)*, 1996.

Two-dimensional ion-imaging of fragment angular distributions after photolysis of state-selected and oriented triatomic molecules

J.M. Teule, M.H. Hilgeman, M.H.M. Janssen, D.W. Chandler*,
C.A. Taatjes* and S. Stolte

*Dept. of Physical and Theoretical Chemistry, Lasercentre Vrije Universiteit, De Boelelaan
1083, 1081 HV Amsterdam, The Netherlands (e-mail adress: teule@chem.vu.nl)*
*Combustion Research Facility, Sandia National Laboratories,
P.O.box 969, Livermore CA 94551*

Abstract. Photodissociation experiments of state-selected and oriented triatomics are presented. Selective ionization using REMPI in combination with two-dimensional ion-imaging allows us to measure both the internal energy and angular distribution of the fragments. The dissociation of N_2O is studied using one laser around 204 nm for both the dissociation of the molecule and the ionization of the fragments. The angular distributions of $O(^1D)$ and $N_2(J)$ are presented and implications of these results on the dissociation dynamics are discussed.

INTRODUCTION

The study of photodissociation dynamics gives direct information on the intramoleculair forces during the breaking of a chemical bond. If the quantum states of the parent molecule prior to dissociation can be selected, and the internal energy and angular distribution of the fragments are measured, one can obtain more details of the dissociation dynamics on the anisotropic excited state potential.

For randomly oriented parent molecules the deviation of the fragment recoil distribution from a spherical angular distribution is described by the anisotropy parameter β (1):

$$I(\theta) \propto [1+\beta P_2(\cos\theta)]$$

where θ is the angle between the polarization of the photolysis laser and the recoil direction, and P_2 the second order Legendre polynomial. In the case of state-selected and oriented molecules higher order terms of the Legendre polynomials are necessary to describe the fragment angular distribution. These terms can elucidate effects of rotation on the timescale of dissociation, the angle between the transition dipole moment and the laser polarization or the dynamics on the potential energy surface.

Linear triatomic molecules with one quantum in the bending vibrational mode can be state-selected using the hexapole focussing technique. The photodissociation of N_2O near 200 nm is especially interesting, because it almost exclusively forms oxygen atoms in the excited (^1D)-state. The N_2 fragment is formed in highly rotationally excited states, which is explained by assuming that the dissociation occurs via a vibronically allowed transition to a bent excited state (2). Though the anisotropy parameter and translational energy distributions of $O(^1D)$ have been studied by several workers (3,4,5), the photodissociation dynamics are still not well understood. Two-dimensional ion-imaging of the fragment angular distributions after photolysis of state-selected and oriented N_2O can possibly elucidate the formation of aligned $O(^1D)$ and the role of an extra dissociation channel in greater detail.

STATE-SELECTION AND ION-IMAGING

N_2O molecules can acquire an angular momentum along the molecular axes by l-type doubling of the bending vibrational mode (6). This makes it possible to select N_2O in a particular (J/M) state by an electrostatic hexapole field. In an N_2O beam that is strongly rotationally cooled by a supersonic expansion, a single (J/M) state can be selected and focussed into the dissociation region. In the laser interaction zone a strong electric field can be applied to orient the state-selected molecules prior to dissociation.

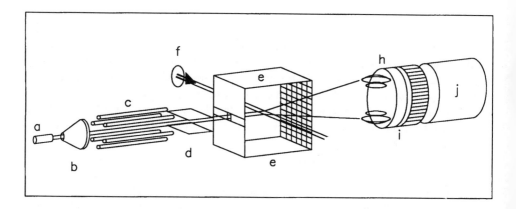

FIGURE 1 Schematic view of the experimental apparatus: a) pulsed nozzle, b) skimmer, c) hexapole, d) guiding field electrodes, e) orientation field electrodes, f) dissociation and ionization laser, h) micro channel plate, i) phosphor screen, j) CCD camera.

A linearly polarized laser beam is used for both the dissociation of the state-selected N_2O molecules and the selective ionization of the fragments. The N_2 fragments are detected with (2+1) REMPI using the $a''^1\Sigma_g^+ \leftarrow X^1\Sigma_g^+$ transition around 203 nm. $O(^1D)$ can be ionized via (2+1) REMPI using the $(^1P_1 \leftarrow {}^1D_2)$ transition at 205.4 nm or via the $(^1F_3 \leftarrow {}^1D_2)$ transition at 203.7 nm. The ions are accelerated by an extraction field perpendicular to the laser beam into a time-of-flight (TOF) tube containing a micro channel plate detector. A phosphor screen converts the signal into light, which is imaged by a CCD-camera. The three-dimensional recoil distribution is thus projected onto a two-dimensional surface. Due to cylindrical symmetry of the original distribution around the laser polarization axis, the three-dimensional distribution can be reconstructed using an Abel-transformation.

RESULTS AND DISCUSSION

The N_2O photodissociation is attributed to the $2^1A'(B^1\Delta) \leftarrow 1^1A'(X^1\Sigma)$ transition (7). A parallel transition via this bent excited state is consistent with the high rotational excitation in the N_2 fragment (2). The photofragments are not produced exclusively along the direction of the laser polarization, as can be seen in figure 2. Two images are shown of the $O(^1D)$ recoil distribution, after photodissociation of state-selected N_2O ($v_2-1,J=1,l=1$) using either 205.4 nm (figure 2a) or 203.7 nm (figure 2b). The laser polarization is directed along the Y-axis. These figures show that detection of $O(^1D)$ via the $(^1F_3)$ state results in a more isotropic image than detection via the $(^1P_1)$ state. After Abel inversion of the images an effective β parameter can be extracted from the angular distributions. Detection via the $(^1P_1)$ state results in a β_{eff} of 1.4, whereas the image detected via the $(^1F_3)$ state gives a smaller β_{eff} of about 0.5.

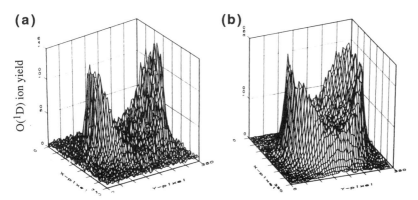

(a) **(b)**

O(^{1}D) ion yield

FIGURE 2 The two-dimensional images of $O(^1D)$ probed (a) via the ^1P state and (b) via the ^1F state are shown. Parent molecules N_2O were state selected but not oriented. The laser polarization is directed along the Y-axis.

The value of $\beta_{eff} \approx 0.5$ observed for the $(^1F_3 \leftarrow {}^1D_2)$ detection is in agreement with previous reports of an anisotropy parameter, where N_2O was dissociated with 193 or 203 nm light (4,5). The difference between the two detection schemes implies a correlation between the alignment of the electronic angular momentum of the $O(^1D)$-atom and the recoil direction of the fragments. In our experimental geometry detection via the $(^1P_1 \leftarrow {}^1D_2)$ transition is only sensitive for $|M|=1$ states (defining the laser polarization axis as quantization axis), whereas using $(^1F_3 \leftarrow {}^1D_2)$ detection mainly $|M|=2$ states are probed. Since we start with N_2O ($l=1$), the initial angular momentum along the quantization axis is $|M|=l=1$. Conservation of angular momentum along this axis during photoabsorption (parallel transition) results in ejection of $O(^1D)$ atoms with $|M=1|$ primarily along the polarization direction. This can explain the relatively high β parameter observed with the $(^1P_1 \leftarrow {}^1D_2)$ detection scheme.

A closer look at the $(^1P_1 \leftarrow {}^1D_2)$ image shows a weak and narrow signal around a scattering angle of 90°, which is associated with molecules having a slightly larger velocity compared to the stronger signal near 0°. This second component was also observed by Suzuki et al. (3), who dissociated non-state-selected N_2O using 205 nm. They determined a β_{eff} of 0.6, which they believe to consist of two components: a slow component with $\beta \approx 1.3$ and a fast component with a very low β. The minor component is ascribed to the contribution of overlapping electronic transitions resulting in a different dissociation pathway.

We are now working on a two-laser experiment, where the N_2O molecule is dissociated using a 193 nm excimer laser, and seperately ionized using either 203 or 205 nm. With the possibility to change the polarization of the probe laser independently, quantitative measurements on the orbital alignment of $O(^1D)$ can be realized. Furthermore, the photodissociation of oriented molecules using two lasers can elucidate more terms of the Legendre polynomials describing the angular distribution.

REFERENCES

1. Zare, R.N., *Mol. Photochem.* **4**, 1-37 (1972).
2. Hanisco, T.F., Kummel A.C., *J. Phys. Chem.* **97**, 7242-7246 (1993).
3. Suzuki, T., Katayanagi, H., Mo, Y., Tonokura, K., *Chem. Phys. Lett.* in press.
4. Springsteen, L.L., Satyapal, S., Matsumi, Y., Dobeck, L.M., Houston, P.L., *J. Phys. Chem.* **97**, 7239-7241 (1993).
5. Felder, P., Haas, B.M., Huber, J.R., *Chem. Phys. Lett.* **186**, 177-182 (1991).
6. Mastenbroek, J.W.G., Taatjes, C.A., Nauta, K., Janssen, M.H.M., Stolte, S., *J. Phys. Chem.* **99**, 4360 (1995).
7. Hopper, D.G., *J. Chem. Phys.* **80**, 4290 (1984).

Laser Ion Source for Selective Production of Short-Lived Exotic Nuclei

Yuri A. Kudryavtsev, Andre Andreyev, Nathalie Bijnens, Jürgen Breitenbach, Serge Franchoo, Johnny Gentens, Mark Huyse, Andreas Piechaczek, Riccardo Raabe, Ils Reusen, Piet Van Duppen, Paul Van Den Bergh, Ludo Vermeeren and Andreas Wöhr.

Instituut voor Kern- en Stralingsfysika, K.U.Leuven
Celestijnenlaan 200 D, B-3001 Leuven, Belgium.

Abstract. An element selective laser ion source has been developed for the production of pure beams of exotic radioactive nuclei. It is based on selective resonant laser ionization of nuclear reaction products thermalized and neutralized in a noble gas at high pressure. The ion source has been installed at the mass separator (LISOL) which is coupled on line to the cyclotron at Louvain-la-Neuve. The ion source has been tested in a wide range of recoil energies going from 1.3 MeV to 90 MeV. Pure beams of ^{54,55}Ni and ^{54}Co, produced in a light-ion induced fusion-evaporation reaction, and of ^{113}Rh and ^{69,71}Ni, produced in proton-induced fission of ^{238}U, were obtained. An efficiency of the ion source of 6.6% for fusion reactions and of 0.22% for fission reactions has been obtained. A selectivity of the ion source of 300 for fusion and 50 for fission reactions has been achieved.

INTRODUCTION

Exotic nuclei far from stability are produced in nuclear reaction in a limited amount. Such nuclei have very short life time and are usually overwhelmed by much more abundant isobars. Laser multistep ionization in combination with a high-pressure gas cell can provide high ionization efficiency, high selectivity and fast extraction of nuclear reaction products from the target-ion source system.

The operational principle of the ion source is based on the selective laser resonant ionization of nuclear reaction products stopped in a high-pressure noble gas [1-3]. A cyclotron beam hits a thin target located in a chamber filled with noble gas. The nuclear reaction products recoiling out of the target are thermalized as neutrals or as 1^+ charged ions and move together with the noble gas in the direction of the exit hole. After a few milliseconds, most of the ions are neutralized due to recombination with plasma electrons created by the primary cyclotron beam. This short ion survival time is the main limiting factor in the efficiency of the standard ion guide ion source [4]. While the atoms move towards the exit hole they pass the laser ionization zone where only selected atoms are ionized by the laser beams. The laser produced ions are then guided by the gas flow through the exit hole, behind which most of the gas is removed by differential pumping. The ions are directed by the skimmer electrode towards the extraction electrode and the analyzing magnet. At the exit of the mass separator an element and isotope pure beam is obtained. The element selectivity is ensured by laser resonance ionization and the isotope selectivity by the mass separator.

THE LASER ION SOURCE FACILITY AT LISOL

The laser ion source has been developed and installed at the front end of the Leuven Isotope Separator On-Line (LISOL), which is coupled on-line to the isochronous cyclotron CYCLONE. Fig. 1 shows the general layout of the laser ion source facility. The cyclotron delivers $^3He^{++}$ beams with an energy of 45 MeV or 27 MeV for a light-ion induced fusion-evaporation reaction and a $^1H^+$ beam with energy 65 MeV for a proton-induced fission reaction.

The body of the ion source has an inner diameter of 2 cm. The cyclotron beam enters the source through a 5 μm Havar window and hits the target(s). High purity helium or argon is used as buffer gas. The diameter of the exit hole is equal to

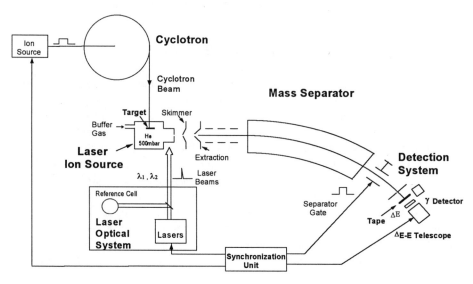

FIGURE 1. General lay-out of the laser ion source facility at Louvain-la-Neuve

0.5 mm. With this diameter, the mean evacuation time of reaction products is equal to 40 ms for He and 120 ms for Ar. A beam of stable ions can be obtained by laser ionization of atoms produced by resistive heating of a filament (Ni, Co or Rh) inside the ion source.

The laser optical system consists of two excimer XeCl lasers (Lambda Physik LPX240i, 400 Hz, 150 mJ, 15 ns) which pump two dye lasers (Lambda Physik Scanmate 2). In order to get light in the region of 230 nm, the radiation from the first step laser is frequency doubled. The laser beams are directed by a set of prisms into the ion source area located at a distance of 15 m from the laser optical system. To tune the laser wavelengths in resonance with the atomic transitions a fraction of the laser beams is deflected into a reference cell. Effective schemes of two-step two-color ionization have been found for Ni, Co and Rh. Table 1 presents wavelengths and energy levels that have been used for the on-line experiments. The chosen atomic transitions can be saturated with the available laser power.

The ions extracted from the ion source are accelerated to an energy of 50 keV. The accelerated ions are then separated according to their mass over charge ratio in a 55^0 dipole magnet with a radius of 1.5 m. After mass separation, the radioactive isotopes are implanted in a movable collector tape. A ΔE-E plastic detector telescope is used for detection of the beta-particles. Gamma-rays are measured by a Ge-detector.

LIGHT ION-INDUCED FUSION EVAPORATION REACTIONS

The on-line performance of the laser ion source has been tested with ^{3}He-induced fusion evaporation reactions on a ^{54}Fe (3 mg/cm^2) target. Table 2 shows the obtained results for yield, efficiency and selectivity. The total efficiency of ion source is determined by the following processes: stopping of reaction products in helium, thermalization in the atomic ground state, survival against formation of molecular ions, transportation towards the exit-hole, photo ionization, acceleration and mass-separation. The contributions of these processes to the total efficiency have been investigated [3].

The production of short living Ni isotopes with a standard target-ion source system is limited by there slow release. The β decay of ^{54}Ni has so far not been

TABLE 1. Transitions used for two step laser ionization of Ni, Co and Rh. λ_1, λ_2 - wavelengths of the first and second step transitions, E_0, E_1, E_2 - energies of the ground intermediate and autoionization levels, f_{01}- absorption oscillator strength.

	First step				Second step		Ion. limit
	λ_1 (nm)	E_0-E_1 (cm^{-1})		f_{01}	λ_2 (nm)	E_2 (cm^{-1})	(cm^{-1})
Ni	232.003	0 - 43090	$^3F_4 \to {}^3G_5^0$	0.69	537.84	61678	61619
Co	230.903	0 - 43295	$^4F_5 \to {}^4F_5^0$	0.045	481.90	63787	63565
Rh	232.258	0 - 43042	$^4F_{9/2} \to$?	?	572.55	60503	60200

181

TABLE 2. Yield, efficiency and selectivity of the laser ion source for different isotopes.

Isotope ($T_{1/2}$)	Reaction	Yield [a] at/pμC (calculation)	Yield [b] ion/pμC (experiment)	Eff. [c] %	Selectivity [d]
^{55}Ni (204ms)	^{54}Fe(^{3}He,2n)^{55}Ni	$3.6\ 10^4$	1650	4.6	280
^{54}Co (193ms)	^{54}Fe(^{3}He,2np)^{54}C	$1.2\ 10^5$	7880	6.6	>380
^{113}Rh (2.7s)	^{238}U(p,f)^{113}Rh	$1.9\ 10^6$	4260	0.22	50

a) number of radioactive atoms recoiling out of the target per pμC of primary cyclotron beam
b) number of radioactive ions in the mass separated beam per μC of primary cyclotron beam
c) [ions/s in mass separated beam] / [atoms/s recoiling out the target] (%),
d) ions/s in mass separated beam (lasers-ON) / ions/s in mass separated beam (lasers-OFF)

studied. The high selectivity (~300) and efficiency (~5%), and short release time (~40 ms) of the laser ion source allowed to measure the decay properties of ^{54}Ni for the first time [5].

PROTON-INDUCED FISSION OF ^{238}U.

The performance of the ion source has also been tested for proton-induced fission on a set of two ^{238}U (10 mg/cm^2) targets, see Table 2. The main difference with the previous case consists in the much higher recoil energy of the reaction products. The fission of uranium atoms gives very wide A and Z distribution of fragments. A production yield of 4260 at/μC was reached for ^{113}Rh (recoil energy 90 MeV) resulting in a total efficiency of 0.22%. This lower value is mainly due to the stopping efficiency of the ^{113}Rh atoms. Only 1.5% of the Rh atoms can be stopped in 500 mbar He, the gas load limit of current apparatus.

The neutron rich nickel isotopes are of interest for nuclear and astrophysics. They can be produced in proton induced fission of ^{238}U. For the first time γ rays were observed from isotopes ^{71}Ni [6].

REFERENCES

1. P. Van Duppen et al., Hyperfine Interactions, **74**, 193-204 (1992).
2. L. Vermeeren et al., Phys. Rev. Let. **73**, 1935-1938 (1994).
3. Yu.A. Kudryavtsev et al., Nucl. Instr. Meth. Phys. Res. **B114**, 350-365 (1996).
4. J. Ärje et al., Nucl. Instr. Meth. Phys. Res., **B26**, 384-393 (1987).
5. I. Reusen et al., 'Precise half-life measurement of ^{55}Ni and first decay properties of ^{54}Ni" in Proceedings of International Conference on Exotic Nuclei and Atomic Masses, 1985, pp. 757-758
6. S. Franchoo et al., 'First observation of neutron rich nickel isotopes with the LISOL laser ion source", Presented at the Belgium Physical Society Meeting, Brussels, June 6-7, 1996.

LASER TECHNOLOGY IN ART CONSERVATION

C. Fotakis, D. Anglos, S. Couris, S. Georgiou, V. Zafiropulos, I. Zergioti

Foundation for Research and Technology-Hellas, Institute of Electronic Structure and Laser
Laser and Applications Division
P.O. Box 1527, 711 10 Heraklion, Crete, Greece

Abstract. Applications of modern laser technology in art conservation are reviewed with emphasis in the conservation of painted artworks. In particular, processes and critical parameters for the cleaning of paintings by excimer lasers are discussed. The influence of photoxidation or free radical formation processes leading to potential damage is examined in experiments involving model and realistic systems. Laser based spectroscopic techniques for non-destructive pigment and media analysis are also presented. A particular issue in this respect is the applicability of such techniques for on-line control of the cleaning process safeguarding from any damage.

INTRODUCTION

Lasers have been employed in art conservation, primarily cleaning, over the last twenty years. One of the most mature applications has been the use of pulsed lasers emitting in the near infrared, for example the Nd:YAG lasers, for the removal of black encrustations from marble, limestones and frescoes (1). In another application, CO_2 lasers have been used for the cleaning of metallic artifacts. Applications of laser techniques for the conservation of painted artworks are only recent and clearly more demanding due to the sensitivity of the painted surface in undergoing laser induced photochemical and thermal changes. Despite these difficulties, there are several examples of successful painting conservation by lasers, including :

1. Laser cleaning of surface layers and/or overpaint removal
2. Laser based spectroscopic techniques for the analysis of binding media and pigments

3. Advanced imaging and laser interferometric techniques for structural diagnostics and
4. Laser based authentication techniques

As in any other laser material processing application, process optimization requires the appropriate interplay among the laser and material parameters involved. In turn, this relies on understanding fundamental aspects of laser matter interaction phenomena occurring in complex systems such as artworks. Here, we shall discuss several of the above applications placing emphasis on the underlying physical and chemical processes involved.

LASER CLEANING OF PAINTED ARTWORKS

Traditional methods of painting conservation rely on mechanical or chemical techniques chosen by a conservator. Because these processes are difficult to control, extensive expertise is necessary to achieve an optimal result. Furthermore, in many cases chemical solvents may penetrate the painting and damage the pigments and media. The conventional analytical techniques applied, although some of them are powerful, they require sampling from the artwork which in most cases is destructive. A major advantage provided by lasers is the degree of control possible which, as has been recently demonstrated, can be applied on-line during the cleaning process safeguarding from potential damages (2).

A typical laser cleaning workstation is shown in Fig. 1. In this version, the painted artwork is mounted on a PC controlled XYZ translator and the cleaning processes can be monitored by recording broad band reflectance spectra and/or emission spectra obtained from the laser produced plasma.

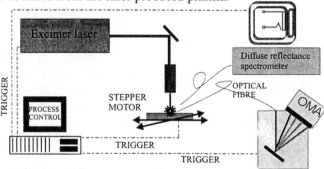

FIGURE 1 : Workstation for excimer laser cleaning of painted artworks.

Laser cleaning is based on the removal of a well defined layer of surface material under fully controlled conditions. It has been shown that in the case of paintings, contaminated surface layers of 20 to 300 μm thickness can be removed by excimer laser ablation with a resolution of 0.3 to 1 μm per pulse (3). This is shown in Fig. 2 for the cleaning of model samples of oxidized varnish (dammar and mastic) by KrF laser emitting at 248 cm. Figure 2a shows ablation rate curves obtained as a function of laser fluence. Considering that a typical varnish (or paint) stroke applies 15 to 20 µm of material and that a typical varnish layer is 50 to 80 µm thick corresponding to 3 to 4 strokes, it is clear that cleaning by excimer laser ablation can be a highly selective process, capable of leaving a thin protective layer of clean varnish over the painted area. From such curves the optimum fluence can be determined for each cleaning case.

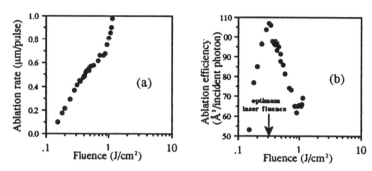

FIGURE 2 : Laser parameter optimization for the removal of dammar varnish film.

STUDY OF PHOTOABLATION EFFECTS

Cleaning by laser ablation may induce three major effects, which are important for the future state of the painted surface. These include thermal interactions, photochemical and photomechanical effects. Although all these processes may operate simultaneously, photomechanical effects become important at higher laser fluences, thermal and photomechanical effects are favored when pulsed infrared lasers are used while photochemical processes leading to photooxidation or free radical formation may be dominant with excimer lasers emitting in the ultraviolet. To assess the interference of any photochemical processes induced during laser cleaning, model systems, based on solid mixtures of a varnish (dammar or mastic) and chlorobenzeme (C_6H_5Cl) were studied under ablation by a KrF laser. C_6H_5Cl has a well known photochemistry in the gas phase and in solution, characterized by a high photodissociation quantum yield at 248 nm. In this way, the study of its photolysis and of the reactivity of the produced fragments provides a sensitive probe of any photochemical processes that occur during laser illumination of such mixtures. To this end, time-of-flight quadrupole mass spectrometry was employed for probing the nature and translational distributions of photoproducts. Figure 3 shows the yield of HCl, as a function of laser fluence for neat films of C_6H_5Cl and in mixtures of dammar, which is a relatively weak absorber and mastic which is a strong absorber at 248 nm.

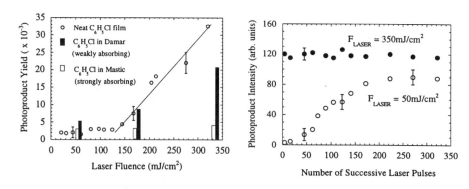

FIGURE 3 : Laser induced photochemical effects studied in model systems.

It is clear that the photofragmentation of C_6H_5Cl is minimal in the case of the strongly absorbing varnish for laser fluences up to 350 mJ/cm^2. In fact, the photolysis yield is lower than that for the gas phase by two orders of magnitude. Furthermore, the mass analysis shows the existense of only HCl and absence of $(C_6H_5)_2$. These observations are compatible with an ablation mechanism, in which the medium retains largely its "structural" integrity preventing phenyl radicals from diffusing and recombining. Furthermore, at low fluences (~50 mJ/cm^2) photoproduct accumulation with successive laser pulses is observed. In striking contrast, at higher fluences, such as those used for laser cleaning (e.g. 300 mJ/cm^2), there is no evidence for accumulation, evidently because of entrainment of the photoproducts in the ejected plume. In conclusion, experiments with these model systems strongly indicate minimal interference from deleterious photochemical effects in the ablated varnish.

LASER BASED ANALYTICAL TECHNIQUES

Parallel to laser cleaning applications, intense interest also exists for the development and application of laser based analytical techniques as tools for artwork diagnostics. Analytical techniques are essential to artwork conservation since they provide important information regarding the physical and chemical structure of artworks. Techniques which are non-destructive and can be performed in situ are highly desirable. To date, a wide variety of physicochemical analytical techniques have been employed to attack quite diverse and often complex problems in art conservation. Recently, laser induced breakdown spectroscopy (LIBS) and laser induced fluorescence (LIF) spectroscopy have been investigated in this respect.

LIBS is a well established, sensitive and highly selective analytical technique appropriate for elemental determination which has found applications in diverse fields (4). As the majority of pigments used in paintings from antiquity to modern times are metal containing substances, LIBS appears to be quite a suitable technique for elemental analysis of pigments. A LIBS apparatus is transportable to the place the artwork is kept and no sample preparation is required. Because the laser beam is focused on a tiny spot (of ~ 30 μm diameter) extremely small quantities of sample material are consumed (of the order of less than a microgram). In addition, spatial resolution is achieved allowing for studies of inhomogeneities in the composition of the sample. The emission signal intensity for most elements of interest is usually high enough so that a single laser pulse is sufficient to provide a good spectrum. Figure 4a shows a typical LIBS spectrum of Ti obtained

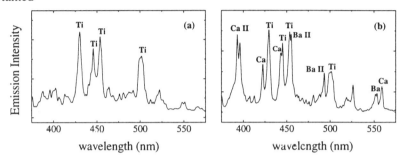

FIGURE 4 : LIBS spectra from (4a) a titanium white sample and (4b) white pigment in an oil painting.

from a titanium white sample. Figure 4b shows a representative LIBS spectrum obtained from white painted areas of a painting attributed to Renoir. The presence of Ti as indicated in Fig. 4b, raises questions regarding the authenticity of this painting since titanium white was not used as a pigment during the times of Renoir. The effectiveness of LIBS for the on-line monitoring of the cleaning process has also been demonstrated (2).

LIF is a versatile, non-destructive analytical technique, which can be performed in situ -on the artwork itself- and provides information which can be directly related to the molecular structure of pigments or binding media, both inorganic and organic. In a typical LIF experiment a low intensity (continuous or pulsed) laser beam is used to excite a small area on the surface of the sample (5). Fluorescent substances in the sample emit broadband fluorescence which is recorded and analysed. Typical fluorescence spectra of pigments are shown in Fig. 5a. In Fig. 5b fluorescence emission from different points of an oil painting is used to differentiate between original and restored areas.

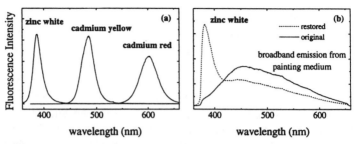

FIGURE 5 : LIF spectra from (5a) pigment samples and (5b) oil painting.

CONCLUDING REMARKS

The transfer and adaptation of established laser material processing and analytical techniques to art conservation appears to be particularly promising. In particular, the appropriate choice of laser parameters permits the controlled cleaning of painted artworks. Laser based spectroscopic techniques are useful for the *in situ* and non-destructive analysis of pigments and media as well as for monitoring the cleaning process. Finally, it should be stressed that the successful implementation of lasers in art conservation is a strongly interdisciplinary field relying on collaboration between art historians, conservators and natural scientists.

AKNOWLEDGEMENTS

The collaboration of Mr. M. Doulgeridis and Ms. A. Fosteridou (National Gallery of Athens) and Mr. Milko den Leeuw (Atelier voor Restauratie & Research van Schilderijen) is aknowledged.

REFERENCES

1. Cooper, M.I., Emmony, D.C., and Larson, J., *Optics and Laser Technology* **27**, 1995, 69.
2. Maravelaki, P.V., Zafiropulos, V., Kylikoglou, V., Kalaitzaki, M., and Fotakis, C., *Spectrochimica Acta*, (to be published, 1996).
3. Fotakis, C., *Optics and Photonics News* **6**, 1995, 30.
4. Cremers, D.A., Barefield II, J.E., Koskelo, A.C., *Appl. Spectrosc.* **49**, 1995, 857.
5. Anglos, D., Solomidou, M., Zergioti, I., Zafiropulos, V., Papazoglou, T.G., and Fotakis, C., *Appl. Spectrosc.* (to be published, 1996).

SESSION VII: RIS
AND LASER DESORPTION

RIS and laser induced fluorescence of Nb from laser-ablated metal surfaces

M. I. K. Santala, H. M. Lauranto, and R. R. E. Salomaa

Helsinki University of Technology, Department of Technical Physics, Rakentajanaukio 2 C, FIN-02150 Espoo, Finland. Email: Marko.Santala@hut.fi

Abstract. Laser ablation was studied as a method for producing atomic and ionic vapor of refractory Nb metal under high-vacuum conditions for RIS and LIF experiments.

INTRODUCTION

We have previously done RIS of Nb with thermal atomization [1,2]. The ultimate goal of our Nb studies is to develop fast neutron dosimetry of nuclear reactor materials by determining their ^{93m}Nb content. Because refractory Nb metal is difficult to evaporate thermally, we have modified our RIS apparatus for use of laser ablation under high-vacuum conditions as an alternative atomization method. We have performed laser-induced-fluorescence (LIF) and RIS experiments in the ablated vapour.

EXPERIMENTAL SETUP

The main component of our experimental setup is a cubic high-vacuum chamber (Fig. 1). The bottom flange is used for sample support, the ablation laser beam enters from the top, and the two side flanges are used for passing the probe laser beams through the chamber in the horizontal direction. The front flange can be used either for ion detection by installing an ion detector or for optical detection by installing a window and suitable optics. The chamber is pumped to 10^{-5}–10^{-6} mbar.

A Nd:YAG laser at 532 nm was used for ablation. The laser beam was first expanded to 2 cm in diameter and then focused into the chamber by a +300-mm lens. The laser intensity was controlled either by attenuating the beam or by varying the distance between the ablation surface and the focusing lens. The maximum pulse energy was 50 mJ, the pulse duration 5 ns, and the repetition rate 20 Hz. The spot size was varied from several mm to below 0.2 mm. The ablation intensities were 10–50 J/cm^2 in the LIF experiments and below 1 J/cm^2 in the RIS experiments. The ablation sample was a commercial Nb foil (Aldrich, 99.8 %).

One (in LIF) or two (in RIS) dye laser beams generated by a XeCl-excimer laser pumped dye laser system were guided through the ablated Nb plume for resonant excitation. The maximum pulse energies of the dye lasers were about 1 mJ, pulse durations about 10 ns, and typical beam diameters in the chamber 2–5 mm. Second harmonic generation was used to generate UV wavelengths below 340 nm with a maximum energy of the order of 100 µJ. The bandwidths were 4–6 GHz or 1–2 GHz with an intracavity etalon.

The Nd:YAG and excimer lasers operate synchronously with an adjustable delay; typical delays were 5 µs (LIF) and 40 µs (RIS). The distance of the dye laser beam from the ablation surface was typically about 4.5 cm in the LIF experiments and about 6.5 cm in the RIS experiments.

In the LIF experiments the fluorescence light was collected by f/8 optics, focused to the entrance slit of a 400 mm monochromator and detected by a photo-multiplier (PM). The PM signal was integrated by a gated integrator and recorded by a PC computer. In the RIS experiments the excited atoms were ionized and then deflected to an ion detector by a high voltage pulse ($\approx$10 kV/cm, 20 ns) triggered by the excimer laser. The signal from the ion detector was recorded as in LIF. For RIS measurements, an electrostatic ion suppressor with a bias voltage of 600 V and a 0.22 µF capacitor was constructed to eliminate the ions in the ablation plume.

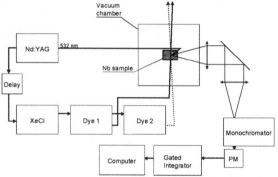

FIGURE 1. Schematic view of the experimental setup in the LIF measurements. In the RIS measurements the optical detection equipment was replaced by an ion detector.

RESULTS

Laser ablation was generally found to produce efficiently atomic or ionic vapour for various spectroscopic experiments. By varying the ablation intensity one can vary the composition and the density of the ablation plume. High ablation intensity produces more ions and denser plumes, which are suitable for LIF experiments with a fairly low inherent sensitivity. Typical ablation velocity was in the range 10–20 km/s. Lower ablation intensity combined with the ion suppressor produces low-density atomic plumes, with velocities down to 1–2 km/s, that are applicable for the sensitive RIS experiments. Figure 2 displays the behaviour of the observed signal when the dye laser(s) was at resonance and the delay between the Nd:YAG and the dye laser(s) was varied. Figure 2a shows LIF signal obtained at high ablation intensity and Fig. 2b the RIS signal at lower intensity. Despite the ion suppressor some ions or highly-excited atoms can enter the interaction region causing unwanted background. This non-resonant signal measured with blocked dye lasers is displayed in Fig. 2c.

Signal stability of the ablation source was not fully satisfactory. Pulse to pulse variation of the signal was over one order of magnitude at worst. Long term variations were occasionally large as well. At large intensities it was possible to burn through the Nb plate ($\approx$1 mm) in a few minutes. The small spot size made the source very sensitive to vibrations: a much larger signal was observed from a fresh surface than from an eroded surface. At small intensities one spot could be utilized for hours with little visible erosion. The composition and the density of the ablation plume, and therefore the observed signal and the non-resonant background, were found to be very sensitive to Nd:YAG intensity. A particularly disturbing feature of the source was sporadic "superpulses", where only large background signal was observed. To eliminate the high-background pulses, an efficient discrimination scheme was developed: another gated integrator detects the ion signal just before the real Nb signal, and the signal data is rejected if too high background is observed.

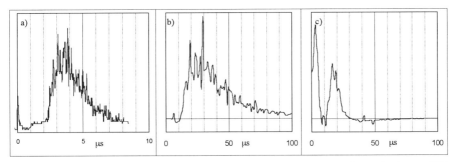

FIGURE 2. Signal vs. Nd:YAG—excimer delay. a) LIF, b) RIS and c) non-RIS background.

In LIF experiments a particularly striking feature of the ablation plume was its high degree of ionization: it was easy to completely saturate the PM by ionic LIF signal while atomic LIF signal was difficult to detect. The degree of ionization could possibly be explained by near-resonant multiphoton ionization [3]. Two photons from a frequency-doubled Nd:YAG correspond to 37577 cm^{-1} and Nb has an excited state at 37578.72 cm^{-1} ($4d^46s$ $^6D_{3/2}$), so that a two-photon transition from ground state ($4d^45s$ $^6D_{1/2}$) could be excited by the wings of the broadband Nd:YAG emission (energy difference 2 cm^{-1}). A third photon would be sufficient for ionization. A typical LIF spectrum obtained by scanning the monochromator is displayed in Fig. 3. The dye laser was tuned to the transition 801—37528 cm^{-1}. The observed peaks correspond to the different decay channels to low-lying states displayed on the state diagram.

To obtain a strong RIS signal requires that both dye lasers are tuned to resonances. As the atom density as well as the non-resonant background were strongly dependent on the ablation intensity, finding the resonances could be tedious, especially as the resonances usable for the second step are not tabulated.

Several different resonance ionization excitation schemes were studied. Hyperfine structures of several $J = 1/2 \to 1/2$ transitions were observed. Previously unknown splitting of the state 23984.87 cm^{-1} ($4d^4 5p\ ^6F_{1/2}$) was measured and found to be about 7 GHz, a surprisingly large value for a state with no s-electron. Fig. 4 shows the RIS signal as a function of the first-step wavelength. The second step wavelength was kept constant.

In conclusion, despite some stability problems, laser ablation was found to be a good alternative to thermal evaporation.

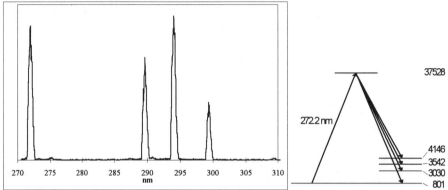

FIGURE 3. Typical LIF emission spectrum (left) and the corresponding level scheme.

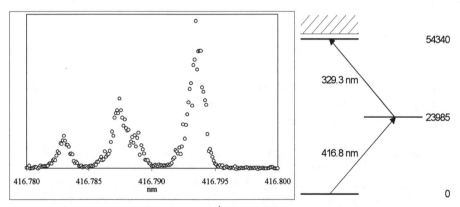

FIGURE 4. Spectrum of 0—23985 cm^{-1} transition and the excitation scheme used.

[1]   H. Lauranto, T. Kajava, T. Lönnroth, R. Salomaa, and J. Tulkki. Inst. Phys. Conf. Ser. 114: RIS-90, 307–311 (1991).
[2]   H. M. Lauranto, T. T. Kajava, and R. R. E. Salomaa. Spectrochimica Acta **51B**, 175–180 (1996).
[3]   C. G. Gill, T. M. Allen, J. E. Anderson, T. N. Taylor, P. B. Kelly, and N.S. Nogar. Appl. Opt. 35, 2069–2082 (1996).

Resonant Laser Ablation:
Mechanisms and Applications

J. E. Anderson, T. M. Allen[†], A. W. Garrett, C. G. Gill, P. H. Hemberger
P. B. Kelly[†] and N. S. Nogar

Chemical Sciences and Technology, MS J565, LANL, Los Alamos,
New Mexico 87545
(505) 665-7279, FAX (505)665-4631, nogar@lanl.gov

[†]*Department of Chemistry, University of California, Davis, California 95616*

Abstract. We will report on aspects of resonant laser ablation (RLA) behavior for a number of sample types: metals, alloys, thin films, zeolites and soil. The versatility of RLA is demonstrated, with results on a variety of samples and in several mass spectrometers. In addition, the application to depth profiling of thin films is described; absolute removal rates and detection limits are also displayed. A discussion of possible mechanisms for low-power ablation are presented.

Introduction

Ever since the first report of laser action, it has been recognized that laser ablation (evaporation/volatilization) may provide a useful sampling mechanism for chemical analysis. In particular, laser ablation is rapidly gaining popularity as a method of sample introduction for mass spectrometry. While most laser ablation/mass spectrometry has been performed with fixed frequency lasers operating at relatively high intensities/fluences ($\geq 10^8$ W/cm^2, ≥ 1 J/cm^2), there has been some recent interest in the use of low-power tunable lasers to ablate and resonantly ionize selected components in the ablation plume. This process has been termed resonant laser ablation (RLA) [1]. Potential advantages of RLA include: 1) simplification of the mass spectrum; 2) improvement of the absolute detection limits; and 3) improvement in relative sensitivity.

This work reports on aspects of RLA behavior for a number of sample types, including metals [2], alloys [3], thin films [4], zeolites [5] and soil. The versatility of RLA for analytical measurements is outlined, with results on a variety of samples and in several mass spectrometers. In addition, the application to depth profiling of thin films is described; absolute removal rates and detection limits are also displayed. An analysis of a variety of fundamental diagnostic experiments and calculations is given, and possible mechanisms for low-power ablation will also be presented.

Experimental

The experiments presented by this work were conducted using either a quadrupole ion trap mass spectrometer (ITMS) or in a linear and/or reflectron Time of Flight (ToF) mass spectrometer. Both experimental arrangements are based upon excimer pumped dye lasers, using a Questek XeCl excimer : Lumonics HD300 dye laser for the ITMS work and a Lumonics XeCl excimer: Lambda Physik FL2002 dye laser for the ToF experiments. The details of the experimental apparatus may be found elsewhere [2,3]. Figure 1 shows a diagram of the experimental apparatus used for each mass spectrometer.

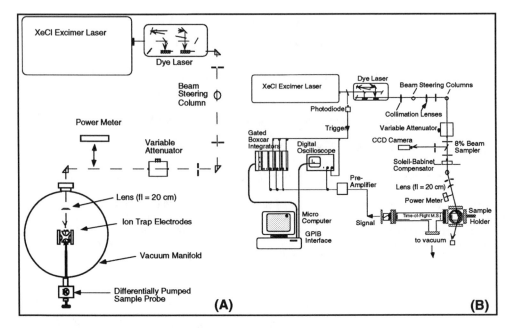

FIGURE 1. Schematic of the ITMS (A) and ToF (B) experimental system.

Samples were prepared as probes for these experiment by machining into a probe tip for metal samples or by bonding thin film wafers to suitable probes with epoxy resin. The powdered samples such as soils or zeolites were bonded to either Si wafers on stainless steel (SS) probes or simply onto the SS probes using a cyanoacrylate ("super glue") based cement.

Mechanistic Studies and Measurements

The use of relatively low fluence laser pulses ($\leq 10^7$ W/cm^2) to sample and selectively ionize analytes from a bulk solid sample has been referred to as resonant laser ablation (RLA) by this and other [6,7] research groups. This method has received considerable attention by this group since this laser based method requires minimal sample preparation (i.e.. no chemical treatments or digests) and affords good discrimination of analyte signal above the background. In RLA, analyte is both desorbed and resonantly ionized by photons from the same laser pulse, with excellent spatial and temporal overlap. Assuming a nominal velocity of 1 X 10^5 cm/s [2] and a 15 nsec pulse duration, the spatial extent of the laser generated plume at the sample surface is ≈ 15 μm, corresponding to a plume density of ≈ 0.1 Torr. The typical ionization fraction for the "2+1" ionization step is $\approx 10^{-3}$, and the typical desorption yield 10^{-6} for the RLA process.

Resonance scans of the various RLA ionization schemes show that the transitions observed are laser linewidth limited at ~0.17 cm^{-1} [3] but can be readily saturated by small increases in laser power since RLA signals have been observed [1] to have a slope ≈ 5 dependence upon laser intensity. By scanning the dye laser across entire transition multiplets for thermometric elements present in a sample and recording spectra, the electronic temperature for RLA can be calculated based upon intensities. For the two photon spectrum of the a^{5}F - e^{5}F and a^{5}D - e^{5}D multiplets of Fe(I), the calculated electronic temperature for RLA was found to be ≈ 1100K.

In addition to the electronic temperature measurements, kinetic energy analysis of the neutrals leaving the sample surface was measured by using a probe laser focused at one of several fixed distances from the sample. By varying the delay between the firing of the ablation laser and the probe laser, velocity distributions were measured. These were then converted to translational temperatures of ≈ 2700 K [2]. This and the previous measurements do not compare well with the calculated surface temperature of ≈ 400 K based upon a one dimensional T-Jump calculation; because the observed desorption yield is higher than anticipated for a strictly thermal desorption process, a non-thermal desorption process is suspected.

The relative signal strength observed for RLA is strongly dependent upon the polarization of the laser light, and also upon the surface quality of the samples. For rough sample surfaces (prepared by 500 grit abrasive paper), the signal intensities from RLA reflect the relative intensity of the residual laser light reflected by the sample (S-polarization gave the strongest reflection and ion signal). In contrast, for highly polished surfaces (100 Å Cu on Si or diamond polished Cu metal), the intensity of the RLA signal was opposite to that observed for light reflected by the sample. This suggested that although the same trend was observed for reflected light as a function of polarization, the different sample surfaces yielded dramatic differences in ion yield as a function of both surface preparation and polarization. These data are presented in figure 2.

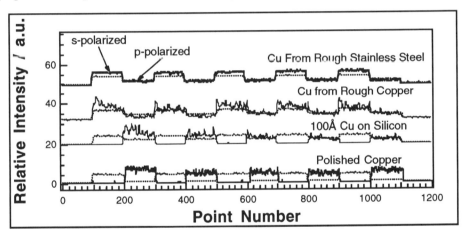

FIGURE 2. Polarization Dependence of RLA signal for different surface qualities. The solid lines represent observed ion signal whereas the dotted lines represent the relative intensity of the laser pulse reflected by the sample surface.

Summarizing the data presented, the material removed from solid samples by the low fluence laser pulses used is much higher than would be anticipated for a strictly thermal process. The wide variation in temperature values determined experimentally from those calculated based upon a thermal desorption mechanism add credibility to a non-thermal desorption hypothesis, and the non-linear intensity dependence of the signal adds further support. The polarization dependence based upon surface preparation suggests that there may be some direct photo physical interaction occurring, such as could be mediated by a surface plasmon interaction. Although surface plasmon interactions are strictly forbidden at planar surfaces, on a microscopic level, the polished surfaces and deposited films have irregularities such as aggregations, kinks and scratches that could act to couple electric field strength from the incident radiation to allow surface plasmon phenomena to occur.

Analytical Applications of Resonant Laser Ablation

A number of analytical applications have been addressed by the application of RLA to chemical analysis. By measuring the time (or number of laser shots) required to penetrate copper films of known thickness [2,3], low fluence removal rates were determined. For typical pulse energies, the average volume removed was 3×10^{-14} cm^3 per pulse. Assuming the same density as for bulk copper, this yields an average of $< 10^9$ atoms/shot, or a depth removal rate of $\approx 10^{-2}$ monolayer/shot and mass removal rate (MMR) of $\approx 10^{-13}$ g/shot. Assuming the MMR is similar for copper films as it is for a bulk copper sample, then the absolute detection limits for trace components in copper is given by the product of MMR and the component's concentration. Lead was easily ionized and detected in a NIST (#494) copper sample at 25 ppm levels, with a S/N ≥ 100, resulting in conservative absolute detection limits in the low attogram range.

The analysis of complex samples by RLA allows the analyst to discriminate analyte signal from those of the bulk matrix, simply by selectively ionizing species of interest. The analysis of trace lead in soil samples was investigated for a NIST (#4355) Peruvian soil, and for lead concentrations of 129 ppm, lead signals were easily observed with high selectivity. This success lead to further investigations in which a series of native soils (Jemez Mts., NM) were spiked with various concentrations of lead. The result of this study was a linear calibration plot for lead over 5 decades of concentration (10 ppb - 1000 ppm), demonstrating the potential for RLA signal quantification.

The analysis of semivolatile (SV) organic compounds is often difficult by mass spectrometry, since these compounds are not readily introduced by either gaseous or direct insertion methods. The use of a selective sorbent material such as a zeolite provides a possible sample handling strategy for laser mass spectrometry on a solid sample support. Because RLA was successful in selectively generating metal ions for gas phase reactions [8], metal (silver) ion exchanged mordenite (a zeolite) was used as a selective sorbent for one of the SV compounds of interest to this group, tri-butyl phosphate (TBP). By implementing RLA tuned to a "2+1" transition for silver at 469.84 nm, silver ions were efficiently made, and the reaction of these silver ions with the TBP desorbed by the same laser pulse gave a strong analytical signal for Ag(TBP) ions[5], the first mass spectral analysis of TBP that yielded a predominant parent species in the mass spectrum.

Acknowledgements. The authors would like to thank Tom Taylor , Bill Earl and Greg Eiden for helpful discussions and investigations that contributed to further understanding of the mechanistic and/or analytical material presented by this work.

REFERENCES

1. G. C. Eiden, J. E. Anderson, and N. S. Nogar, Microchem. J. **50**, 289-300 (1994).
2. C. G. Gill, T. M. Allen, J. E. Anderson *et al.*, Appl. Opt. **35**, 2069-85 (1996).
3. C. G. Gill, A. W. Garrett, P. H. Hemberger *et al.*, Spectrochim. Acta., Part B **in press** (1996).
4. T. M. Allen, P. B. Kelly, J. E. Anderson *et al.*, Appl. Phys. A **61**, 221-225 (1995).
5. C. G. Gill, A. W. Garrett, W. L. Earl *et al.*, JASMS **submitted** (1996).
6. G. Krier, F. Verdun, and J. F. Muller, Fresenius' Z. Anal. Chem. **322** (4), 379-82 (1985).
7. C. J. McLean, J. H. Marsh, A. P. Land *et al.*, Int. J. Mass Spectrom. Ion Process **96** (1), R1-R7 (1990).
8. C. G. Gill, A. W. Garrett, P. H. Hemberger *et al.*, JASMS **7**, 664-7 (1996).

SESSION VIII: RIS
AND ULTRATRACE ANALYSIS

Resonance Ionization and Mass Selection for Molecular Laser Spectroscopy and Environmental Trace Analysis

Ulrich Boesl

Institut für Physikalische und Theoretische Chemie, Technische Universität München
Lichtenbergstraße 4, 85747 Garching , Germany

Abstract. Laser mass spectroscopic methods applied and developed in our group for laser spectroscopy of ionized or neutral molecules and for environmental analysis are reviewed.

Many molecular species are not available as pure substances, but only in complex mixtures and, in addition, often with very low concentrations. Therefore, they are hardly accessible for conventional spectroscopic methods. This restricts fundamental molecular research as well as application to environmental and trace analysis. Typical species are positively or negatively charged molecular systems as well as radicals and short lived intermediate species or extremely rarified, but nevertheless dangerous traces of pollutants. An important tool to cope with these problems is additional mass selection. When using pulsed lasers for spectroscopy, time-of-flight mass analyzers with their many options and features are the ideal instrument for such purposes. Involving multiphoton- and secondary laser excitation even enhances the possibilities of laser time-of-flight spectrometry.

In my group, we developed and used these methods of mass selected laser spectroscopy involving processes such as ionization, electron detachment or dissociation with resonant intermediate states (1,2). Thus, we succeeded in measuring highly resolved UV-spectra of molecular radical cations (for some species even with rotational resolution) (3-6), highly resolved ZEKE (zero kinetic energy electrons) spectra of molecular anions (providing information about neutrals other-wise often unaccessible) (7,8), or the first highly resolved UV-spectra of several neutral molecules (such as chlorinated or polycyclic aromatics) (9,10).

We found strong coupling effects (giving rise to femtosecond relaxation) of two nearly degenerate electronic states of the benzene radical cation (5), studied vibronic dynamics in the acetylene radical cation (4), investigated model systems for small catalytic reaction intermediates (e.g. metal + acetylene) (7,8) and small halogen water clusters (8) and, last not least, started to develop techniques for fast trace analysis of pollutants in automobile exhaust emissions (11) and for isomer selective and fast detection of traces of polychlorinated aromatics in the ppb and sub-ppb concentration region (10). In the following, a summary of experimental details rather than a description of the various experiments (see references) will be given.

In figure 1 a general experimental scheme is represented consisting of a central part (ion source, mass separator, ion detector) and several options for a variety of spectroscopic experiments. The neutral inlet system or source may be a narrow tubing (2) for an effusive molecular beam or a pulsed nozzle for supersonic molecular beams (12). The advantage of the former is a very simple and inexpensive construction. With the end of the tubing being placed between the electrodes of the extracting ion optics high gas densities may be achieved at the position of laser ionization without a too high load of the vacuum. The advantage of the latter is the possibility of efficient cooling the internal degrees of freedom of molecular motion. This allows a decongestion of spectra of even larger molecules (e.g. polycyclic or substituted aromatics (13,9)) giving access to rich spectroscopic information or enabling highly selective excitation. In addition large involatile neutral molecules may be transferred into the gas phase by laser desorption (14) as mentioned above.

There exist a number of laser induced techniques to ionize these neutrals such as resonance enhanced multiphoton ionization (in ion source B or C, figure 1), laser induced vacuum UV ionization, electron attachment (in ion source B) or electron ionization (in ion source C). Electrons may be supplied by laser induced photoelectron emission from metal surfaces, preferably thin wires made out of material with low work function. In addition, desorption of molecular ions from solid surfaces is possible, which, however, is not subject of our work. Ions formed in source B drift into the ion optics and can be extracted by a pulsed electric field while ionization in source C may be performed within a static electric field with instantaneous ion extraction. Extracted ions enter the field free drift region of a time-of-flight mass separator. Time-of-flight analyzers have turned out to be the ideal mass selection tools for pulsed laser induced ionization. A first mass selective detection may be performed in the so called space focus of the ion source. Even for very short field free drift regions (e.g. 10 cm) a mass resolution of 200 to 300 is possible in routine operation; a maximum resolution of 800 has been reached (1). A considerably higher mass resolution is achieved by adding a special ion reflector with further drift regions (15). In such a "reflectron" the ion cloud in the space focus SF is imaged onto the ion detector. This preserves the flight time distribution Δt, but extends the total flight time t thus enhancing the mass resolution $R_{50\%} = 1/2\ t/\Delta t$.

In figure 1, optional photoelectron spectrometers are included. These are preferably electric and magnetic field free drift regions allowing the analysis of kinetic electron energies by measuring electron time-of-flights. They may be combined with ion source C for analyzing photoelectrons emitted at resonance enhanced laser ionization; in this case spectroscopic information about cation ground states and intermediate neutral states is supplied. Such a photoelectron time-of-flight spectrometer can also be placed at the space focus (SF in figure 1). Electrons due to photodetachment of anions may then be analyzed containing information about neutral molecular ground states. Since this detachment takes place at the space focus with its intrinsic mass selection this anion photoelectron spectroscopy is mass selective. For neutral-cation photoelectron spectroscopy at ion source C mass selection is possible by photoion/photoelectron coincidence techniques (16). Resolution is strongly enhanced by selective monitoring "zero-kinetic-energy" electrons (ZEKE-spectroscopy (17)) as a function of exciting laser wavelength, which now has to be tuned. For cation←neutral ZEKE, field ionization of high Rydberg states is involved making high resolution (<0.1 cm^{-1}) as well as mass selectivity possible. In our group, neutral←anion ZEKE in the space focus (intrinsic mass selection) is performed. Since Rydberg-like states do not exist for anions free electrons of very low kinetic energy have to be monitored which makes high resolution difficult (8).

In figure 1, several laser beams are supposed to be used in combination or separately. By laser beam L1 resonance enhanced multiphoton ionization is to be performed. By laser beam L2 molecular cations can be excited giving rise to molecular fragments and fragment ions. Mass selective analysis of the manifold of secondary fragments results in tandem mass spectrometry (1). Recording the intensity of one special fragment ion as function of laser L2 wavelength allows cation UV/VIS spectroscopy (3). Both experiments (tandem MS or cation spectroscopy) are also possible with laser beam L3 coupled into the space focus. By the latter experimental arrangement considerably higher secondary mass selectivity is achieved, while the former arrangement may supply the better sensitivity. In addition, with laser L3 mass selective neutral←anion photoelectron and ZEKE spectroscopy can be performed as mentioned already. Finally, laser beam L4 is used for photoelectron emission from a wire for anion formation by electron attachment or for desorption of involatile neutral molecules into a supersonic molecular gas beam (14,18).

1. Boesl,U., Weinkauf,R., Schlag,E.W., Int.J.Mass Spectrom.Ion Proc., **112**, 121 (1992).
2. Boesl,U., Weinkauf,R., Weickhardt,C., Schlag,E.W., Int.J.Mass Spectrom., **131**, 87 (1994).
3. Weinkauf,R., Walter,K., Boesl,U., Schlag,E.W., J.Chem.Phys., **98**,1914 (1988).
4. Cha,Ch., Weinkauf,R., Boesl,U. , J.Chem.Phys., **103**, 5224 (1995)
5. Walter,K., Weinkauf.R., Boesl,U., Schlag,E.W., Chem.Phys. Lett., **155**, 8 (1989).
6. Weinkauf,R., Boesl,U. , J.Chem.Phys., **98**, 4459 (1993).
7. Drechsler,G., Bäßmann,C., Boesl,U., Schlag,E.W. , J.Molec.Struct., 348, 337 (1995).
8. see Bäßmann,C. et al., RIS 96, Poster Session 2
9. Weickhardt,C., Zimmermann,R., Boesl,U., Schlag,E.W. , Rapid Commun.Mass Spectrom., **7**, 183 (1993); ibid., **8**, 381 (1994).
10. see Zimmermann,R. et al. Ris 96, session 4, postersession 2 and 3.
11. Boesl,U., Nagel.H., Weickhardt,C., Frey,R., Schlag,E.W. 1996/7, The Encyclopedia of Environmental Analysis and Remediation, John Wiley&Sons, to be publishe (1996/7).
12. Levy,D.H. , Science, **214**, 263 (1982).
13. Hayes,J.M. , Chem.Rev., **87**, 745 (1987); Stiller, S.W., Johnston,M.V., Anal.Chem., **59**, 567 (1987); Lubman,D.M. , Anal.Chem., **59**, 31A (1987)
14. von Weyssenhoff,H., Selzle,H.L., Schlag,E.W. , Z.Naturforsch. **A40**, 674 (1985); Grotemeyer,J., Boesl,U., Walter,K., Schlag,E.W. , Org. Mass Spectrom., **21**, 645 (1986); de Vries,M.S., Elloway,D.J., Wendt,R., Hunziker,H.E., Rev.Sci.Instrum. **63**, 3321 (1992); Trembreuill,R., Lubman,D.M., Anal.Chem., **58**, 1299 (1986).
15. Mamyrin,B.A., Karataev,V.I., Shmikk,D.V., Zagulin,V.A., Sov.Phys.-JETP, **37**, 45 (1973).
16. Booze,J.A., Baer,T., J.Chem.Phys., **98**, 186 (1993).
17. Müller-Dethlefs,K., Schlag,E.W., Annu.Rev.Phys. Chem., **42**, 109 (1991).
18. Meijer,G., de Vries,M.S., Hunziker,H.E., Wendt,H.R., J.Chem.Phys., **92**, 7625 (1990).

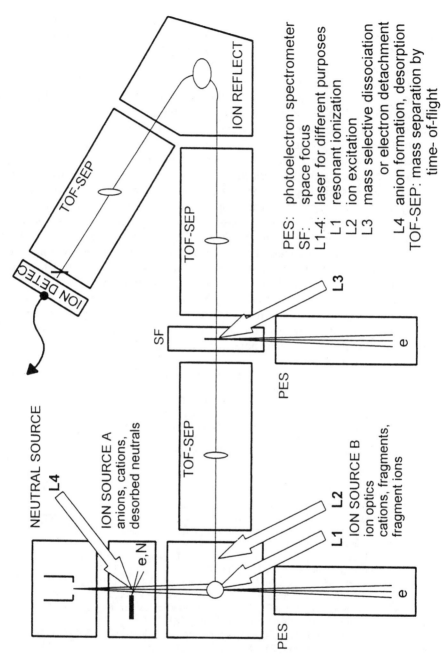

FIGURE 1 General view of a laser mass spectroscopic apparatus with ion source, time-of-flight mass separator, ion detector and several experimental otions, e.g. different types of ionization, mass separation, secondary laser excitation, photoelectron spectrometers.

Trace Analysis of Plutonium in Environmental Samples by Resonance Ionization Mass Spectroscopy (RIMS)

N. Erdmann*, G. Herrmann*, G. Hubert†, S. Köhler*, J.V. Kratz*,
A. Mansel*, M. Nunnemann†, G. Passler†, N. Trautmann*,
and A. Waldek*

*Institut für Kernchemie and †Institut für Physik,
Universität Mainz, D-55099 Mainz, Germany

Abstract. Trace amounts of plutonium in the environment can be detected by resonance ionization mass spectroscopy (RIMS). An atomic beam of plutonium is produced after its chemical separation and deposition on a filament. The atoms are ionized by a three-step excitation using pulsed dye-lasers. The ions are mass-selectively detected in a time-of-flight (TOF) mass spectrometer. With this setup a detection limit of $1 \cdot 10^6$ atoms of plutonium has been achieved. Furthermore, the isotopic composition can be determined. Different samples, including soil from the Chernobyl area, IAEA-certified sediments from the Mururoa Atoll and urine, have been investigated.

INTRODUCTION

Trace amounts of plutonium are present in the environment as a result of global fallout from nuclear weapons tests, satellite and reactor accidents and releases from nuclear facilities. Monitoring plutonium contaminations in the environment as well as studying its ecological behavior requires sensitive and element-selective detection methods. The measurement of the isotopic abundances allows the determination of the origin of the plutonium samples. RIMS meets all these requirements due to its high sensitivity and good element and isotope selectivity.

EXPERIMENTAL

Sample preparation

Prior to the RIMS measurement plutonium is chemically separated from the environmental samples. The chemical procedure is based on a coprecipitation step with $Fe(OH)_3$ and the subsequent separation of plutonium on an anion exchange column, TEVA-SPEC, 50-100 μm. Then plutonium is electrolytically deposited in form of its hydroxide on a tantalum backing (1) and covered with a thin titanium layer produced by sputtering. By heating such a sandwich filament the hydroxide is converted to the oxide, which is reduced to the metallic state during diffusion through the titanium layer. In this way an atomic beam of plutonium is evaporated from the surface of the filament. In general, ^{236}Pu ($T_{1/2}$=2.85 a) is used as tracer for the chemical yield and ^{244}Pu (98% enriched) to monitor the RIMS efficiency.

Experimental setup for RIMS

The experimental setup for RIMS consists of three tunable dye lasers pumped simultaneously by two copper vapor lasers, operating at a repetition rate of 6.5 kHz. The light of the dye lasers is coupled into a quartz fibre and focused into the interaction region of a time-of-flight (TOF) mass spectrometer perpendicular to the created atomic beam. The atoms are photoionized in a three-step, three-colour resonant excitation with the third step leading to an autoionizing state. All plutonium measurements have been performed with the excitation scheme λ_1=586.49 nm, λ_2=665.57 nm and λ_3=577.28 nm. The ions produced are accelerated in electric fields and mass-analyzed in a TOF mass spectrometer. A detailed description of the experimental setup is given in (2,3).

RESULTS

The applicability of RIMS for trace analysis of plutonium was first studied with synthetic samples. A detection limit of $1 \cdot 10^6$ atoms has been achieved and samples containing as little as $3.6 \cdot 10^6$ atoms of ^{236}Pu were measured unambiguously. In order to test the reproducibility of the isotopic analysis synthetic samples containing the isotopes ^{238}Pu through ^{244}Pu were investigated. For the measurement of isotopic compositions λ_1 and λ_3 have to be tuned over the resonances of the isotopes. The results obtained are in good agreement with mass spectrometric data.

Investigation of soil samples and sediments

Soil samples from the village Masany close to the Chernobyl reactor were chemically treated and measured with RIMS. Among the samples a hot particle was investigated separately. It contained 45 mBq of $^{239/240}$Pu. The isotopic composition results are shown in table 1 together with the data calculated from the reactor parameters. As can be seen there is an excellent agreement.
The RIMS spectrum of the hot particle is shown in figure 1(a). The isotopes ^{238}Pu through ^{242}Pu can be seen together with ^{236}Pu which was used to determine the chemical yield as well as the RIMS efficiency.

TABLE 1. Isotopic composition of plutonium in soil samples from the Chernobyl area. The errors are 3σ-errors for the counting statistics. The discrepancies for the hot particle are due to a reduced scanning region of λ_1 and λ_3 in order to enhance the detection effiency.

isotope	calc. data [%]	Hot Particle [%]	Masany I [%]	Masany II [%]
238	0.22	0.9(3)	0.25(2)	0.21(4)
239	64.6	70(3)	66.9(4)	70.5(7)
240	27.6	23(2)	25.3(2)	22.0(4)
241	5.5	4.7(8)	5.9(1)	5.8(2)
242	2.2	1.2(4)	1.6(1)	1.4(1)

Figure 1(b) shows the RIMS spectrum of an IAEA-certified sample of Pacific Ocean sediments collected at the Mururoa Atoll. Here ^{244}Pu was used for monitoring the RIMS efficiency. The difference in isotopic composition between the two samples is quite obvious. The composition of the Chernobyl plutonium is typical for reactor plutonium, whereas that of the Mururoa plutonium for weapon plutonium. Several samples, each prepared from one gramm of the IAEA material, were measured with respect to the total content of plutonium (31 mBq/g $^{239/240}$Pu) and its isotopic composition (97% ^{239}Pu, 3% ^{240}Pu). The results are in agreement with the certified data.

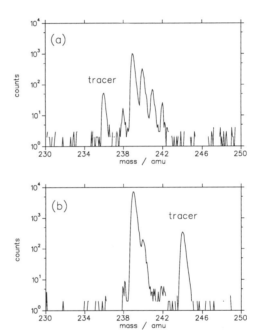

FIGURE 1. RIMS spectra of plutonium from two environmental samples: (a) hot particle from the Chernobyl area, (b) Pacific Ocean sediment from the Mururoa Atoll.

Investigation of urine samples

In collaboration with the National Radiological Protection Board (NRPB) in Great Britain uptake and urinary excretion of plutonium in human volunteers (4) will be studied. As a test a five day pooled urine sample from a volunteer

207

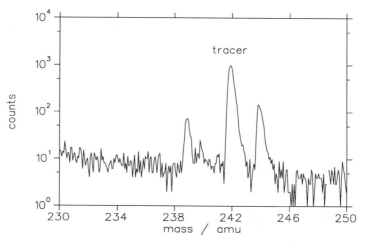

Figure 2. RIMS spectrum of a urine sample from a volunteer pooled over five days.

taken 1111 days after injection of $1.9 \cdot 10^{12}$ atoms of ^{244}Pu (98%) was investigated. Plutonium was chemically separated from the urine and ^{242}Pu was added as tracer at NRPB. From this solution a filament for RIMS was prepared and the ^{242}Pu/^{244}Pu ratio was measured. The RIMS spectrum is shown in figure 2. A content of $4.2 \cdot 10^8$ atoms of ^{244}Pu was determined which corresponds to a urinary output of $8.4 \cdot 10^7$ atoms per day (0.00437% of excretion per day after injection) and is in good agreement with data obtained by mass spectrometric analysis of urine from other volunteers (5). The ^{239}Pu peak in the RIMS spectrum is due to a contamination resulting from the chemical procedure.

ACKNOWLEDGMENTS

This work has been funded by the "Bundesministerium für Bildung, Wissenschaft, Forschung und Technologie" and the "Zentrum für Umweltforschung der Universität Mainz".

REFERENCES

1. Eberhardt, K., et al., *AIP Conf. Proc.* **329**, 503-506 (1994).
2. Ruster, W., et al., *Nucl. Instr. Meth. Phys. Res. A* **281**, 547-558 (1989).
3. Urban, F.-J., et al., *Inst. Phys. Conf. Ser.* **128**, 233-236 (1992).
4. Popplewell, D.S., et al., *Radiat. Prot. Dosim.* **53**, 241-244 (1994).
5. Ham, G.J., private communication (1996).

Application of
Resonance Ionization Mass Spectrometry for
Trace Analysis and in Fundamental Research

G. Passler

Institut für Physik, Johannes Gutenberg-Universität Mainz, D-55099 Mainz, Germany

Abstract. Resonance ionization mass spectrometry (RIMS) has been used for ultra-trace analysis on long-lived radioisotopes like Pu, Tc and ^{89,90}Sr in various environmental samples. The experimental approaches cover pulsed laser spectroscopy on a thermal atomic beam and subsequent time-of-flight mass analysis, a pulsed laser ion source combined with conventional mass spectrometry, and collinear resonance ionization on a mass-separated fast atomic beam. The high sensitivity of RIMS also enables atomic spectroscopy on rare isotopes. For the first time experimental values for the ionization potential of actinides up to Cf have been determined. The paper reviews the dependency of the different experimental approaches on the analytical problem.

INTRODUCTION

As a highly selective and very sensitive method resonance ionization mass spectrometry (RIMS) is an ideal tool for ultra-trace analysis of contaminations in environmental, biological or technological samples as well as for atomic spectroscopy studies on very rare elements.

Trace determination of radioisotopes, occuring naturally in the environment or released by nuclear weapons tests, nuclear power plants and other technical or medical applications, is usually done by radiometric methods. However, amazingly, for some of the most important and hazardous radioisotopes radiometry faces various problems. For very long-lived α- and β-emitters (e.g. $T_{1/2}[^{239}Pu] = 2.41 \cdot 10^4$ a, $T_{1/2}[^{99g}Tc] = 2.1 \cdot 10^5$ a) the measuring time is very long, and despite the high sensitivity the detection limits are impaired by background. Some isotopes, e.g. the pairs $^{239}Pu/^{240}Pu$ and $^{238}Pu/^{241}Am$ cannot be distinguished by α-spectroscopy due to very similar energies of the α-lines. As the continous energy spectra of the β-emitters ^{89}Sr and ^{90}Sr overlap and usually cannot be measured separately by β-spectroscopy, ^{90}Sr ($T_{1/2} = 28.5$ a) is detected via the daughter isotope ^{90}Y. This includes a waiting time of about 10-14 days.

However, for surveillance of the environment and a fast risk assessment in case of a nuclear accident, detection limits of $10^6 - 10^7$ atoms at a short measuring time are desirable for those hazardous radioisotopes, in order to detect an increase above the omnipresent background contamination as soon as possible. Isotope selective detection contains additional information on the source of a contamination. Studies of the migration behaviour of those radionuclides or, for example, personnel dose monitoring in nuclear facilities require low detection limits and isotope selectivity as well.

The detection limits of non-radiative detection methods, like the various types of mass spectrometry, do not depend on the halflife and the decay type of the isotope. In some cases non-radiative detection methods face another problem though, that

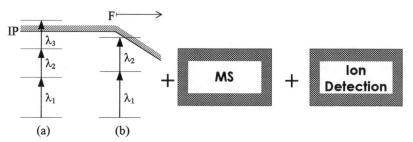

◊ high element selectivity   ◊ high isotope selectivity   ◊ high detection efficiency
◊ (high isotope selectivity)   ◊ background suppression
◊ high ionization efficiency

FIGURE 1. Main components of a RIMS set-up and their contribution to its features. On the left side two resonant ionization schemes out of a number of possible ones are shown: a) three-step resonant excitation of an auto-ionizing state, b) excitation of a Rydberg state and subsequent field ionization.

is not known in radiometry: interferences from isobars and from stable isotopes of the same element, that may be highly abundant in the environment, can strongly affect the detection limit. In contrast to other non-radiative detection methods, RIMS provides high element selectivity by multiple resonant excitation and ionization of the respective element, thus minimizing isobaric interferences. The mass selective component ensures isotope selectivity and significantly reduces background. High sensitivity is guaranteed by a high ionization probability due to the high optical cross sections as well as by the ion detection. If necessary, very high additional isotope selectivity can be achieved by making use of the isotope shift in the optical excitation. Figure 1 shows a schematic diagram of RIMS and the contributions of the single components.

The extraordinary sensitivity of RIMS also enables atomic spectroscopy studies – like isotope shift measurements, Rydberg or autoionizing state spectroscopy, or the determination of the ionization potential – of elements that are available in amounts of only a few picogram or cannot be handled in greater amounts due to their high radioactivity.

EXPERIMENTAL APPROACHES

In the following the RIMS set-ups, realized for trace analysis and spectroscopy of plutonium and other actinides, technetium, and strontium ($^{89/90}$Sr) are described, and it is shown how the experimental approach is determined by the respective requirements.

Time-of-Flight Mass Spectrometer

Trace Determination of Plutonium

For trace analysis of plutonium the isotopic composition of a sample is of great interest, as it contains information on the source of a contamination. For this purpose an isotope selectivity in the order of $S \approx 10^3$ is needed, which is easily obtained by the mass selective component. Therefore pulsed broad-band high power dye-lasers are used. Mass analysis is performed by a time-of-flight spectrometer that allows the simultaneous measurement of several isotopes. The experimental set-up is described in detail by (1, 2). As the sample is evaporated as continous thermal atomic beam, high-repetition rate lasers (Cu-vapour pump lasers at $\nu_{rep} = 6.5$ kHz) are required in order to get sufficient temporal overlap of the sample atoms and the laser pulses. A quite difficult task was the development of an efficient atomic beam source, due to the complex chemical properties of Pu and other actinides. In cooperation with B. Eichler (PSI, Villigen, Switzerland) a sandwich filament has been developed where the sample is electrolytically deposited on a tantalum backing and subsequently coated by a titanium layer (3).

With this set-up the detection limit (DL) has been determined to $DL_{RIMS} = 1 \cdot 10^6$ atoms (signal/noise ≥ 3 σ) for a single isotope (in comparison to $DL_{radiometric} \geq 4 \cdot 10^8$ atoms of ^{239}Pu). Though broad-band dye-lasers are used ($\Delta\nu \approx 6$ Ghz), the first and third excitation step must be scanned due to the isotope shift in the optical transitions if the isotopic composition is measured. Therefore 10^7 atoms of the least abundant isotope are necessary for a complete information on the isotopic composition with an accuracy of about 10 % of the actual values, due to losses while scanning the lasers (4).

Ionization Potential of Actinides

The sensitivity of the set-up described above predestines it for atomic spectroscopy studies on rare elements. The first ionization potential (IP), a fundamental quantity for understanding the atomic spectra and the chemical properties of an element, had not been determined experimentally so far for any element heavier than plutonium, due to the large samples needed with the usual methods. As the actinides beyond Pu are available in small amounts only and are difficult to handle due to their high radiotoxicity, only semi-empirical calculations had been published. The time-of-flight spectrometer is an ideal tool for the determination of the ionization potential of these heavy elements (5, 6). According to the saddle point model the ionization threshold is lowered in the presence of an external electric field. The well-known and homogenous acceleration field in the interaction region is used for this purpose. If the laser is scanned across the threshold, a sudden onset of the ion signal is detected. By variation of the electric field and determination of the ionization threshold as a function of the field strength, the ionization potential is easily determined by extrapolation to zero field. In this way the IP of the actinides Am, Cm, Bk, and Cf have been experimentally determined for the first time, using samples of 10^{12} atoms each. The results show an accuracy of some 10^{-4} eV, while the values diverge from the semi-empirical extrapolations up to $3 \cdot 10^{-2}$ eV.

Laser Ion Source

The sensitivity of the time-of-flight spectrometer set-up is restricted by the spatial and temporal overlap of the thermal atomic beam and the pulsed lasers ($\varepsilon_{overlap}$ = 0.26 % at 6.5 kHz repetition rate and 3 mm beam waist). Evaporating the sample inside a small heated chamber, where it is resonantly ionized by the laser beams that enter the chamber through a small hole, this overlap can be considerably increased. The ions are extracted by an electric field via the same hole or a second one, while the neutral atoms are „stored" inside the chamber by multiple collisions with the wall for a period of a few laser shots. In this way ionization efficiencies of about 10 % are attainable. In order to make use of the high element and isotope selectivity of RIMS, surface ionization of sample atoms as well as interfering elements in the hot chamber must be suppressed. Therefore a proper choice of the chamber material is important, depending on the analytical problem to be treated. The mass selective detection is performed by a conventional mass spectrometer as the temporal structure of the extracted ion pulse lasts about 15 μs, i.e. too long for a time-of-flight analysis.

Plutonium

As the ionization potential of Pu is rather low (IP_{Pu} = 6.026 eV) the evaporation temperature as well as the work function of the chamber material should be as low as possible. Tantalum or titanium, already known from the sandwich filaments, are well suited with respect to the adsorption enthalpy and the work function. As titanium is too soft even at temperatures of only 1200 K, tantalum is the better choice. It has been shown experimentally that in such a chamber the resonant ionization of

Pu is the dominant process, and surface ionization can be neglected. With this set-up a photo-ionization efficiency in the chamber of $\varepsilon_{photo} = 2\%$ has been achieved for Pu. This value is derived from the overall detection efficiency $\varepsilon_{Det} = 10^{-4}$, taking into account that the ion beam extracted from the chamber does not match the mass spectrometer entrance slit resulting in a low transmission of only 0.5 %. Nevertheless a detection limit (for a single isotope) of $DL_{Pu} = 5 \cdot 10^5$ atoms is found, which can be lowered by at least one order of magnitude if the transmission of the mass spectrometer is improved.

Technetium

In the case of Tc the thermal ionization of the sample atoms can be neglected, due to the higher ionization potential of $IP_{Tc} = 7.28$ eV. But in this case another problem occurs: For an experiment on the solar neutrino flux the ratio $^{97,98}Tc/^{99}Tc$ is to be determined (7). The sample that is extracted from 10,000 t of molybdenum ore contains about 10^8 atoms of ^{97}Tc and ^{98}Tc, 10^{11}-10^{12} atoms ^{99}Tc and more than 10^{12} atoms of molybdenum. Therefore surface ionization of molybdenum must be avoided in order to suppress isobaric interferences stemming from $^{97,98}Mo$. Furthermore the chamber material must not be contaminated by Mo. The same is true for trace determination of ^{99}Tc in environmental or biological samples where the tracer ^{95m}Tc is measured simultaneously in order to get quantitative results. In the latter case an interference with thermally ionized ^{95}Mo must be avoided. However, all high temperature metals contain molybdenum in the ppm regime. Therefore ultrapure pyrolithic graphite is chosen as chamber material. In this way a detection limit for a single isotope of $DL_{Tc} = 10^5$ atoms has been reached, corresponding to a photo-ionization efficiency of $\varepsilon_{photo} = 8\%$. The molybdenum background from the graphite chamber can be neglected, and the selectivity against molybdenum contained in the sample is measured to $S_{Tc:Mo} \geq 10^5$ at T = 2200 K and far higher for lower temperatures.

Laser ion sources are also used to an increasing degree at on-line mass separators (8). This application also aims at the efficient suppression of isobars or interfering neighbouring isotopes.

Collinear Resonance Ionization on a Fast Atomic Beam

The trace determination of the radioactive isotopes $^{89,90}Sr$ requires a completely different approach (9). As less than 10^8 atoms of these radioisotopes are to be detected in presence of up to 10^{18} atoms of stable strontium (mainly ^{88}Sr), the isotope selectivity required is $S > 10^{10}$. Conventional mass spectrometry is not able to achieve such a low abundance sensitivity. Therefore optical isotope selectivity must be added. However, unfortunately, the isotope shift in the Sr isotopes is by far too low for this purpose. For this reason an "artificial isotope shift" is introduced by making use of the Doppler shift in collinear excitation of a fast atomic beam. As all isotopes are accelerated to the same energy, their velocity is different according to the different masses. This leads to an "isotope shift" of several GHz per isotope.

The strontium sample is introduced into a conventional ion source of a 33 keV mass separator, where most of the stable Sr is removed from the beam. Then the ion beam is neutralized in a charge exchange cell, the fast atoms are excited to a Rydberg state by a collinearly superimposed laser beam, and subsequently field ionized. These resonantly produced ions are deflected from the atomic beam and detected with a channeltron. With this set-up the detection limits for ^{90}Sr and ^{89}Sr are DL $^{90}_{Sr}$ = $3 \cdot 10^6$ atoms and DL $^{89}_{Sr}$ = $5 \cdot 10^7$ atoms, respectively. The selectivity is $S \geq 10^{10}$, corresponding to the demands (10, 11).

OUTLOOK

Using RIMS for trace determination very low detection limits and extremely high isotope selectivity can be achieved, provided that the experimental approach is optimized for the respective analytical problem. For some selected elements, like e.g. plutonium, RIMS may become a standard analytical method, if the set-up is simplified and automized. The use of laser diodes, like demonstrated for the isotope selective trace detection of calcium (12), surely points the way.

ACKNOWLEDGEMENTS

The work presented here was carried out by F. Albus, F. Ames, B. A. Bushaw, K. Eberhardt, N. Erdmann, H. Funk, H.-U. Hasse, G. Herrmann, G. Huber, H.-J. Kluge, S. Köhler, J.-V. Kratz, J. Lantzsch, A. Mansel, L. Monz, P. Müller, W. Nörtershäuser, M. Nunnemann, E.W. Otten, P. M. Rao, J. Riegel, J. Stenner, N. Takahashi, N. Trautmann, F.-J. Urban, A. Waldek, and K. Wendt.

REFERENCES

1. Ruster, W., et al., *Nucl. Instr. Meth. Phys. Res.* **A281**, 547 (1989)
2. Urban, F.-J., et al., *Inst. Phys. Conf. Ser.* **128**, 233 (1992)
3. Eberhardt, K., et al., *AIP Conf. Proc.* **329**, 503 (1994)
4. Erdmann, N., et al., *these conference proceedings*
5. Köhler, S., et al., *AIP Conf. Proc.* **329**, 377 (1994)
6. Nunnemann, M., et al., *these conference proceedings*
7. Hasse, H.-U., et al., *AIP Conf. Proc.* **329**, 499 (1994)
8. Janas, Z., et al., *Phys. Scr.* **T56**, 262 (1995)
9. Monz, L., et al., *Inst. Phys. Conf. Ser.* **128**, 225 (1992)
10. Lantzsch, J., et al., *AIP Conf. Proc.* **329**, 251 (1994)
11. Wendt, K., et al., *these conference proceedings*
12. Bushaw, B. A., et al., *these conference proceedings*

Three-Color Resonance Ionization Spectroscopy of Zr in Si

C. S. Hansen,[1] W. F. Calaway,[1] M. J. Pellin,[1] R. C. Wiens,[2] and D. S. Burnett[2]

[1]*Materials Science, Chemistry, and Chemical Technology Divisions*
Argonne National Laboratory, Argonne, Illinois 60439

[2]*Divisions of Geology and Planetary Science*
California Institute of Technology, Pasadena, California 91125

Abstract. It has been proposed that the composition of the solar wind could be measured directly by transporting ultrapure collectors into space, exposing them to the solar wind, and returning them to earth for analysis. In a study to help assess the applicability of present and future postionization secondary neutral mass spectrometers for measuring solar wind implanted samples, measurements of Zr in Si were performed. A three-color resonant ionization scheme proved to be efficient while producing a background count rate limited by secondary ion signal (5×10^{-4} counts/laser pulse) This lowered the detection limit for these measurements to below 500 ppt for 450,000 averages. Unexpectedly, the Zr concentration in the Si was measured to be over 4 ppb, well above the detection limit of the analysis. This high concentration is thought to result from contamination during sample preparation, since a series of tests were performed that rule out memory effects during the analysis.

1. INTRODUCTION

Resonance ionization spectroscopy (RIS) is characterized by high ionization efficiency and atomic selectivity making it well suited for low-level elemental analysis. It has successfully been used for selective postionization of sputtered neutral species, providing powerful ultratrace analysis capabilities below atomic fractions of 10^{-9}[1-3] while removing only a few monolayers of the substrate. Results obtained using RIS with secondary neutral mass spectrometry (SNMS) are also readily quantifiable.[4]

As continued progress is made to lower detection limits and as a wider variety of systems are studied, increasingly difficult challenges are presented that test the limits of RIS/SNMS capabilities. One such challenge is presented by the needs of the planetary science community to study the composition of the solar wind. It has been proposed that this can be done directly by sending ultrapure collectors into space and exposing them to the solar wind for roughly two years before returning them to earth for analysis.[5] Predictions of the flux and energy of solar wind ions indicate that collectors would receive doses from 10^5-10^{11} ions/cm^2 for elements heavier than Fe, implanted in a shallow distribution less than 100 nm deep.[6] This corresponds to concentrations from 1 ppm to below 1 ppt.

A RIS/SNMS instrument (SARISA IV) developed at Argonne National Laboratory[7,8] is part of a study to determine the applicability of the technique to the solar wind analysis problem. The measurement of Zr in a pure Si sample was used for a test. Si was chosen because it is one of the purest materials available, and is thoroughly characterized by other methods. Zr is an element of interest as a test of solar wind models and has a proven three-color ionization scheme,[9] which can be exploited to minimize background levels.

Results obtained from initial experiments on the Zr/Si system are presented here. A new lower detection limit is reported for the instrument, resulting from improved suppression of secondary ions. A second, surprising, result was the presence of Zr in the near-surface region well above the detection limit.

2. EXPERIMENTAL

Sputtered neutrals, produced from 2500 V Ar+ ion bombardment in 800 ns pulses, are postionized in the gas phase ~ 0.5 mm from the target surface from a volume of approximately 10 mm^3. A double hemisphere, energy and angle refocusing time-of-flight (EARTOF) mass spectrometer provides discrimination against secondary ions and other noise sources. Secondary ions are suppressed by applying a voltage to the target during the Ar+ ion bombardment that imparts sufficient energy to the secondary ions such that the EARTOF prevents their transmission to the detector. The target voltage is pulsed down to the appropriate level just before the postionizing lasers are fired. Atomic selectivity is achieved in Zr by the three-color ionization scheme shown in Fig. 1.[9] To distinguish the resonant photoion signal from background levels, measurements were made with the first laser detuned 0.1 nm to the red of the resonance wavelength. Silicon wafers provided by two independent manufacturers were used as the target materials.

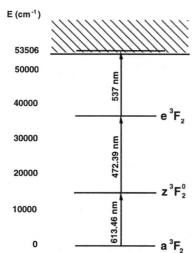

FIGURE 1. Three-color scheme used to resonantly ionize Zr.

Three depth profiles of Zr concentration in Si were measured on two Si wafers by interleaving analyses with ion beam milling of the sample. The milled area (4 mm^2) was much greater than the sampled area (0.03 mm^2) to ensure that crater-wall effects did not bias the analyses. Each spot was analyzed to a depth of 100 nm. Typically, 25 to 50 measurements were made over the 100 nm range, with each measurement consisting of 10,000 laser shots (5000 signal and 5000 background). Signal from a pure Zr metal target was used to calibrate the instrument, allowing quantitative analyses to be made.[4]

3. RESULTS AND DISCUSSION

Using the measured sensitivity and background from the three analyses, an average detection limit for Zr in Si was determined. The top curve in Fig. 2 shows this detection limit as a function of the number of averages. This curve is based on a background count rate of 5×10^{-4} counts/average, obtained by integrating the off-resonance signal over the main Zr isotopes (90-94 Dalton), and assuming a signal-to-noise ratio of unity. The main noise sources were found to be secondary ions traversing the time-of-flight path and scattered ions reaching the detector directly. If nonresonantly ionized species contributed to the background level, their effect was not discernible above these other background sources.

The measured background is more than two orders of magnitude above the dark count rate of the multichannel plate detector. While it is anticipated that some instrument modifications can be made to reduce the total background, the dark counts alone keep the detection limit above 20 ppt in 10^6 averages using the present primary ion source. By installing a higher current source, however, measurements down to the 1 ppt level are feasible, as is demonstrated by the lower family of curves in Fig. 2. The

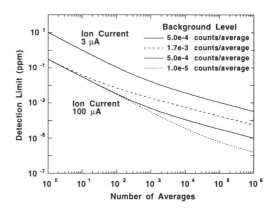

FIGURE 2. Detection limit (for Zr in Si) of the SARISA instrument for the present ion source (top curve) and a future ion source (bottom 3 curves). The lowest curve is the limit imposed by the detector dark count rate.

dashed curve corresponds to the present background level scaled linearly with increased ion current, while the solid line is based on a background level independent of source current. Tests conducted to determine how the background will scale with ion current indicated that the actual detection limit will fall somewhere between the two lines. The dotted line in Fig. 2 shows the predicted detection limit if instrument modifications are made that reduce the background level to the detector dark count rate. As can be seen in Fig. 2, the detection limit drops to 1 ppt in 10^6 averages in this case--near the estimated limits required by the solar wind experiment.

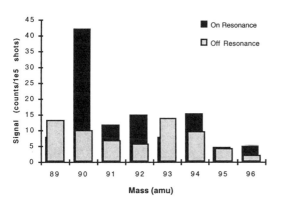

FIGURE 3. Mass spectrum demonstrating the presence of Zr in the Si sample analysis. On resonance signal is greater than off resonance signal for masses corresponding to Zr isotopes (90-92, 94 Dalton).

A quite unexpected result from this study was the unquestionable detection of Zr in the Si samples. Each of the depth profiles revealed a uniform distribution of Zr over the 100 nm depth that was measured. A mass spectrum of the cumulative signal from one of the depth profiles (2.4×10^5 averages) is seen in Fig. 3, where the isotopic signature for Zr is clearly reproduced in the measured mass spectrum. The on-resonance signal is well above background at mass 90, the major isotope of Zr (51%). For the minor isotopes of Zr at masses 91, 92, and 94, the net signal is positive in each case though not statistically above the background. The total integrated Zr signal is 4.4×10^{-4} counts/average, corresponding to a concentration of 4 ppb. Similar results were obtained for the two depth profiles on the other sample, although the measured overall Zr concentration was a factor of two lower than for the first target.

The accuracy of the SARISA instrument, when using a pure metal foil for calibration (due to the required dynamic range), is good to within a factor of three,[4] so measured Zr concentrations for these two samples are equal within the analysis uncertainty. Since these high Zr levels were found on samples from different manufacturers, it is believed that the Si wafers were contaminated during the preparation or analysis, not during the manufacturing process. During preparation, wafers were not handled in cleanroom conditions, though precautions were taken so they were not brought into contact with contaminated surfaces.

Contamination during the measurements (memory effect) has been ruled out by several tests performed during the analyses. Typically, a spare piece of pure Si was sputtered for an hour prior to collecting data to cover any Zr in the vacuum system resulting from the calibration. This covering step was eliminated before the second analysis on the second sample, but no change in Zr signal was observed. Secondly, in each analysis the Zr concentration was found to be constant over the measured depth. This would not be expected if Zr was being deposited onto the sample. In addition, at the end of the depth profile, no increase in Zr signal was observed when the sample surface was reanalysed at a new location. All of these results indicate that Zr on nearby surfaces was well covered and not being redeposited onto the Si sample. Therefore, it is believed that the contamination of the Si samples is occurring prior to their introduction into the SARISA instrument.

ACKNOWLEDGMENTS

This work is supported by the U.S. Department of Energy, BES-Material Sciences, under Contract W-31-109-ENG-38 and by NASA Grant NAGW-4182.

REFERENCES

1. M. H. Pellin, C. E. Young, W. F. Calaway, J. E. Whitten, D. M. Gruen, H. D. Blum, I. D. Hutcheon, and G. J. Wasserburg, *Phil. Trans. R. Soc. Lond. A* **333**, 133 (1990).
2. N. Thonnard, J. E. Parks, R. D. Willis, L. J. Moore, and H. F. Arlinghaus, *Surf. and Interface Anal.* **14**, 751 (1989).
3. D. L. Pappas, D. M. Hrubowchak, M. H. Erwin, and N. Winograd, *Science* **243**, 64 (1989).
4. W. F. Calaway, S. R. Coon, M. J. Pellin, C. E. Young, J. E. Whitten, R. C. Wiens, D. M. Gruen, G. Stingeder, V. Penka, M. Grasserbauer, and D. S. Burnett, *Inst. Phys. Conf. Ser.* **128**, 271 (1992).
5. D. Rapp, F. Naderi, M. Neugebauer, D. Sevilla, D. Sweetnam, D. Burnett, R. Wiens, N. Smith, B. Clark, D. McComas, and E. Stansbery, *Astronautica Acta*, submitted (1996).
6. E. Anders and N. Grevesse, *Geochim Cosmochim Acta* **53**, 197 (1989).
7. M. J. Pellin, C. E. Young, W. F. Calaway, J. W. Burnett, B. Jørgensen, E. L. Schweitzer, and D. M. Gruen, *Nucl. Instrum. Methods Phys. Res.* **B18**, 446 (1987).
8. M. J. Pellin, C. E. Young, and D. M. Gruen, *Scanning Microsc.* **2**, 1353 (1988).
9. D. R. Spiegel, W. F. Calaway, G. A. Curlee, A. M. Davis, R. S. Lewis, M. J. Pellin, D. M. Gruen, and R. N. Clayton, *Anal. Chem.* **66**, 2647 (1994).

Trace CO Detection by REMPI at 230 nm

W. X. Peng*, K. W. D. Ledingham and R. P. Singhal

Department of Physics and Astronomy, the University of Glasgow, Glasgow G12 8QQ, Scotland UK
*Permernant address: Department of Physics, Jilin University, Changchun 130023 PR China

Abstract. A laser based procedure, which combines the REMPI technique and ion chamber detection to give a very high sensitivity, has been developed at Glasgow to detect urban air pollutants NOx (x=1,2) and CO. Several REMPI schemes have been tested for CO monitoring. Among them, the (2+1) process at 230nm, i.e. the two photon transition $B^1\Sigma^+\leftarrow X^1\Sigma^+$ (0,0) and one photon ionisation is the best one for trace detection. A sensitivitiy of 0.6 ppm is reached by this procedure which is far below the standard recommended by the authorised UK expert panel: 10 ppm. Street air samples have been measured and compared with standard gas samples. The results showed that the CO concentration in the air at rush hour in Glasgow is about 2.5 ppm. This value is in accordance with the data announced by the British Atmospheric Data Centre which says the CO concentration in Glasgow is 3 ppm. Other measurements, e.g. laser power dependence of CO ion yields, are also included in this paper.

INTRODUCTION

The detection and identification of environmentally sensitive gases at trace levels in the atmosphere is an issue of great importance. Ozone, nitrogen oxides (NOx) sulphur dioxide (SO_2), carbon monoxide (CO), PM10, benzene, 1,3 butadiene and 25 hydrocarbon species are treated as harmful air pollutants which should be measured for air quality monitoring purpose. Of all the air pollutant gases, CO is one of the most dangerous since it can cause death. Carbon monoxide is emitted into the atmosphere as a result of combustion processes and from the oxidation of hydrocarbons and other compounds. Several laser spectroscopic techniques have been used to measure the concentration of CO in the atmosphere, e.g. Laser induced fluorescence (LIF)[1], Infrared absorption spectroscopy (IAS)[2] and tunable diode laser absoption spectroscopy (TDLAS)[3]. Since the detection of resonantly enhanced multiphoton ionization (REMPI) can be more sensitive than fluorescence detection, we therefore built a laser-based procedure to detect the trace CO in the urban air. This technique has already been used to detect trace NOx (x=1,2) in Glasgow.[4] Using this approach, the limit of detection can be made down to background level and even lower than that.

To find a spectroscopic fingerprint of CO for the use of trace detection, one needs to investigate its REMPI laser spectroscopy. A number of REMPI schemes have been used to study CO molecules, e. g. transitions (A $^1\Pi\leftarrow$X $^1\Sigma^+$)[5], (B $^1\Sigma^+\leftarrow$X $^1\Sigma^+$)[6], (C $^1\Pi\leftarrow$X $^1\Sigma^+$)[7] and (E $^1\Pi\leftarrow$X $^1\Sigma^+$)[8]. We chose B $^1\Sigma^+\leftarrow$X $^1\Sigma^+$ two photon for excitation and the third photon for ionization at 230 nm to conduct CO the sensitivity detection experiments, because this process can yield strong ionization signals and give a sharp peak. The two photon electric dipole selection rules for a $\Sigma\rightarrow\Sigma$ transition predicts that $\Delta J=0, \pm2$, i.e. there should be Q, S and O branches with P and R being forbidden.

EXPERIMENTAL

The experiment set-up has been published in reference[4], and only a simple description is given here. A Lumonics excimer-pumped dye laser system (Models TEM 860-M and EPD 330) was operated with dye Coumarin 47 (LC 4700) to give tunable laser radiation in the wavelength range 220-242 nm after frequency doubling with an Inrad autotracker II system. The resulting UV beam was separated from the fundamental beam by using an Inrad Harmonic Separator. The UV laser was scaned around 230nm to record the CO spectra. A 30 cm lens was used to focus the 230 nm radiation into ion detector. The nomal laser pulse energy at 230 nm was 300 μJ. An attenuator was used to reduce the laser power. The order of magnitude of the laser fluence is about 10^{10} W/cm^2.

The ion detector was a diecast aluminium alloy box with two parallel stainless steel plates, 15x25 mm, and quartz windows for laser access constituted the detector. Ports for gas flow were added to allow dilution experiments to be performed.

The data acquisition system is based on a Stanford Research System Gated Integrator and Boxcar Averagers (SR 250) with LabView data acquisition software and controlled by a Macintosh IIfx computer via an SR 245 computer interface. Both ion yields and the laser pulse energy can be recorded. The signals from the ion chamber and Molectron Model J3-09 pyroelectric joulemeter can be monitored by the LeCroy 9410 digital oscilloscope simultaneously. Sample concentrations can be varied by on-line mixing using a gas flowing system . The ratio between the sample gas flow rate and the balance gas flow rate can be changed by adjusting the controlling valves on the flowing meters.

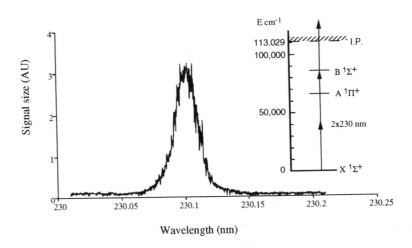

Figure 1. A Recorded Spectrum of (2+1) REMPI Process of CO at 230 nm

RESULTS AND DISCUSSION

1. The shape of the two photon transition spectrum of CO molecules at 230 nm.

Figure 1 shows the two photon excitation and one photon ionization spectrum of $B^1\Sigma^+ \leftarrow X\ ^1\Sigma^+$ (0,0) of CO gas which is seeded into air with concentration of 20 ppm. This spectrum consists of only one symmetric unresolved band centred at 230.1 nm which corresponds to Q-branch of the transition. The most striking feature of the spectrum is the complete absence of the O and S branches. This observation is in agreement with DiMauro et al[9]. The detail rotational structure cann't be seen, because this branch is very congested. Due to the similarity of the potential curves of the B and X state all lines of Q-branch are piled up in a single bandhead with a width of $\Delta\lambda = 0.02$ nm, which could not be resolved into individual rotational lines.

2. Concentration measurements

Figure 2 shows a good linear relation between ion signal size and CO concentrations within the experimental error ranges. The measurements were taken with a laser pulse energy of 160 µJ. At this condition, the limit of detection of CO in air is 0.6 ppm. Figure 3 shows a spectrum of sample gas taken from the urban air at Glasgow. Comparing this spectrum with the spectra generated by the standard CO gas (from Linder Gas Ltd UK), it turns out that the CO concentration of the urban air sample is about 2.5 ppm, this amount is below the safety standard of 10 ppm[10].

CONCLUSION

We have investigated the (2+1) REMPI process at 230 nm for the electronic transition of $B\ ^1\Sigma^+ \leftarrow X\ ^1\Sigma^+$ of CO molecules. A series of spectral measurements showed a very good linear relation between ion yields and CO concentration in the air and a sensitivity of 0.6 ppm for trace CO detection can be reached with the UV laser pulse energy of 160 µJ. Both the CO experiments and the pervious NOx

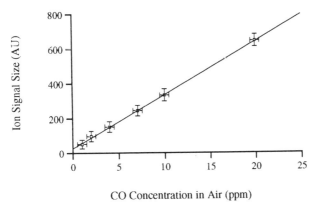

Figure 2. Ion Signal Size versus CO Concentration in air

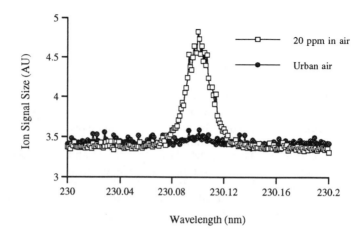

Figure 3. Spectra of Urban Air and Standard Gas

experiments proved that this laser based procedure can be used to carry out air pollution monitoring. This approch has the advantage of easy calibration, suitable for multi-components and in-situ monitoring.

ACKNOWLEDGEMENT

WXP wishes to thank Glasgow University for a scholarship.

REFERENCES

1.Pfab, J., Edi: Clark, R.J.H. and Hester, R.E.,*Spectroscopy in Environmental Sciences*, New York, USA, John Wiley and Sons, Inc., 1995, pp.149-218.

2.Hanst, P L and Hanst, S. T., Edi. Markus W. Sigrist, *Air monitoring by spectroscopic techniques*, New York, USA, John Wiley and Sons, Inc., 1994, pp.335-470

3.Schiff.H.I., Mackey, G.I.and Bechara, Edi. Markus W. Sigrist, J., *Air monitoring by spectroscopic techniques* New York, USA, John Wiley and Sons, Inc., pp.239-334, 1994

4. Peng, W. X., Ledingham, K.W.D., Marshall, A., and Singhal, R. P., *Analyst*, **120**, 1995, 2537-2542

5. Jiang, B., Sha, G., Sun, W., Zhang, C., He, J., Xu. S., and Zhang, C., *J. Chem. Phys.* **97**(7), 1992, 4697-4703

6. Koehoven, S.M., Buma, W.J. and Lange, C. A., *J. Chem. Phys.* **99**(7), 1993, 5061-5070

7. Hines, M.A., Michelson, H.A., and Zare, R. N., *J. Chem. Phys.* **93**(12), 1992, 8557-8564

8. Zacharias, H. and Rottke, H., *Optics comm.*, **55** (2), 1985, 87-90

9. DiMauro, L.F. and Miller, T.A., *Chem. Phys. Lett.*, **138**(2,3), 1987, 175-180

10. Department of the Environment, UK. Expert Panel on Air Quality Standards: Carbon Monoxide, 1994

Using an Atom Interferometer to Take the *Gedanken* Out of Feynman's *Gedankenexperiment*

David E. Pritchard, Troy D. Hammond, Alan Lenef,
Jörg Schmiedmayer*, Richard A. Rubenstein, Edward T. Smith,
and Michael S. Chapman

Department of Physics and Research Laboratory of Electronics, Massachusetts Institute of Technology, Cambridge, Massachusetts 02139, USA

*Institut für Experimentalphysik, Universität Innsbruck, A-6020 Innsubruck Austria

Abstract. We give a description of two experiments performed in an atom interferometer at MIT. By scattering a single photon off of the atom as it passes through the interferometer, we perform a version of a classic *gedankenexperiment*, a demonstration of a Feynman light microscope. As path information about the atom Is gained, contrast in the atom fringes (coherence) is lost. The lost coherence is then recovered by observing only atoms which scatter photons into a particular final direction. This paper reflects the main emphasis of D. E. Pritchard's talk at the RIS meeting. Information about other topics covered in that talk, as well as a review of all of the published work performed with the MIT atom/molecule interferometer, is available on the world wide web at http://coffee.mit.edu/.

In the *Feynman Lectures on Physics*, the famous Caltech physicist remarked that the quantum mechanical interference of particles observed in matter wave interferometers is "a phenomenon which is impossible, *absolutely* impossible to explain in any classical way, and which has in it the heart of quantum mechanics. In reality it contains the *only* mystery." (1) He is referring of course to the phenomenon of wave-particle duality, in which matter at the atomic scale sometimes behaves like particles and sometimes behaves like waves. However, observation of both wave and particle behavior in the same measurement is somehow inevitably forbidden by the laws of quantum mechanics.

Wave-particle duality, along with Heisenberg's uncertainty principle (which sets a lower limit of precision on the simultaneous measurement of a particle's position and momentum) are the two most pervasive examples of the quantum mechanical principle of complementarity. A stumbling block for the lay person's comprehension, complementarity is "not a problem" for professional physicists who have learned to accept these restrictions on the observable. These physicists probably forget to mention that, before deciding that this duality was not a problem, they had to mull over certain *gedankenexperiments* (thought experiments) in which their intuition was re-educated about the peculiarities of quantum behavior.

Many of these *gedankenexperiments* involve particle interferometers, because they so beautifully exemplify wave-particle duality. In order to predict the interference fringes that are observed at the detector in these interferometers, one is forced to represent the particle by a de Broglie wave that travels over both of two separate paths at the same time and then recombines with itself, producing interference. However, our classical intuition insists that the particle must have traveled one path or the other, but not both. In order to re-educate our intuition, a host of "which-way" *gedankenexperiments* have been thought up in which one attempts to measure which path the particle traversed ("which-way") while still preserving the interference pattern recorded by the detector (2). In all such *gedankenexperiments*, however, it can be shown that the which-way measurement invariably reduces the visibility of the observed interference fringes — in fact, the more certainly the measurement determines which side of the interferometer the particle traversed, the more the fringe contrast is reduced (*i.e.* the less the particle exhibits wavelike behavior).

Recent developments in atom optics and interferometry have allowed us to perform one of the most famous of these *gedankenexperiments*: Feynman's suggestion that Heisenberg's light microscope be used to see which side of the interferometer the particle traversed (3). His which-way measurement scheme uses an optical microscope to observe which path the particles take as they pass through a Young's two-slit experiment. Of course, the particles must be illuminated with a light source in order to be imaged, and Feynman claims that it is just the act of scattering photons from the particles that destroys their interference fringes. This is true even if we arrange to scatter only a single photon from each particle passing through the apparatus. He notes that we can minimize the impact of the photon on the interference pattern if we use light of longer wavelength or reduce the slit separation (it is only the ratio that matters), but only at the expense of resultant uncertainty in the which-way information obtained. Indeed, complementarity would suggest that fringe contrast must disappear when the slit separation is larger than half the

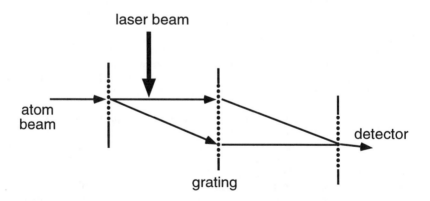

FIGURE 1. A schematic of the experiment showing the three grating Mach-Zehnder atom interferometer and the laser beam used to illuminate the atoms. The grating period was 200 nm and the separation between gratings was 60 cm.

wavelength of the scattered light (λ / 2), since at this point, we could resolve which path the particle took.

In our experiment (see Fig. 1), we used a three grating Mach-Zehnder atom interferometer instead of a two slit experiment (4). A laser beam was used to scatter a single photon from each atom passing through the interferometer, and we measured the contrast of the atomic interference fringes for different path separations at the point of scattering (thus, this experiment is equivalent to a Young's two-slit experiment with a variable slit separation). Not surprisingly (to professional physicists) the principle of complementarity is upheld: the fringe contrast is high when the separation of the paths is much less than (λ / 2) but decreases to zero at about this separation (see Fig. 2).

At larger separations, Fig. 2 reveals not only the general suppression of the fringe contrast expected from complementarity, but also several subsequent revivals of the fringe contrast. Applying the complementarity principle again, these contrast revivals must reflect the imperfect spatial localization provided by the scattered photon (5). This imperfect localization results from the diffraction rings in the image (even a perfect lens has diffraction). If a photon from an atom in one arm of the interferometer is imaged, forming a diffraction ring, it is possible for one of the bright fringes of the diffraction ring to coincide with the position of the other arm of the interferometer. This case is indistinguishable from the one in which the roles of

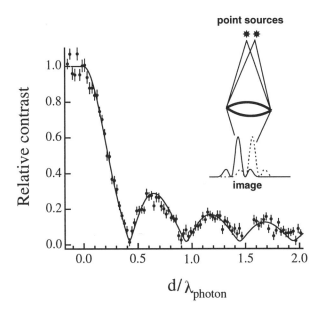

FIGURE 2. Loss of atom interference fringe contrast vs. separation (*d*) of the two atom paths at the intersection of the laser beam (the laser beam was translated along the atom beam direction to scatter at different separations). The inset illustrates the ambiguity of optically resolving two closely spaced point sources which leads to uncertainty in which path the atom took.

the first and second arms are reversed (see inset in Fig. 2). Under these circumstances there is significant uncertainty about which side the atom that scattered the photon really traversed; consequently the fringe contrast can be (and is) greater than zero. The several small recurrences of the contrast shown in Fig. 2 correspond to the several diffraction rings successively aligning themselves with the other arm of the interferometer.

Our experiment also addresses another fundamental question: where is the coherence lost to and how may it be regained? This is a more troublesome problem for the professional physicist because the elastic scattering of a photon is not a dissipative process *per se* and may be treated with Schrödinger's equation without any *ad hoc* dissipative term. Such an analysis preserves the full quantum coherence and results in an atomic wave function that is "entangled," or intertwined with that of the scattered photon. That is, if one knows the final scattering direction of the photon, then the atomic interference fringes are not destroyed but instead acquire a phase shift correlated with the scattering direction. (This does not violate complementarity because a measurement of the exact direction of the photon requires a lens with a small aperture, which lacks sufficient resolution to determine which side of the interferometer the photon scattered from.) The atom-photon entanglement provides a quantitative explanation of the disappearance of the fringes

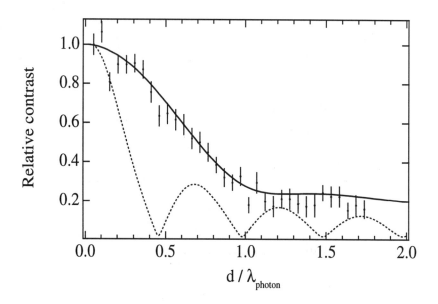

FIGURE 3. Same data as in Figure 2, except that this time we detected only atoms that scattered photons in a particular direction. Fringe contrast is recovered only at the expense of which-way information since the spatial resolution of a measurement made with the scattered photons is reduced along with the uncertainty in their final direction. The dashed line is for comparison, and shows the contrast if all atoms are observed.

when the direction of the photon is not measured. The key concept is that photons scattered in different directions impart different phase shifts to the atom fringes; when these phase shifts are large enough to cause the peaks of some of the fringes to line up with the valleys of others, the result is cancellation of the fringes in the total intensity pattern observed. In fact, the phase shifts increase according to the separation of the paths at the point of scattering, which explains why the fringes are generally more strongly suppressed for larger path separations.

To verify the entanglement between the photon scattering direction and the atom fringes explicitly, we observed the atom fringes formed only by those atoms that had scattered photons in the same general final direction. As expected, the atom fringe contrast was higher in this case. In addition, we found that it shifted in phase according to which final scattering direction we chose. Hence we were able to recover the atomic coherence lost when the scattering direction of the photons was not restricted (see Fig. 3).

Entanglement of one system (say, a particle) due to interaction with another system (a photon, or some other measuring device) is an important issue in contemporary quantum mechanics, particularly with regards to Einstein-Podolsky-Rosen (EPR) type correlations and to understanding the measurement process in general and specifically the loss of coherence which occurs in moving from the quantum to the classical regime (6). In our experiment, we have studied this at the most fundamental level: a single particle undergoing a single scattering event. Thus this experiment not only brings an old *gedankenexperiment* into reality, but also probes deeply enough to reveal its relevance to issues of contemporary interest in quantum mechanics.

ACKNOWLEDGMENTS

The atom gratings used in this work were made in collaboration with Mike Rooks at the Cornell Nanofabrication Facility at Cornell University. We are grateful for the existence and assistance of this facility, and also for help with nanofabrication from Hank Smith and Mark Schattenburg at M.I.T. This work was made possible by support from the Army Research Office contracts DAAL03-89-K-0082 and ASSERT 29970-PH-AAS, the Office of Naval Research contract N00014-89-J-1207, NSF contract 9222768-PHY, and the Joint Services Electronics Program contract DAAL03-89-C-0001. TDH and ETS acknowledge the support of National Science Foundation graduate fellowships. JS acknowledges the support of an Erwin Schrödinger Fellowship of the Fond zur Förderung der Wissenschaftilchen Forschung in Austria and an APART fellowship of the Austrian Academy of Sciences.

REFERENCES

1. Feynman, R.P., Leighton, R.B., and Sands, M., *The Feynman Lectures on Physics*, Reading, MA: Addison-Wesley, 1965, vol. III, p. 1-1.

2. Bohr, N. in *A. Einstein: Philosopher - Scientist*, edited by P.A. Schilpp, Evanston, IL: Library of Living Philosophers, 1949, pp. 200-241.

3. Chapman, M. S., Hammond, T. D., Lenef, A., Schmiedmayer, J., Rubenstein, R.A., Smith, E., and Pritchard, D.E., *Phys. Rev. Lett.* **75**, 3783-3787 (1995).

4. Keith, D.W., Ekstrom, C.R., Turchette, Q.A., and Pritchard, D.E., *Phys. Rev. Lett.* **66**, 2693-2696 (1991).
    Schmiedmayer, J., Chapman, M. S., Ekstrom, C.R., Hammond, T. D., Wehinger, S., and Pritchard, D.E., *Phys. Rev. Lett.* **74**, 1043-1047 (1995).

5. Tan, S.M., and Walls, D.F., *Phys. Rev. A* **47**, 4663-4676 (1993).
    Sleator, T., Carnal, O., Pfau, T., Faulstich, A., Takuma, H., and Mlynek, J., in *Laser Spectroscopy X*, edited by M. Ducloy, E. Giacobino and G. Camy, Singapore: World Scientific, 1991, p. 264z.

6. *Quantum Theory and Measurement,* edited by J.A. Wheeler and W.H. Zurek, Princeton: Princeton University Press, 1983.
    Zurek, W.H., *Physics Today* **44**, 36-44 (1991), and references therein.

SESSION IX: RIS
AND MOLECULAR CLUSTERS

Charge Resonance and Charge Transfer Interactions
in Naphthalene Homo- and Hetero-Dimers

Y. Inokuchi, M. Matsumoto, K. Ohashi, and N. Nishi

Department of Chemistry, Faculty of Science, Kyushu University, Fukuoka 812-81

and Institute for Molecular Science. Myodaiji, Okazaki 444, Japan

Abstract   Charge resonance interaction in naphthalene homo- and hetero-dimer cations is studied by photodissociation spectroscopy of the charge resonance and the local excitation transitions.   The resonance interaction in naphthalene dimer cation is slightly weaker than that of a benzene dimer cation because of partial overlapping of the respective aromatic rings. A local excitation band of the benzene cation chromophore is observed in the spectrum of a naphthalene-benzene hetero-dimer cation at nearly the same position as that of the benzene dimer cation.   This indicates that in spite of its higher ionization potential the positive charge stays on the benzene molecule in some probability. On the basis of the band position of the charge resonance transition as well as the intensity of the local excitation band, the probability is analyzed to be approximately 9 %. This means 91 % is localized on the naphthalene chromophore in this hetero-dimer.

CHARGE RESONANCE INTERACTION IN CLUSTERS

A cation molecule in contact with neutral molecule(s) has a significant effect on the electrons in higher occupied orbitals of the neutral species.   In particular, aromatic molecules with low ionization potentials could show some specific behavior in molecular cluster cations. A model of charge delocalization over all molecules in a cluster has been proposed for benzene cation clusters, $(C_6H_6)_n^+$, on the basis of this situation.[1]   However, not only the local excitation(LE) bands but also the charge resonance(CR) bands of $(C_6H_6)_n^+$ with n=3-6 showed the spectral positions and widths quite similar to those of the dimer cation, $(C_6H_6)_2^+$, although the trimer cation showed the largest spectral shift of the LE(π, σ) band.

[2,3] Interestingly the spectrum of the hexamer cation, $(C_6H_6)_6^+$, exhibited nearly the same pattern as that of the dimer. These results clearly demonstrate that the charge is localized on the dimer unit. This means that other neutral molecule(s) interact with the dimer cation rather weakly just like solvent molecule(s).

Similar ion core structures were found in $(CO_2)_n^+$ [4] and Ar_n^+ [5] clusters. Ionization potentials of benzene(9.2437 eV) and naphthalene(8.1442 eV) are, however, very much smaller than those of CO_2(13.773 eV) and Ar(15.7596 eV), suggesting higher possibilities of electron hopping to the cation species induced by inter-molecular motions of neutral and cation molecules. The charge resonance interaction in a cation homo-dimer provides the ground state wavefunction:

$\Psi_+ = \sqrt{1/2}\,\psi(B_1^+)\psi(B_2) + \sqrt{1/2}\,\psi(B_1)\psi(B_2^+)$. This interaction produces also the

antisymmetric state: $\Psi_- = \sqrt{1/2}\,\psi(B_1^+)\psi(B_2) - \sqrt{1/2}\,\psi(B_1)\psi(B_2^+)$. The CR transition

takes place from the ground symmetric state to the antisymmetric one. The upper state is unstable because of the repulsive interaction. In both states of the homo-dimer, the B_1 and B_2 molecules have an equal probability of carrying a charge. $(C_6H_6)_2^+$ shows the CR band at 920 nm. The CR transition moves the charge from B_1 to B_2 and B_2 to B_1 with the frequency of a photon energy. The transition probability is fairly large compared to the local excitations in this case.

The resonance interaction is expected also in the hetero-dimers composed of molecules with similar ionization potentials. We have studied the photodissociation spectrum of the benzene-toluene hetero-dimer cation.[6] The spectrum revealed a stong CR band located at 1170 nm. This observation suggested that the probability of the charge residing in toluene is approximately 70 % and in benzene 30 %. The difference in the ionization potential is 0.42 eV. This energy is smaller than the stabilization energy(0.674 eV) of $(C_6H_6)_2^+$. Thus, the appearance of the CR band is reasonable.

The ionization potential of naphthalene is 8.144 eV, 1.10 eV lower than benzene. The difference in the ionization potentials is now much larger than the resonance interaction energy in $(C_6H_6)_2^+$. Therefore, the positive charge in the naphthalene-benzene dimer cation$(C_{10}H_8\text{-}C_6H_6)^+$ is expected to be localized on the naphthalene chromophore. Saigusa and Lim[7] reported the photo-dissociation spectra of naphthalene cation clusters. Their band positions were not in agreement with our spectrum of the naphthalene dimer cation, $(C_{10}H_8)_2^+$.[8]

EXPERIMENTAL

The apparatus used in this work has been described elsewhere.[8]   The experiment was carried out using an octopole ion trap with two quadrupole mass filters. The parent ions were prepared using the laser-induced plasma technique. Cation clusters were formed by accretion of neutral molecule(s) to the ionized one and were successively cooled in the expansion. Naphthalene or naphthalene-benzene dimer cations were selected by a quadrupole mass filter.   After 90° deflection the ion beam was introduced into the octopole trap where the photodissociation laser beam was propagated coaxially, exciting the trapped ions. The photoexcitation induced the fragmentation of $(C_{10}H_8)_2{}^+$ or $(C_{10}H_8\text{-}C_6H_6)^+$ producing $C_{10}H_8{}^+$ fragments. Yields of the fragment ions were measured as a function of photon energy and laser output.   The output of an optical parametric oscillator (Spectra-Physics MOPO-730) pumped with a Nd:YAG laser (Spectra-Physics GCR-250) was utilized in the regions of 460-680nm and 740-1200nm. A dye laser (Spectra Physics PDL-3) was used for the wavelength range 680-740nm.

PHOTODISSOCIATION SPECTRA

Figure 1 shows the photodissociation spectra of $(C_{10}H_8)_2{}^+$ (top) and $(C_{10}H_8\text{-}C_6H_6)^+$ (middle) as well as that of $(C_6H_6)_2{}^+$ (bottom). Relative intensities of the three spectra in Figure 1 are just arbitrary.   They are adjusted for the comparison of the band positions. The charge resonance band of $(C_{10}H_8)_2{}^+$ appears at 1180 nm corresponding to a transition energy of 1.05 eV.   On the basis of a simple interaction model[3], this value gives an interaction energy of 0.525 eV (cf. 0.674 eV for $(C_6H_6)_2{}^+$).   Inokuchi et al. proposed a partial overlap structure for $(C_{10}H_8)_2{}^+$ based on the position of the LE band at 580 nm[8]. A partial overlap structure is in conformity with an interaction energy weaker than that of $(C_6H_6)_2{}^+$.

In a hetero-dimer cation, the position of the CR band is related to the difference in the ionization potentials.   A larger difference locates the excited state at a higher energy, while the interaction energy decreases inversely proportional to the square of the difference.   The CR band of $(C_{10}H_8 \text{-} C_6H_6)^+$ appears at 920 nm, the same position as that of $(C_6H_6)_2{}^+$.   There is no reason for the CR band positions to coincide.

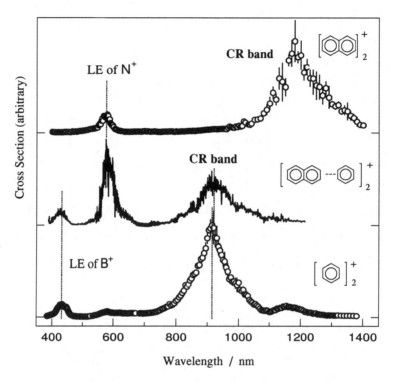

Figure 1.   Photodissociation spectra of naphthalene dimer cation(top), naphthalene-benzene hetero-dimer cation(middle), and benzene dimer cation(bottom).   The middle spectrum was taken through the naphthalene cation window of the mass spectrometer. The data were corrected for the laser intensity distribution.

The local excitation band at 580 nm in the spectrum of $(C_{10}H_8 \text{-} C_6H_6)^+$ appears stronger than the CR band.   The relative cross sections of the 580 nm bands in the two spectra could be compared by measuring the photodepletion yields of the parent signals in a time-of-flight mass spectrum.   It was found that the peak intensity of the LE band of $(C_{10}H_8 \text{-} C_6H_6)^+$ at 580 nm is 0.6 ($\pm$0.2) times as strong as that of $(C_{10}H_8)_2{}^+$.   Therefore the vertical intensity scale of the middle spectrum should be reduced to 14% of the scale of the top spectrum.   The peak intensity of the CR band of the hetero-dimer is then only 7 % of that of the homo-dimer.

PRESENCE OF THE "LE" BAND OF ($C_6H_6^+$) IN ($C_{10}H_8$ - C_6H_6)$^+$

In the spectrum of (C_6H_6)$_2^+$, the LE band of the monomer cation of (C_6H_6)$^+$ appeared at 430 nm. The hetero-dimer cation also weakly showed this band at the same position. In Figure 2, an expanded view of the LE bands of the two cation dimers is given on the same intensity scale. The vertical scale is a measure of intensities relative to the peak of the CR band of (C_6H_6)$_2^+$.

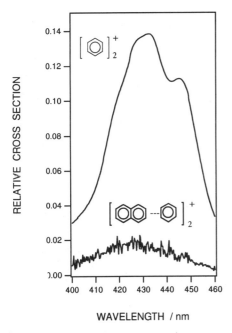

Figure 2. The second LE bands of (C_6H_6)$_2^+$ and ($C_{10}H_8$ - C_6H_6)$^+$.

Although the signal is rather noisy, the spectrum of the hetero-dimer shows a profile quite similar to that of the local transition of ($C_6H_6^+$) in (C_6H_6)$_2^+$. Moreover, the peak positions also coincide. In this region, there exists a very broad band originating from ($C_{10}H_8$)$^+$. This background band is almost flat in this region contributing approximately 40% of the observed intensity. Therefore, the relative cross section of the LE transition of the hetero-dimer is estimated to be 8 (±2)% of that of (C_6H_6)$_2^+$. This fact strongly suggests that the charge is localized on the benzene with a probability of approximately 10 %. This value has to be varified using a more reliable method.

NEAR RESONANCE INTERACTION IN HETERO-DIMERS

As demonstrated in the case of benzene-toluene cations[6], hetero-dimer cations can show intervalence transitions from the ground (A^+ + B) state to the excited (A + B^+) state, which is a charge transfer transition. The presence of this transition also demonstrates that the ground state is stabilized by the resonance interaction between the two states. The degree of stabilization is related to the difference in the ionization potentials of A and B molecules. The CR transition of $(C_{10}H_8- C_6H_6)^+$ was observed at 1.35 eV. The difference in the ionization potentials is 1.10 eV. Since locally excited states may not strongly couple with these local ground states, a simple perturbation theory can be applied to estimate the mixing coefficients based on these values. Figure 3 shows a result of the analysis. One numerical analysis results in a ground state wavefuntion with an admixture of the $\psi(N)\ \psi(B^+)$ configuration of 9 %. This value is in accord with that obtained in the previous section.

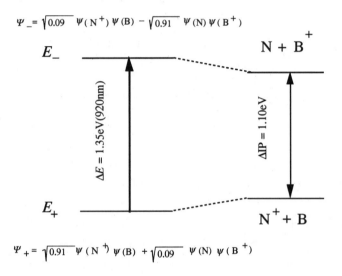

$$\Psi_- = \sqrt{0.09}\ \Psi(N^+)\Psi(B) - \sqrt{0.91}\ \Psi(N)\Psi(B^+)$$

$$\Psi_+ = \sqrt{0.91}\ \Psi(N^+)\Psi(B) + \sqrt{0.09}\ \Psi(N)\Psi(B^+)$$

Figure 3.   Interaction diagram for naphthalene(N)- benzene(B) dimer cation.

REFERENCES

1. Krause, H., Ernstberger, B., and Neusser, H.J., *Chem. Phys. Lett.*, **184**, 411-417(1991)
2. Ohashi, K., and Nishi, N., *J. Phys. Chem.*, **96**, 2931-2932(1992)
3. Ohashi, K., Nakai, Y., and Nishi, N., *Laser Chem.*, **15**, 93-111 (1995)
4. Johnson, M.A., Alexander, M.L., Lineberger, W.C., *Chem. Phys. Lett.*, **112**, 285-290(1984)
5. Deluca, M.J., and Johnson, M.A., *Chem. Phys. Lett.*, **162**, 445-448(1989)
6. Ohashi, K., Inokuchi, Y., Shibata, T., Nakai, Y., and Nishi, N., Structures and Dynamics of Clusters, Frontiers Science Ser. **16**, 357-362 (1996)
7. Saigusa, H., and Lim., E.C., *J. Phys. Chem.*, **98**, 13470-13474(1994)
8. Inokuchi, Y., Ohashi, Y., Matsumoto, M., and Nishi, N., *J. Phys. Chem.*, **99**, 3416-3418(1995)

Ionization Detected Vibrational Spectroscopy of Size-Selected Hydrogen-Bonding Clusters of Phenol

Naohiko Mikami, Takayuki Ebata, and Asuka Fujii

Department of Chemistry, Graduate School of Science,
Tohoku University, Sendai 980-77 JAPAN

Abstract. Infrared (IR) and Raman spectroscopies involving ionization detection methods were applied for observing OH vibrations of size-selected phenol clusters. Also IR spectra of size-selected phenol cluster cations were observed by using IR multiphoton dissociation spectroscopy combined with an ion trapping technique. The cyclic form of the neutral phenol trimer was confirmed with experimental evidences. In the trimer cation, on the other hand, the IR spectrum shows that the cyclic form is no longer stable, but a chain form is feasible, in which all the phenyl moieties are bound by a single chain of hydrogen bonds. The drastic structure change induced upon ionization of the neutral trimer is discussed.

INTRODUCTION

Clusters in which a molecule is surrounded by other molecules are thought to be microscopic models for solute-solvent systems in condensed phases. Recently, vibrational spectroscopies of size-selected clusters have become popular for investigations of intermolecular binding structures of clusters (1-4). Hydrogen-bonded clusters of organic acids, such as phenol and tropolone, provide us with new materials for a microscopic view of acidic nature in condensed phases. Very recently, we have reported vibrational spectroscopic studies of OH stretching vibrations of clusters of these acids solvated with water molecules (5,6). The vibrational spectra are quite effective to characterize hydrogen-bonding structures in the clusters, which are expected to be closely related to local structures among solute-solvent molecules in their aqueous solutions. In this paper, we will present vibrational spectroscopic studies of phenol clusters, $(phenol)_n$, and their cations, $(phenol)_n^+$, by using ionization detected infrared (IDIR) and ionization detected stimulated Raman (IDSR) methods as well as infrared multiphoton dissociation spectroscopy in combination with an ion trapping technique (7).

EXPERIMENTAL

Fig.1(a) and (b) show excitation schemes used in IDIR and IDSR spectroscopy, respectively (1,5). The ion signal was generated by resonance enhanced multiphoton ionization (REMPI) via the 0-0 band of the S_1-S_0 transition of a particular cluster, with which the size-selection can be accomplished. The ion signal represents a measure of the S_0 population of the selected cluster species. The S_0

population is reduced when vibrational transitions are induced by $h\nu_{IR}$ or by the stimulated Raman process. The vibrational spectra of a particular cluster were obtained as ion-depletion spectra. Fig.1(c) shows a scheme for the trapped ion photodissociation (TIP) spectroscopy (7). Cluster ions generated by REMPI were introduced into a quadrupole ion trap cell, and the size-selection was carried out by selecting an appropriate condition for electric potentials of the cell. Details of the ion trapping method were described elsewhere (7). The vibrational spectra of particular species were obtained as yield spectra of the fragment ion generated by IR-multiphoton dissociation. Hydrogen-bonded clusters were prepared by supersonic expansion of the phenol vapor seeded in helium gas at about 3 atm.

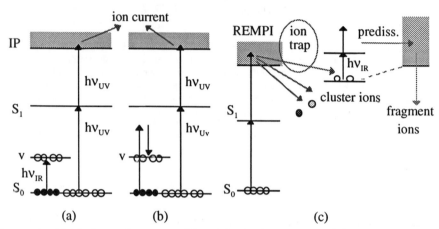

(a) (b) (c)

Figure 1. Excitation schemes used in (a) ionization detected infrared (IDIR) and (b) ionization detected stimulated Raman (IDSR) spectroscopy, and (c) for infrared multiphoton dissociation spectroscopy combined with an ion trapping technique.

RESULTS AND DISCUSSION

IR and Raman spectra of neutral (phenol)ₙ

Fig.2 shows the OH stretching region of IR and Raman spectra of (a) bare phenol, (b) phenol dimer, and phenol trimer (c), for which the frequencies of the ionization laser $h\nu_{UV}$ were tuned at 36348, 36044, and 36202 cm^{-1}, respectively.

In the dimer IR spectrum, two OH bands at 3654 and 3530 cm^{-1} are seen, which correspond well to the Raman bands observed with IDSR spectroscopy. Since two OH oscillators of the dimer are active in both IR and Raman spectra, an asymmetric form is expected for the dimer, in which the two phenol moieties are not equivalent, that is, one acts as a proton donor and the other as an acceptor. It has been established that a small frequency shift occurs for OH oscillators of proton accepting sites, while a large low-frequency shift takes place for proton donating OH oscillators. It is noticed that the band at 3654 cm^{-1} is very close to the OH band of bare phenol at 3657 cm^{-1}. In this respect, therefore, the band at 3654 cm^{-1} is assigned to the OH oscillator of the accepting site and the other band to that of the donating site. The schematic dimer form is illustrated in the figure.

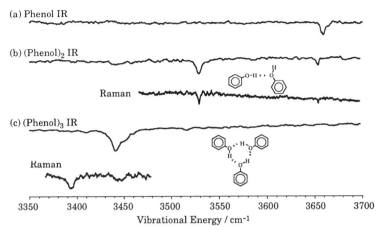

Figure 2. IR and Raman spectra of the OH stretching region of (a) bare phenol, (b) phenol dimer, and (c) phenol trimer (8).

In the trimer IR spectrum, on the other hand, an intense band occurs at 3441 cm^{-1} with a shoulder at 3449 cm^{-1}, while no band is seen in the frequency region of OH oscillators of accepting sites. Thus, the spectrum indicates that the trimer has no OH bond free from hydrogen-bonding. In the Raman spectrum, in contrast, two bands are observed; an intense band occurs at 3394 cm^{-1} and a weak band at 3441cm^{-1}. It must be noted that the former is missing in the IR spectrum. Such an IR/Raman propensity suggests that the trimer appears to be a cyclic structure with a highly symmetric hydrogen-bond. As illustrated in the figure, the cyclic trimer with C_3 symmetry is proposed, in which three OH stretching vibrations will split into two groups: one is totally symmetric and is Raman active but IR forbidden, the other is doubly degenerate and IR active non-totally symmetric vibrations. Depolarization ratio measurements of the trimer Raman spectrum have confirmed that the cyclic form is realistic (8).

IR spectra of cations, (phenol)$_n^+$

Fig.3 shows IR multiphoton dissociation spectra of (phenol)$_n^+$ for n=2-4, obtained with the trapped ion photodissociation (TIP) method (9,10). In each spectrum, an intense band is dominant, appearing at 3620, 3627 and 3632 cm^{-1}, for n=2, 3, and 4, respectively. No sharp band occurs in the range 3000-3600 cm^{-1}, where only broad absorptions are observed. It was found that these prominent bands are very close to the OH frequency of bare phenol in its neutral state (3657 cm^{-1}), but that they do not correspond well to the OH band of the phenol ion (3535 cm^{-1}). In IR spectra of many hydrogen-bonded neutral clusters (5,6), it has been established that OH bands occurring in the range 3600-3700 cm^{-1} are responsible for OH oscillators at hydrogen-accepting sites, in which the OH bond is not directly involved in hydrogen bonding. Such OH bonds at accepting sites may be called "dangling" bonds in hydrogen bonding systems. In this respect, therefore, there must exist, at least, one dangling OH bond in every species of (phenol)$_n^+$. Our result indicates that every (phenol)$_n^+$ has a neutral phenol moiety with a dangling OH bond.

239

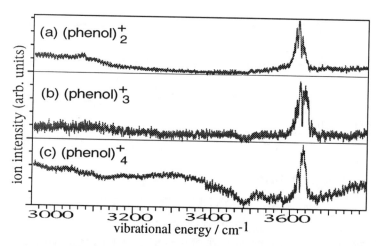

Figure 3. IR spectra of (a) (phenol)$_2^+$, (b) (phenol)$_3^+$, and (c) (phenol)$_4^+$. Sharp dips between 3610-20 cm^{-1} are due to water vapor absorptions.

Consequently, the dimer ion must have an asymmetric form, in which the phenol ion site appears as a proton donor and the neutral phenol site acts as an acceptor. Since the neutral dimer exhibits the asymmetric hydrogen-bonded structure, no drastic structure change in intermolecular binding is induced upon ionization. In (phenol)$_3^+$, on the other hand, a substantial structure change is revealed from the analysis of its IR spectrum, in which at least one dangling bond must be involved. The presence of a dangling bond indicates that the trimer ion appears to have an asymmetric form with respect to the intermolecular hydrogen bonding. In the electronic spectrum of the trimer ion in the wavelength range 400-1600 nm, no charge resonance transition was observed, indicating that the charge resonance interaction among phenyl rings is insignificant (10). It is concluded that all the phenyl moieties in the trimer ion are bound by a linear chain of OH--O hydrogen bonds, while a symmetrically cyclic form is evident for the neutral trimer. In small-size clusters, it should be noticed that the ring formation leads to more stabilization energy for the cluster formation, but results in some stress of the OH--O bond-angles. Such a substantial structural change upon ionization can be explained by a balance between the destabilization due to the stress and the hydrogen-bonding energy associated with the ring formation.

REFERENCES

1. Hartland,G.V.,Benson,B.F.,Venturo,V.A., and Felker,P.M. *J. Phys. Chem.* **96**, 1164 (1992).
2. Tanabe, S., Ebata, T., Fujii, M., and Mikami, N., *Chem. Phys. Letters* **215**, 347 (1993).
3. Pribble, R. N., Zwier, T. S., *Faraday Discuss.* **97**, 229 (1994).
4. Mikami, N., *Bull. Chem. Soc. Jpn.* **68**, 683 (1995), and references therein.
5. Watanabe, T., Ebata, T., Tanabe, S., and Mikami, N. *J. Chem. Phys.* **105**, (1996) in press.
6. Mitsuzuka, A., Fujii, A., Ebata, T., and Mikami, N. *J. Chem. Phys.* **105**, (1996) in press.
7. Sato, S. and Mikami, N. *J. Phys. Chem.* **100**, 4765 (1996), and references therein.
8. Ebata, T., Watanabe, T., and Mikami, N. *J. Phys. Chem.* **99**, 5761 (1995).
9. Sawamura,T., Fujii,A., Sato,S., Ebata,T., and Mikami, N. *J. Phys. Chem.* **100**, 8131 (1996).
10. Fujii, A., Iwasaki, A., Yoshida, K., Ebata, T., and Mikami, N. *J. Phys. Chem.* submitted.

Observation of Atomic-like Resonance in Multiphoton Ionization Spectra of the Metal-rich Magnesium-oxide Calcium-oxide and Barium-oxide Clusters

M. Foltin, H. Sakurai, and A.W. Castleman, Jr.

Department of Chemistry, The Pennsylvania State University, 152 Davey Lab., University Park, PA 16802, USA

Abstract. We observe that the $(CaO)_n Ca_m$ and $(BaO)_n Ba_m$ clusters containing few excess Ca and Ba atoms (i.e., $6 \leq m < n$) can be resonantly ionized via the $(ns^2)^1 S_0 \rightarrow (nsnp)^1 P_1$ atomic resonance of the isolated Ca and Ba atoms. The resonance is attributed to monomer atoms evaporated from the cluster undergoing excitation transfer to the cluster. In $(MgO)_n Mg_m$ clusters containing many excess Mg atoms (i.e., m is large) we observe a depletion of the $(MgO)_n Mg^{++}$ doubly-charged cluster ion intensity in a resonance which is red shifted by 35 meV from the $(3s^2)^1 S_0 \rightarrow (3s3p)^1 P_1$ atomic resonance. This different behaviour is interpreted in terms of different strength of cluster-monomer interaction and different rate of electron-to-vibrational relaxation in both cases.

INTRODUCTION

In recent years considerable attention has been directed to studies of metallization of alkali-halide clusters. Previous theoretical (1), photoelectron spectroscopy (2) and mass spectrometric (3) studies on $(NaCl)_n Na_m$ clusters and their ions have shown that the valence electrons in these clusters are delocalized and form closed shell structures like in pure Na_m clusters. Regarding the analogous alkaline earth oxide systems, considerably less is known about the evolution of their electronic properties with increasing metallization with cluster size. Better understanding of these properties may be of practical importance since these oxides are used as catalysts. Herein, the photoionization dynamics of the $(CaO)_n Ca_m$ and $(BaO)_n Ba_m$ clusters containing very few excess metal atoms (i.e., $6 \leq m < n$ in calcium case) and the $(MgO)_n Mg_m$ clusters containing many excess Mg atoms (i.e., m is large) is studied, which were found to resonantly absorb photons with energies close to the $(ns^2)^1 S_0 \rightarrow (nsnp)^1 P_1$ atomic resonance of the isolated Ca, Ba and Mg atoms, respectively.

EXPERIMENTAL

The apparatus used in these experiments has been described elsewhere (4). Briefly, in a liquid nitrogen-cooled source chamber, Ca, Ba or Mg metal is evaporated from a resistively heated alumina crucible. The metal vapor is mixed with flowing cold He (3000 sccm flow rate) and N_2O (0.8 to 5 sccm flow rate), whereupon it is oxidized, cools and undergoes concomitant clustering reactions. Clusters are carried by He flow in a flow tube where at the end a small fraction of the flow passes through 1.5 mm orifice in a sampling cone and forms a molecular beam weakly expanding in a vacuum (pressure 10^{-4} Torr). The beam enters the ion source where it is irradiated by tuneable dye-laser beam(s) or by XeCl (308 nm) excimer laser. The ions formed by photoionization of the molecular beam are mass-analyzed in a time-of-flight mass spectrometer equipped with a reflectron.

RESULTS AND DISCUSSION

The composition of the neutral $(CaO)_nCa_m$ calcium-oxide clusters studied in this work has been probed with XeCl excimer laser (4.03 eV photons) at low laser fluence, revealing that the excess metal content is somewhat larger than m=6 and possibly increases with the cluster size n, but the m/n ratio is probably less than one. The excess Ca atoms are evaporated when increasing the laser intensity, i.e., at intensities above 5 mJ/cm^2 the $(CaO)_nCa^+$ cluster ions dominate the mass spectrum. When the photon energy is tuned to 2.95 eV, no cluster ions are detected at laser fluence below approx. 5 mJ/cm^2. However, as the dye-laser is tuned to the $(4s^2)^1S_0 \rightarrow (4s4p)^1P_1$ resonance of the Ca atom at 422.67 nm (2.933 eV), the $(CaO)_nCa^+$ cluster ions appear in large abundance even at low laser fluence (see Fig. 1). Surprisingly, the cluster resonance is centered to within $\pm$ 1 meV exactly at the position of the Ca atomic resonance, it has a line-width of less than 2.0 meV and its position does not vary with the cluster size n. We would expect broadening and cluster size dependent spectral shift of the resonance caused by interactions between the Ca ligands and the $(CaO)_n$ ionic core. We could exclude possibilities that this resonant ionization is caused by reactions of clusters with excited Ca* atoms or Ca$^+$ ions formed by resonant ionization of Ca atoms in the beam or in the background (5). We observed analogous cluster resonance in the $(BaO)_nBa_m$ clusters at 553.55 nm and we found that for small clusters (n<9) this resonance is quenched when a light pulse from a Nd-YAG laser (532 nm) is applied 100 ns before the resonant pulse, whereas no quenching is observed when the Nd-YAG pulse is applied 100 ns after the resonant pulse. The ionization mechanism consistent with these observations is as follows: Absorption of the first photon occurs on the cluster ladder (wavelength-independently), leading to vibronic excitation of the cluster and subsequent evaporation of one or more rather weakly bound Ca (or Ba) ligands. The Ca atom very efficiently absorbs light

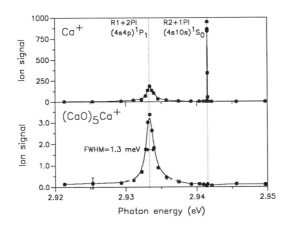

Figure 1. Dependence of ion signal on the photon energy compared for the Ca^+ atomic ion and the $(CaO)_8Ca^+$ cluster ions. Energies of the $(4s4p)^1P_1$ and $(4s10s)^1S_0$ Ca atomic resonances are marked with vertical dashed lines.

resonant with the $(4s^2)^1S_0 \rightarrow (4s4p)^1P_1$ transition, i.e., from the population rate equations we calculate that at the laser fluence of 5 mJ/cm^2, more than 10% of all Ca atoms will be excited to the $(4s4p)^1P_1$ state during 1 ps of their exposure to the laser light. Assuming that the evaporated Ca ligand carries away about 0.2 eV of kinetic energy, in 1 ps it will separate from the cluster by less than 10Å. Qualitative arguments show that the potential surface of the $Ca^*(4s4p)\text{-}(CaO)_nCa_{m-1}$ state is crossed by that of the $Ca^+\text{-}(CaO)_nCa_{m-1}^-$ ion-pair state at ~10Å separation. Thus, it is reasonable to assume that the evaporated Ca ligand excited to the $(4s4p)^1P_1$ state can undergo excitation transfer to the cluster by a "harpooning mechanism" where the $Ca^+\text{-}(CaO)_nCa_{m-1}^-$ ion-pair state plays the role of an intermediate between the two excited states. In the case of small $(BaO)_nBa_m$ clusters (n<9), most of the excess Ba atoms are apparently desorbed when the Nd-YAG laser beam is applied 100 ns before the resonant ionization laser beam, which explains the quenching of the resonance.

The case of magnesium oxide clusters is quite different in that besides the singly-charged cluster ions $(MgO)_nMg^+$ and $(MgO)_n^+$ we also observe the doubly-charged ions $(MgO)_nMg^{++}$ when the photon energy is scanned around the Mg $(3s^2)^1S_0 \rightarrow (3s3p)^1P_1$ atomic resonance. Moreover, we observe a photo-depletion of the $(MgO)_nMg^{++}$ cluster ion signal in a band which is red-shifted by 35 meV from the position of the atomic resonance and has a full width at half maximum of approx. 120 meV (Fig. 2). Based on the present findings that the magnesium oxide clusters studied in this work are more metal rich than the calcium oxide clusters (5), we speculate that unlike in the calcium-oxide clusters where the evaporating Ca ligands are the photoabsorbing chromophores, in the more metal-rich magnesium-oxide clusters some weakly bound Mg ligands are the chromophores.

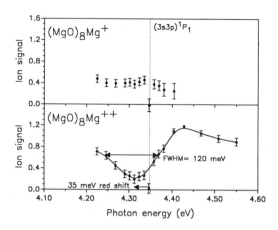

Figure 2. Ion signal as a function of photon energy compared for the doubly-charged $(MgO)_8Mg^{++}$ (on the bottom) and the singly-charged $(MgO)_8Mg^+$ (on the top) cluster ions. Energy of the atomic resonance is marked by dashed line.

The photodepletion seen in the lower spectrum in Fig. 2 most likely occurs because the cluster is stripped of most of its metal ligands after excitation transfer from the chromophore to the cluster, therefore inhibiting multiple ionization. The depletion of the doubly-charged cluster ions is accompanied by no enhancement in the singly-charged cluster ion intensity. Hence the electronic-to-vibrational relaxation of the excited state in magnesium-oxide clusters is apparently faster than in calcium-oxide clusters. These observations provide evidence of a transition of electronic properties of alkaline-earth oxide clusters with increasing metallization, which, however, occurs at m>>6, i.e., it is much slower than in alkali oxides and halides.

ACKNOWLEDGEMENTS

Financial support by the Department of Energy, Division of Chemical Sciences, Office of Basic Energy Sciences, Office of Energy Research, Grant No. DE-FGO2-92ER14258, is gratefully acknowledged.

REFERENCES

1. Rajagopal, G., Barnett, R. N., and Landman, U., *Phys. Rev. Lett.* **67**, 727 (1991)
2. Xia, P., and Bloomfield, L. A., *Phys. Rev. Lett.* **72**, 2577 (1994)
3. Pollack, S., Wang, C. R. C., and Kappes, M. M., *Z. Phys.* **D12**, 241 (1989)
4. Farley, R. W., Ziemann, P., and Castleman, Jr., A. W., *Z. Phys.* **D14**, 353 (1989)
5. Foltin, M., Sakurai, H., and Castleman, Jr., A. W., *Chem. Phys. Lett.*, to be submitted

Photoionization Spectroscopy of Small Alkali and Superalkali Clusters

D. T. Vituccio, O. Golonzka, R. F. W. Herrmann,[1]
S. Rakowsky,[2] and W. E. Ernst

Department of Physics, The Pennsylvania State University, University Park, PA 16802, USA
[1]*Max Born Institut, Rudower Chausee 6, D-12489 Berlin, Germany*
[2]*Max-Planck-Institut für Strömungsforschung, Bunsenstrasse 10, D-37073 Göttingen, Germany*

Abstract. Alkali clusters and nonstoichiometric alkali-halide and alkali-oxide complexes, often referred to as superalkalis, have been studied spectroscopically in molecular beam expansions. Continuous wave resonant two-photon ionization spectroscopy including optical-optical double resonance at high resolution provided insight into the vibronic coupling in alkali trimers, while pulsed laser ionization spectroscopy of Na_2F and Na_3O was used to derive ionization thresholds.

INTRODUCTION

With only one valence electron per atom, alkali clusters represent the simplest small metallic system and offer an excellent opportunity to study the development of structural and electronic properties as a function of cluster size. On the other hand, ionic alkali-halide clusters can be thought of as a simple, finite-size, insulating system. By producing nonstoichiometric, halogen-deficient clusters, one can systematically study the transition from an insulating to a metallic state. Both halogen-deficient alkali-halide clusters M_nX_{n-m} and oxygen-deficient aggregates of the type $M_{2n}O_{n-m}$ possess lower ionization potentials than pure alkalis and, for this reason, are sometimes called superalkalis. Alkali and superalkali clusters are produced in molecular beams which contain a mixture of molecular species of different sizes. For achieving size selectivity in spectroscopic studies, ionization spectroscopy with subsequent mass selective detection of ions is the experimental method of choice.

R2PI OF ALKALI TRIMERS AT ROTATIONAL RESOLUTION

In both clusters and solids, the nuclei are often not arranged in the geometrical structure of highest symmetry which may alter their electronic behavior and chemical properties in unexpected ways. Jahn-Teller (JT) or pseudo-Jahn-Teller (PJT) coupling of the electronic and nuclear degrees of freedom, a breakdown of the Born-Oppenheimer approximation, can cause interesting structural deviations. Triatomic species are the smallest ones in which the JT or PJT effect can be observed and are still simple enough for a detailed investigation of the underlying physics.

Among alkali clusters, the sodium trimer has been the subject of many spectroscopic studies since it can be produced in reasonable abundance and several electronic states can be accessed with visible lasers (1). Of particular interest is the B state of Na_3, in which the molecule has no rigid geometrical shape but periodically alters its geometry. It performs a so-called pseudorotational motion that can be described by a superposition of the two nonsymmetric vibrational modes (2). During one period of this pseudorotation the geometry changes through different obtuse and acute isosceles shapes. The vibronic wavefunction contains a phase factor $e^{ij\phi}$ with j being the quantum number of the vibronic angular momentum of the pseudorotation and ϕ is the respective angle. For a long time, it was believed that the B state of Na_3 represented an example of the wavefunction changing its sign upon one cycle through ϕ; i.e., acquiring a phase change of π upon going from $\phi = 0$ to $\phi = 2\pi$ due to the occurrence of a degeneracy on the potential surface. This behavior would represent a case of Berry's phase in a molecule (3) and cause a half-odd integer quantization of the quantum number j since it would take two cycles through ϕ to return to the original value of $e^{ij\phi}$. Recently, we were able to unambiguously assign integer values j = 0, 1, 2, 3, ... by performing high resolution resonant two-photon ionization spectroscopy (R2PI) and by analyzing the rotational structure and Coriolis interaction in pseudorotational bands (4).

Sodium clusters are produced by seeding argon, at pressures up to 15 bar, with sodium vapor and expanding the mixture through a 30 to 50 μm nozzle heated to 750° C. Details about the molecular beam apparatus are described elsewhere (5). For the spectroscopic studies, the cluster beam is crossed by several laser beams. Sodium trimers are resonantly excited with a single-mode, continuous-wave dye laser and subsequently ionized with 514 nm light from an argon-ion laser. Ions are detected through a quadrupole mass filter. In this way, optical spectra are recorded with 0.005 cm^{-1} linewidths and can be assigned to clusters of a particular size due to the resonance ionization and ion detection

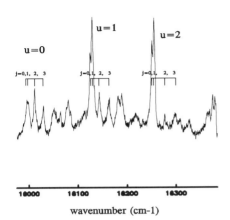

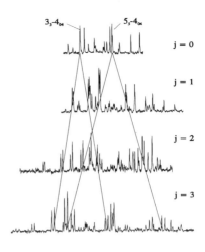

FIGURE 1. Resonant two-photon ionization spectrum of the *B* state of Na$_3$ taken at low resolution. Vibrational bands are denoted by the quantum numbers u and j.

FIGURE 2. Coriolis splitting in the B state of Na$_3$. Two transitions are labeled in the top scan and their counterparts are traced throughout higher j bands. The straight lines indicate that the splitting is linear in j while the absence of splitting in the top scan proves that j is integer quantized.

processes. Although a rather low rotational temperature of about 5 K is achieved in the supersonic beam expansion and the spectra are rotationally resolved, the line density approaches 200 lines per cm^{-1}. To simplify the spectra, optical-optical double resonance (OODR) spectroscopy, in which a second single-mode, cw dye laser crosses the molecular beam upstream from the ionization region, is employed. The second dye laser is scanned throughout an electronic system and a decrease in the ion current is seen when it induces a transition emanating from the same ground state rotational level as the resonant transition in the ionization region. Thus, only transitions sharing the ground state rotational level with the resonant (pump line) transition are recorded (4, 5). Figure 1 shows a low resolution survey spectrum of the Na$_3$ *B-X* system around 620 nm. A series of OODR spectra is depicted in Fig. 2. In these recordings, one cw dye laser is used to pump a narrow line in the dense spectrum while the second one is scanned across the pseudorotational bands of j = 0, 1, 2, and 3. Here, the observed lines all derive from the same ground state levels, and we were able to follow changes in the rotational structure through all these bands. In particular, interaction of the rotation of the molecule with the pseudorotation shows up in a

Coriolis splitting of each rotational level. The magnitude of this splitting depends on the size of the pseudorotational angular momentum (4). In Fig. 2, straight lines connect the transitions which share both upper and lower state rotational quantum numbers and, in this way, show the linear dependence of the Coriolis splitting on j. Our experiments provided the first quantitative proof that $\mathbf{j}$ is integer quantized and, therefore, that the B state of Na_3 is not a case of Berry's phase (4). Furthermore, they revealed new details about the coupling of the unpaired electron spin in this molecule with the nuclear motion (6) and led to new developments in the theory of pseudorotating molecular states (7).

IONIZATION SPECTROSCOPY OF Na_2F AND Na_3O

By reacting the sodium cluster beam with gases like SF_6 or N_2O outside the oven source, in front of the nozzle, mixed sodium-fluoride or sodium-oxide clusters are produced. The electronic structure of nonstoichiometric complexes is most interesting. For example, the unpaired electron in a molecule like Na_2F or Na_3O should give rise to an optical absorption spectrum analogous to that of an F-center in a crystal. Much less is known about these species than about pure alkali clusters. Therefore, we started our investigation with the precise measurement of ionization potentials.

Light from a pulsed dye laser is used to ionize molecules in the ionization region of the quadrupole mass spectrometer. We covered a range of photon energies between 3 and 4.5 eV by applying the output of an excimer-pumped UV dye laser or of a frequency doubled dye laser. In Fig. 3, a laser scan around the

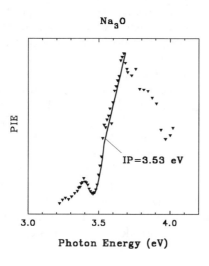

Na_3O

PIE

IP=3.53 eV

3.0   3.5   4.0

Photon Energy (eV)

FIGURE 3. Photoionization efficiency (PIE) vs. photon energy for Na_3O. The PIE is defined as the ratio of the measured ion yield to the photon flux density--assuming no saturation effects are present. The line through the data points is the calculated PIE which has been fit to the data.

photoionization threshold of Na_3O is depicted. Thresholds of 3.81(2) and 3.53(5) eV for Na_2F and Na_3O, respectively, were determined. Details of these findings are reported in a forthcoming paper (8). In the same work, the vibrational temperatures of the two species were estimated to be in the vicinity of 650 K. The high temperature is caused by the exothermicity of the reaction of the sodium clusters with SF_6 or N_2O and presents us with a serious obstacle to the successful measurement of vibrationally resolved R2PI spectra (8).

New Approach for the Production of Ultracold Aggregates

The generation of metal-halides and oxides through a gas phase reaction yields clusters of rather high internal temperature thereby preventing the measurement of well-resolved spectra. A traditional method for producing and studying cold aggregates has been matrix isolation spectroscopy. Unfortunately, the interaction of the aggregate of interest with the matrix environment often leads to spectral line shifts and broadening. Very recently, large helium clusters with an internal temperature of 0.4 K have been doped with alkali atoms (9). The alkali atoms remain on the surface of the helium clusters and react to form dimers and trimers for which the vibrational bands are barely shifted compared to the gas-phase species. By adding an oxidant gas to the alkali doped He_n, the desired aggregates could be generated at very low vibrational temperatures. The helium produces very little perturbation and thus the "cluster isolation" technique could become an ideal method for the spectroscopy of superalkali clusters.

ACKNOWLEDGEMENT
This work was supported by the National Science Foundation.

REFERENCES

1. Broyer, M., Delacrétaz, G., Labastie, P., Whetten, R. L., Wolf, J. P., and Wöste, L., Z. Phys. D. **3**, 131 (1986).
2. Herzberg, G., Molecular Spectra and Molecular Structure, Vol. III, Malabar FL: Krieger Publishing Co., 1991.
3. Geometric Phases in Physics, edited by Shapere, A., and Wilczek, F., Singapore: World Scientific, 1989.
4. Ernst, W. E., and Rakowsky, S., Phys. Rev. Lett. **74**, 58 (1995).
5. Ernst, W. E., and Rakowsky, S., Ber. Bunsenges. Phys. Chem. **99**, 441 (1995).
6. Vituccio, D. T., Golonzka, O., and Ernst, W. E., to be published.
7. Ohashi, N., Tsuura, M., and Hougen, J. T., J. Mol. Spec. **173**, 79 (1995).
8. Vituccio, D. T., Herrmann, R. F. W., Golonzka, O., and Ernst, W. E., to be published.
9. Stienkemeier, F., Higgins, J., Ernst, W. E., and Scoles, G., Phys. Rev. Lett. **78**, 3592 (1995).

SESSION X: RIS APPLIED TO NUCLEAR AND PARTICLE PHYSICS

Application of RIS to Search for
Double Beta Decay of ^{136}Xe

M.Miyajima, S.Sasaki, H.Tawara and E.Shibamura*

National Laboratory for High Energy Physics
Oho 1-1, Tsukuba, Ibaraki, 305 Japan
*Saitama College of Health
Kami-okubo 519, Urawa, Saitama, 338 Japan

Abstract. In order to search for the double beta decay of ^{136}Xe, its decay daughters of ^{136}Ba are being collected with positive ion collectors in xenon gas and liquid. A time-of-flight mass spectrometer in which Ba atoms are selectively ionized by RIS is being developed to quantitatively measure the number of Ba atoms. A new method is proposed to study the neutrinoless double beta decay of ^{136}Xe.

Introduction

Nuclear double beta ($\beta\beta$) decay is one of the rarest processes in nature. The $\beta\beta$ decay is a process with two main decay modes, the two neutrino (2ν) mode and the neutrinoless (0ν) mode. In the 2$\nu\beta\beta$ mode, the nucleus decays emitting two neutrinos together with two electrons and in the 0$\nu\beta\beta$ mode, the nucleus emits only two electrons without neutrinos. The 2$\nu\beta\beta$ decay is expected to take place by second order process of the standard weak interaction and has no connection with the nature of the neutrino(1). On the other hand, the 0$\nu\beta\beta$ decay violates a lepton number conservation and is forbidden in the standard electroweak theory. However, this process is expected to occur if the neutrino is a Majorana particle and massive(1).

Many theoretical and experimental studies have been so far made to search for the 0$\nu\beta\beta$ decays in many decay candidates together with those 2$\nu\beta\beta$ decays during the recent decades to obtain evidence of the nature of the neutrino (1). However, the 0$\nu\beta\beta$ decay has not been found yet, and the lower limits of half-life are established for 10 candidates at present by a counting method. On the other hand, the 2$\nu\beta\beta$ decay has been measured for nine candidates with one of the geochemical, radiochemical or counting method (2). The counting method is sensitive to the emission of neutrinos, but is required to largely suppress background radiation (background problem) such as alpha-, beta- and gamma-rays from radioactive contamination in the experimental apparatus and also from naturally existing radioactive nuclei in its environment, cosmic rays and radiation from nuclear reactions induced by cosmic rays. In order to overcome the background problem, one possible method is to identify the existence of a decay daughter at the decay position. In 1977, Hurst et al. demonstrated that an atom of Cs could be observed by a method of resonance ionization spectroscopy (RIS) (3). If this technique could be possible to incorporate into the $\beta\beta$ decay experiment, measurements may be almost free from the background problem. As one of applications of RIS, we are presently developing a time-of-flight mass spectrometer in which decay daughters of ^{136}Ba are selectively ionized by RIS in order to search for the $\beta\beta$ decay of ^{136}Xe (4,5,6). Xenon is selected from a view point that its liquid is an excellent media for radiation detection (7). Furthermore, two isotopes of ^{134}Xe and ^{136}Xe are candidates for the $\beta\beta$ decay and their isotopic abundance are 10.4% and 8.9%, respectively in the natural mixture. The Q-values of the $\beta\beta$ decay are 0.847 MeV and 2.479 MeV for ^{134}Xe and ^{136}Xe, respectively. So the isotope of ^{136}Xe has a shorter life-time and is suitable to search for its $\beta\beta$ decay. However, our goal is to search for the $0\nu\beta\beta$ decay of ^{136}Xe under the condition free from backgrounds by incorporating a system for identification of decay daughters into a liquid xenon ionization drift chamber (5).

In this paper, we describe our method to measure the $2\nu\beta\beta$ mode of ^{136}Xe in liquid and also gaseous xenon. A test apparatus is proposed to search for the $0\nu\beta\beta$ decay of ^{136}Xe together with its daughter in real time.

Experimental Principle in Searching for the $2\nu\beta\beta$ Decays of ^{136}Xe

The $\beta\beta$ decay of ^{136}Xe leaves its daughter ^{136}Ba^{++} in gaseous or liquid xenon. The ionization potential of XeI is 12.1 eV and that of BaI and BaII is 5.21 eV and 10.0 eV, so Ba^{++} can not be neutralized by getting electrons from the neighboring xenon atoms in gaseous xenon. However, Ba^{++} may be possibly converted to Ba^{+} in liquid xenon, since a band structure may locally exist in liquid xenon with the almost same band gap energy of 9.9 eV as that of solid xenon. In either phase of xenon, Ba ions are stable and move slowly under an electric field, while free electrons ionized by two beta rays emitted at a $\beta\beta$ decay are rapidly swept away from their original positions. Thus, it is possible to collect the decay daughters of ^{136}Xe under an electric field in either phase of xenon with a positive-ion-collector which is described later. The positively charged Ba ions move along the electric field to the direction of an collector electrode (collector) in the positive-ion-collector and attach to its surface. The number of Ba ions which can be collected on the collector surface during a period of T, is $\lambda N(^{136}Xe)TF_c$, where λ is the decay constant of the $\beta\beta$ decay of ^{136}Xe, $N(^{136}Xe)$ the number of ^{136}Xe in the positive-ion-collector, and F_c the collection efficiency of Ba ions in gaseous or liquid xenon. The collector is transferred to a time-of-flight mass spectrometer (TOFMS) without exposing to air and incorporated into the center of a repeller electrode. The number of ^{136}Ba atoms are quantitatively measured with TOFMS in which RIS is applied to

Table 1. Number of ^{136}Xe atoms in PIC and expected number of collected Ba per year.

	PIC-I	PIC-II	PIC-III
Phase	liquid	gas (6 atm)	gas (11atm)
Number of ^{136}Xe ($\times 10^{24}$)	3.7	5.9	13.2
Operation (y)	0.5	0.36	in preparation
Number of ^{136}Ba (y^{-1})	2600	4100	9100

selectively ionize Ba atoms. The number of Ba atoms counted by TOFMS is $\lambda N(^{136}Xe)TF_cF_m$, where F_m is the overall detection efficiency for the Ba atoms in TOFMS. Thus, the life-time of the $2\nu\beta\beta$ decay of ^{136}Xe is measured by absolutely determining the efficiencies F_c and F_m.

Positive-Ion-Collector for ^{136}Ba in Gaseous and Liquid Xenon

The positive-ion-collector (PIC) consists of a collector electrode (collector) and a vessel, which is filled with gaseous or liquid xenon. The collector is fixed at the center of PIC and negatively biased to collect Ba ions. The shape of collector is an octahedron of stainless steel of which a cross section is hexagon with a side of 4mm long and its length is 8 mm in order to fix into the repeller electrode in TOFMS. There are presently two operated PICs and a PIC in preparation as listed in Table 1. The expected number of ^{136}Ba which we can collect in a year with each PIC is also shown in Table 1, assuming the half-life of 10^{22} y and F_c is unity.

In order to determine F_c, it is necessary to observe the number of ^{136}Ba detected by TOFMS as a function of biasing voltage in PIC. We are presently measuring the saturation curves and drift velocities in xenon using positive ions of ^{208}Tl instead of Ba ions. The ^{208}Tl ions are recoiled with a kinetic energy of 120 keV into gaseous or liquid xenon by alpha decays of ^{212}Bi. The recoiled ^{208}Tl ions produce many electron-ion pairs along its path (maybe roughly 40 μm in xenon gas of 1 atm and about three orders of magnitude shorter in liquid) and are neutralized by recombination with ionized electrons. However, those lose a chance to be neutralized if an electric field sweeps out the ionized electron swarms, because the ionization potentials are 6.11 eV for TlI and 20.4 eV for TlII, respectively. The situation is the same for Ba ions in xenon. The ^{208}Tl ions were collected on a thin Al sheet and the beta-rays with a maximum energy of 1.80 MeV were detected with a PIN Si photodiode (18×18 mm^2) which is fixed to the backside of the Al sheet. Positive identification was made with the half-life of 3.053 min. An typical example of beta ray counts per every 10 s in liquid xenon is plotted as a function of time as shown in Fig.1 (left). The counts per every 10 s rapidly increase just after the bias voltage was applied between two electrodes, almost saturate and decrease with the half-life of 3 min just after the bias voltage was cut. Also, the saturated counts rate were measured as a function of applied voltage as shown in Fig.1 (right). Saturation occurs above the field of 500 V/cm. However, measured points largely scatter at the lower field. It depends on impurities in gaseous xenon such as O_2, H_2O and etc.. In the case of liquid, survival of ^{208}Tl ions is more sensitive to impurities, and the saturation could not be observed below the field of 6 kV/cm. In the case of the $\beta\beta$ decay, the energy of recoiled ^{136}Ba ions is at most 30 eV, and only free electrons ionized by two beta rays exist surrounding a ^{136}Ba ion. The specific ionization of a recoiled Tl ion is much larger than that of electrons emitted at a ββ decay both in xenon gas and liquid. Then, no recombination of Ba ions with the ionized electrons may exist, and the saturation at the collection of Ba ions should occur at a lower electric field both in gaseous and liquid xenon. The F_c may be expected to be unity in gaseous xenon, but we need further studies about saturation in liquid xenon.

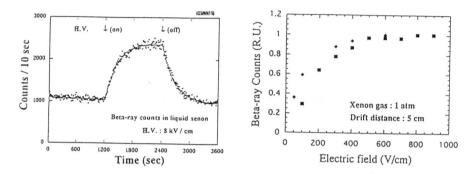

Figure 1. Beta-ray counts in liquid xenon (left) and saturation of beta-ray counts in xenon gas (right).

Time-of-Fight Mass Spectrometer

The Ba atoms on the collector surface are quantitatively analyzed with a TOFMS. The efficiency F_m is given by

$$F_m = F_{lib} \times F_{sam} \times F_{RIS} \times F_{tra} \times F_{MCP},$$

where F_{lib} is the efficiency for liberation of the Ba atoms from the collector surface by a laser shot, F_{sam} the sampling efficiency for liberated Ba atoms by laser shot, F_{RIS} the efficiency for RIS of Ba atoms, F_{tra} the transmission efficiency of the ionized Ba ions and F_{MCP} the single ion detection efficiency of MCP. In order to determine the half-life of the ββ decay of ^{136}Xe, it is essential to experimentally determine each efficiency absolutely. We have so far attempted to determine F_{tra} (7) and F_{MCP}(9) to our TOFMS filled with xenon gas at a pressure of 1×10^{-6} to 15×10^{-6} Torr. The F_{tra} was determined with two sets of well calibrated charge sensitive amplifiers, one of which is connected to a repeller electrode and the other is connected to a collector electrode used in stead of the MCP. The Xe atoms were ionized through an excited level 6p[3/2]₂ by three photons of 252.484 nm from a dye laser. The F_{tra}, simply defined as a ratio of the signals from the repeller to those from the collector, was measured by changing several parameters. Experimental details will be presented in other section of this conference. The F_{tra} was 0.73 ± 0.7 in our TOFMS. Its relatively large error mainly originates from a slight dependence of F_{tra} on positions of the laser beam in the ionization region.

In order to study the single ion detection technique and measure F_{MCP}, we have been developing an apparatus to produce at most one ion in the region of interest with alpha-particles from a source of ^{241}Am. It consists of an alpha-particle source, a Si detector for defining the alpha-particle trajectory, two electrodes for acceleration of ions, and a detector assembly of MCP in a chamber. The chamber is filled with a rare gas at a pressure of 5×10^{-6} Torr. The ion is accelerated to several kV and is measured with a detector assembly of MCP. Measurements are made by a coincidence method such that we count an ion by the MCP only when we detect an alpha-particle at the Si detector. We are interested in the pulse height spectrum measured with the MCP due to single ions and background to determine the absolute detection efficiency. The distribution was so wide and its width was 120% at FWHM. The background, which originates from dark current in the MCP, was observed to be 0.5 cps/cm^2. The F_{MCP} was estimated to be 0.59 ± 0.09 for Ar ions based on the alpha-particle ionization cross section of argon. In this measurement, systematic errors of +40% and -10% should be added from uncertainties in gas pressure, cross section and counting losses of multiply charged ions. However, the present F_{MCP} is almost equal to the open-area-ratio of MCP.

Ionization Drift Chamber with Single Ion Identification System

In order to solve the background problem in the counting method and to search for the $0\nu\beta\beta$ decay of ^{136}Xe, we are presently developing a liquid xenon ionization drift chamber equipped with an identification system of the daughters. The schematic drawing of the apparatus is shown in Fig.2. The effective volume is about 400 cm^3. In liquid xenon, we can observe ionization and scintillation at the same time. Furthermore, daughter ions of ^{136}Ba is stable and move slowly under an electric field in liquid xenon as mentioned above. Assuming a decay in the $0\nu\beta\beta$ mode occurred in the effective volume, E/W_s scintillation photons are emitted under zero electric field and about one-third of those under an electric field, and also E/W ionized electrons are produced, where E (2.48 MeV) is the summed energy of emitted two beta rays, and Ws (16.3 eV) and W (15.6 eV) are the average energies expended per scintillation photon and electron-ion pair both in liquid xenon. The scintillation photons are observed by photo-multipliers with a quartz window and a photo-cathode of Cs-As (10). A swarm of ionized electrons drifts to the outer region under an

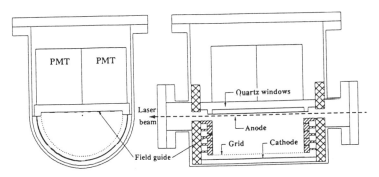

Figure 2. A test apparatus for searching the $0\nu\beta\beta$ decay of ^{136}Xe together with its daughter.

electric field, and is detected by one of collector electrodes after passing through a grid. The measured number of electrons is proportional to E, while the observed number of photons by the PMTs strongly depends on positions where events occur. However, it is possible to use scintillation for triggering the successive phenomena such as observation of ionization at the collector and so on, since the decay time of scintillation is so fast (2 to 5 ns) in liquid xenon. The drift velocity of an electron swarm is constant (about 3×10^5 cm/s) above an electric field of 3 kV/cm. Then, a time from observation of scintillation to that of ionization gives a distance of the event from the collector, namely a position of a Ba ion. The Ba ion slowly moves to the anode wire with a velocity, which are presently not known exactly. When the Ba ion is transported near the anode wire, it will be shot by laser beams several times and might be identified by its fluorescence. The fluorescence is observed with the same PMT as scintillation is observed.

Discussions

In development of the TOFMS, F_{lib} and F_{sam} are still left to be determined. We are presently building a Ba source in TOFMS to make an ultra thin layer of Ba atoms on the collector surface. The Ba source with a known surface density is considered to be essential for determining F_{lib} and F_{sam}.

In the counting method, the resolution of total energy of two beta-rays is the most important factor to separate the $0v$ mode events from a long tail of the $2v$ mode events as well as to resolve the background problem. The identification of decay daughters seems to be one of solutions for the background problem. However, there can not be presently found any techniques to identify a single atom in liquid. The fluorescence method seems to be worthwhile to try to identify the Ba daughters.

References

1. For review see Tretyak, V.I. and Zdesenko, Yu. G., Atomic Data and Nuclear Data Tables, 61, 43(1995). References there.

2. Moe,M and Vogel.P, Ann. Rev. Nucl. Sci., 44, 247(1994).
   Moe, M, Nucl. Phys., B (Proc. Suppl.) 38, 36(1995).

3. Hurst,G.S., Nayfeh. M.H., and Young, J.P., Appl. Phys. Lett. 30, 229(1977).

4. Miyajima, M., Sasaki, S. and Tawara, H., KEK-Proceedings 91-5, 19(1991).

5. Miyajima, M., Sasaki, S. and Tawara, H., Hyperfine Interactions, 74, 159(1992).

6. Miyajima, M., Sasaki, S. and Tawara, H.,IEEE Trans. Nucl. Sci. 41,835(1994).

7. Miyajima, M. and Sasaki. S., AIP Conf. Proc., 32, 211(1994).

8. Doke, T., Portugal Phys. 54, 9(1981).

9. Tawara. H., Sasaki, S. Shibamura, E. and Miyajima, M., KEK-Proceedings. to be published.

10. Miyajima, M., Sasaki, S. and Shibamura. E., Nucl. Instr. & Meth.. B63, 297(1992).

Resonant Ionization of Thermal Muonium Produced by 500-MeV Protons

Y. Miyake[1], K. Shimomura[1], A.P. Mills, Jr.[1,2], and K. Nagamine[1,3]

[1] Meson Science Laboratory, University of Tokyo (UT-MSL), Hongo, Bunkyo-ku, Tokyo, Japan
[2] Bell Laboratories, Lucent Technologies, Murray Hill, NJ 07974, U.S.A.
[3] Muon Science Laboratory, Institute of Physics and Chemical Research (RIKEN), Wako, Saitama 351-01, Japan

Abstract. Generation of the ultra slow muon was realized by developing solid-state laser systems to generate Lyman-α photons for muonium atoms. We have also shown that all of the hydrogen isotopes (H, D and T) which are generated by the nuclear reactions of 500 MeV proton can be resonantly ionized after thermarization and evaporated into vacuum. Therefore, any hydrogen isotope can be extracted without changing any experimental condition but the Lyman-α wavelength.

INTRODUCTION

At UT-MSL we have been pursuing the "Ultra Slow Muon Project", in which thermal Muonium atoms (designated as **Mu**; consisting of a μ^+ and an e^-) are generated from the surface of a hot tungsten foil (1), placed at the primary 500 MeV proton beam line and resonantly ionized by intense laser pulses synchronized with the emission of **Mu**, obtaining a source of the ultra slow muons. Recently, by developing OPO (Optical Parametric Oscillator) and Ti-Sapphire (TiS) laser systems to generate Lyman-α photons and employing a 6N-tungsten target, we have succeeded in the generation of ultra slow muons by the laser resonant ionization method (2). In this report, recent developments of the ultra slow muon project at UT-MSL are reported.

EXPERIMENTAL ARRANGEMENT

We performed our experiments using ~ 5 μA of pulsed 500-MeV protons (50 ns width and 20 Hz) available at the P4 line of UT-MSL located at KEK (National Laboratory for High Energy Physics). In the proton beam line, a tungsten foil (99.9999%) of 50 μm thickness was placed inside an UHV target chamber right behind a BN (Boron Nitride) slab of 2 mm thickness in which π^+ production, $\pi^+ \rightarrow \mu^+$ decay and slowing down of μ^+ occur. Charged particles ionized by the lasers were extracted by an immersion lens with an acceleration voltage of 9.0 kV, and transported to a micro-channel plate (MCP) detector located 8 m from the target. Any ions produced at the tungsten target region can be identified either through the mass/charge (Q) ratio by setting the bending magnet in the ion optics, or through the time-of-flight (TOF) spectrum. A detailed description of the system is reported in ref. (3). For ionizing **Mu** atoms efficiently, we adopted a resonant ionization scheme through the *1S→2P→unbound* transition. VUV generation was achieved by a Sum-Difference Frequency Mixing method using two photons of 212.55 nm for two-photon resonant excitation of the $4P^5 5P[1/2,0]$ state in krypton (4), subtracted by a photon of a tunable difference wavelength. A single mode 850 nm light with a band width of 0.5-1.0 GHz which is obtained either from an OPO system (Continuum Mirrage800) or a narrow band TiS laser (STI-HRL100), is amplified by a TiS crystal (18 mm in diameter and 15 mm long) which is pumped by two synchronized frequency-doubled Nd-YAG lasers from the both surfaces. The output energy of the single mode 850 nm is as high as 300-600 mJ/p with use of the input energy of 50-70 mJ/p. The amplified 850 nm lights are quadrupled by using two β-Ba$_2$BO$_4$ crystals, generating 212.55 nm of an intensity of 8-12 mJ/p with a ring pattern. The difference wavelength between the $4P^5 5P[1/2,0]$ state of krypton and the desired Lyman-α wavelength was generated by a broad-band TiS laser system (STI-LRL).

RESULTS AND DISCUSSION

Resonant ionization of **Mu, D** and **T** was attempted by scanning the Mass/Q in the region of **Mu, D** and **T** with the laser tuned to their respective Lyman-α wavelengths (122.09, 121.53 and 121.52 nm), where the laser pulses are delayed by Δt_{laser} relative to the proton pulse to allow time for Mu, H, D or T to evolve from the target. In Fig.1 the two-dimensional spectra obtained by scanning the mass region of **Mu, D** and **T** are summarized. Significant events counted by the MCP were obtained at the position of the correct mass and correct TOF for **Mu, D** and **T**. In the case of **D**, a H_2^+ or D^+ signal at a timing that would correspond to $\Delta t_{laser}=0$ is seen, which is probably associated with a low energy tail of direct recoil emissions due to proton bombardment. As a further confirmation of the resonant ionization signal, the VUV wavelength was detuned so as to obtain resonance spectra for **H, D, T** and **Mu** as is shown in Fig. 2. The observed resonance peaks agree with the values calculated. The width of the **Mu** resonance is consistent with the FWHM expected due to Doppler broadening and a **Mu** temperature of 2000 K. The H resonance was taken as a cold run where residual hydrogen in the ultra high vacuum (3 × 10⁻¹⁰mbar) can be extracted.

In the next step, time-evolution measurements of the neutral atoms **T** or **Mu** produced by nuclear reactions at 2000 K were carried out by changing the Δt_{laser}. Fig. 3 shows the results of the time evolution of **T** and **Mu** atoms at 2000 K. No significant events were obtained before the time of the proton beam arrival (t_0)

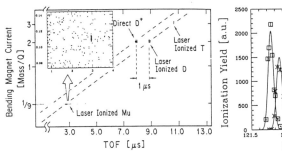
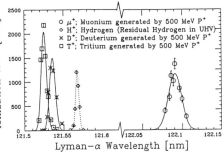

Figure1. Two dimensional Mass-TOF plot of the MCP signal exhibiting the generation of ultra slow μ⁺ by the RIS method Δt_{laser}=1.0 μs.

Figure 2. The resonance curve of the resonantly. ionized **Mu**, **H**, **D** and **T**.

until some threshold delay time. The yield started increasing after the threshold delay time, then had a peak at around a 3 μs for **T** and 0.8 μs delay for **Mu**, and then decreased with increasing time delay. The time evolution curve for **Mu** has an earlier peak because of its lighter mass and faster decay because of its intrinsic lifetime of 2.2 μs, compared with the case of **T**. It can be understood from the picture of the time-dependent profile of evaporated neutral **T** or **Mu** atoms at the location where Lyman-α and 355 nm are passing; beforehand tritons and muons produced by nuclear reactions are thermalized and diffused in hot tungsten, and are subsequently neutralized at the tungsten surface. The solid lines in Fig. 3 are fitted using a formula describing a one-dimensional diffusion-limited time-dependent flux of particles in the target and a beam Maxwellian time-of-flight distribution from the surface including a loss rate of the particle flux due to the vacancy or impurity trapping λ_l. λ_l was able to be fitted well with 0.0036(0.0001) μs⁻¹ for **T** and 0.017(0.07) μs⁻¹ for **Mu**, which is showing that λ_l is proportional to $1/\sqrt{\text{mass}}$. This is consistent with the classical diffusion model in the bulk.

In order to examine mechanism of evaporation of thermal **H** isotope in more detail, we made investigations on the temperature dependence of **T** and **Mu** evaporation in separated runs. Fig. 4 shows the total yield of **T** and **Mu** against temperature which is obtained through the fitting of the time evolution experiments at each temperature between R.T. and 2150 K. Both the **T** and **Mu** evaporation were negligible at the temperature below 1200 K and started to increase above 1200 K, and increased monotonically with temperature and leveled off at around 2000 K. A least square fit of an Arrhenius-type activation curve to the data gives E_μ= 1.64 (4) eV and E_T. = 1.86(3) eV. Hydrogen atoms are bound to the W surface by about 3 eV, and to the solid by about 1.5 eV (1). Therefore we may conclude that both the heaviest and lightest hydrogen isotope **T** and **Mu**

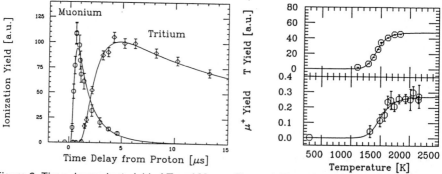

Figure 3. Time-dependent yield of **T** and **Mu** laser resonantly ionized, as a function of Δt_{laser}.

Figure 4. Temperature dependence of the resonantly ionized **T** and **Mu** atoms.

atoms are thermionically emitted from the bulk. As a conclusion of the present experiment,in which we used a laser resonant ionization method for detecting neutrals, we have demonstrated, a clear evidence of thermal neutral **T** and **Mu** production from a hot tungsten surface originating from high-energy nuclear reactions. We can extract any hydrogen isotope by only changing the Lyman-α wavelength without changing any other experimental conditions, even on the order of μs. Therefore our set-up has a potential for many experiments particularly for systematically examining hydrogen dynamics on surface etc. in the order of μs. Finally, it should be emphasized that the present experiments were realized only by the use of the pulsed proton beam available at a rapid-cycling synchrotron.

ACKNOWLEDGMENTS

The authors would like to acknowledge Drs. J.P. Marangos, P. Birrer and T. Kuga, Drs. P. Jones, Messrs. T. Shiraishi, I. Ishida, and S. Matsuo for their contributions to the helpful discussion and creation of the laser system.

REFERENCES

1. A.P. Mills,Jr., J. Imazato, S. Saito, A. Uedono, Y. Kawashima,and K. Nagamine; Phys. Rev. Lett. 56, 1463 (1986).

2. K. Nagamine, Y. Miyake, K. Shimomura, P. Birrer, J. P. Marangos, M. Iwasaki, P. Strasser and T. Kuga ; Phys. Rev. Lett. 74, 4811 (1995).

3. Y. Miyake, J. P. Marangos, K. Shimomura, P. Birrer and K. Nagamine, Nucl. Instr. and Meth. B95, 265 (1995).

4. J.P. Marangos, N. Shen, H. Ma, M.H.R. Hutchinson and J.P. Connerade; J. Opt. Soc. Am. B7, 1254 (1990).

Efficient Detection of Photons
Emitted from Fast Moving Atoms

Bernhard Lehmann, Harald Quintel, Andrea Ludin*,
Thomas Tschannen

Physics Institute, University of Bern, Sidlerstr.5, CH-3012 Bern, Switzerland
present address : Lamont-Doherty Earth Observatory, Palisades, NY 10964, USA

Abstract. Metastable atoms of krypton and photons from a tunable cw infrared diode laser at 812 nm meet in counterpropagating beams. A photomultiplier mounted perpendicular to the beams detects photons reemitted from the passing atoms. Multiple diffuse reflections from a thermoplastics tube are used to achieve the high collection efficiency necessary for photon burst detection.

INTRODUCTION

Single atom counting by Resonance Ionization Spectroscopy with pulsed tunable lasers strongly relies on time gating the detectors for background reduction (1). In an alternative approach for detecting very rare noble gas isotopes (^{81}Kr, ^{85}Kr) metastable atoms in a fast beam are collinearly excited by tunable narrowband cw infrared diode lasers. Large isotope shifts in the absorption spectra caused by the Doppler effect enables the selective excitation of atoms of a selected noble gas isotope (2). Extremely efficient background reduction can be achieved by taking advantage of photon burst detection if several photons can be detected from each atom of interest. One key requirement in such a technique is the ability to guide photons emitted along a straight line in random directions to a photon multiplier. Our approach makes use of multiple diffuse reflections in a high-reflectance thermoplastics tube.

COLLINEAR BEAM SPECTROSCOPY

In Fig.1 the optical set-up of our experiment is presented. Frequency stabilization of a single-mode diode laser is realized by monitoring shifts relative to a stabilized HeNe laser in a scanning confocal Fabry-Perot interferometer with a free spectral range of 2 GHz and a finesse of ~300 in both the 633 nm and the 812 nm region. About a third of the laser power is guided to the atomic beam line through a single-mode optical fiber.

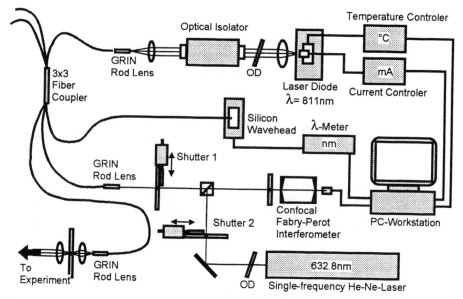

Figure 1 : Schematic of the Optical Lay-Out

RECYCLING OF ATOMS

One liter of modern water contains 10^{-4} cm^3 STP of Kr with only 1400 atoms of ^{81}Kr. It is therefore essential to be able to work with very small Kr gas samples in order for an optical technique to be useful for e.g. groundwater dating in environmental sciences. In Fig.2 we present our gas recycling system. With a gas filling of $3 \cdot 10^{-3}$ cm^3 STP of Kr we can maintain a steady state Kr ion beam of ~170 nA for several hours (3). A flux of 2000 ^{81}Kr ions per hour results in a modern sample.

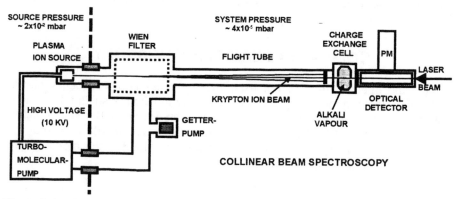

Figure 2. Ion Beam Line and Detector Arrangement

DETECTOR GEOMETRY

A rectangular stainless steel tube is welded between two standard CF63 vacuum flanges. The inner walls are covered with plates of a thermoplastics material (Spectralon) as shown in Fig.3. This material has a reflectance of 99% at a wavelength of 810 nm. An RCA 8852 photomultiplier (active area 17 cm^2) is mounted at right angle to the beam line just outside a window in order to minimize the distance to the beams to about 2 cm. The laser and the atomic beam enter through two 2mm diameter apertures at both ends with typical beam diameters of 1 mm. Three similar detectors of different lenghts with one photomultiplier each can be mounted in series to an overall detector length of 1 meter. At 10 keV beam energy the transit time of a Kr atom through such a long detector is 6.5 µs; at a planned beam energy of 800 eV the time will be 23 µs.

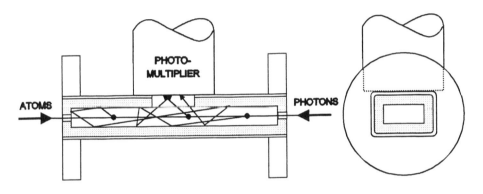

Figure 3 : Schematic of the Detector

Based on our experience with a prototype version where a photon collection efficiency of 7% was estimated with the PM positioned at a distance of 16 cm from the beam axis (2) we expect to increase this factor to >50% with the new design.

PHOTON BURST DETECTION

The potential for very high sensitivity measurements by photon burst mass spectrometry in collinear beam spectroscopy was outlined in detail by Fairbank in 1987 (4). The optical transition at 811.93 nm (for [81]Kr at 10 keV with counterpropagating beams) has an estimated cycling time of 60 ns. The counting electronics therefore has to be capable of recognizing groups of n photons within the transittime of an atom through the detector with a temporal resolution of at least 15 MHz. PM signals pass an amplifier and a discriminator before entering a "logic box" that generates the timing sequence. A master scaler determines the burst length and 16 slave scalers accumulate the number of bursts of a given length n. In Fig.4 an example of a burst distribution for photomultiplier dark counts at room temperature over a period of 4 hours is presented as an illustration (5). This dark count rate can easily be lowered by cooling the detector, however, stray-light from the laser causes a background of about this magnitude.

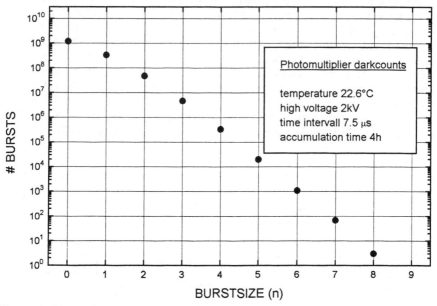

Figure 4 : Photon Burst Counting Statistics

An overall system test combining all the optical, electronic and mechanical elements presented in this short conference contribution is planned for the near future using ^{81}Kr atoms from a commercial ^{81}Rb/^{81}Kr^{m} generator used in medical lung diagnostics.

ACKNOWLEDGEMENTS

The authors thank H.P.Moret, H.Riesen and K.Grossenbacher for their skillfull technical work. The project is supported by the Swiss National Science Foundation and the State of Bern.

REFERENCES

1. Thonnard N., Wright M.C., Davis W.A. and Willis R.D., Inst.Phys.Conf.Ser. No 128, 27-30, 1992.
2. Ludin A., Lehmann B.E. : Appl.Phys.B 61, 461-465 (1995)
3. B.E.Lehmann, D.F.Rauber, N.Thonnard, R.D.Willis : Nuclear Instruments and Methods B28, 571-574 (1987)
4. Fairbank W.M. Jr., Nucl.Instr.Meth. B29, 407, 1987.
5. Th.Tschannen : Elektronische Verarbeitung schneller Impulsfolgen von einem "Photon Burst Detector", Physics Diploma Thesis, University of Bern , unpublished (1995)

Determination of the First Ionization Potential of Berkelium and Californium by Resonance Ionization Mass Spectroscopy

M. Nunnemann*, K. Eberhardt†, N. Erdmann†,
G. Herrmann†, G. Huber*, S. Köhler†, J. V. Kratz†,
A. Nähler†, G. Passler* and N. Trautmann†

†Institut für Kernchemie and *Institut für Physik,
Universität Mainz, D-55099 Mainz, Germany

Abstract. Resonance ionization mass spectroscopy (RIMS) is used for the precise determination of the first ionization potential (IP) of transuranium elements. Small amounts of material ($\approx$ 0.4 ng) are sufficient for these measurements due to the high sensitivity of RIMS enabling the investigation of the actinides beyond plutonium, which are accessible only in limited amounts and difficult to handle due to their high radioactivity. The method presented takes advantage of the dependence of the ionization threshold on an external static electric field. With samples of 10^{12} atoms of ^{249}Bk and ^{249}Cf experimental values for the first ionization potentials of $IP_{Bk} = 49989(2)$ cm^{-1} and $IP_{Cf} = 50665(2)$ cm^{-1} were obtained.

INTRODUCTION

The first ionization potential (IP) of an element is a fundamental property in physics and chemistry. Precise experimental values serve as a crucial test of multi-configuration Dirac-Fock (MCDF) calculations, at present the most successful theoretical treatment for heavy multi-electron atoms [1]. MCDF calculations are also important for a better understanding of relativistic effects in heavy elements [2]. Until very recently for the actinides beyond plutonium no experimental values for the IP were known. This is due to the small quantities available of the havier actinides and the difficulty in handling these strongly radioactive elements. Compared to measurements via surface ionization [3], electron impact ionization [4] or the Born-Haber cycles [5] laser spectroscopy provides the most accurate value of the IP obtained by the convergence of Rydberg state series [6-9]. In many cases, the Rydberg series are distorted by configuration interaction. Therefore a large amount of data is required for a correct interpretation of the spectra requiring larger samples.

Our method takes advantage of the dependence of the ionization threshold on an external static electric field and enables the determination of the IP with sample sizes of $\approx 10^{12}$ atoms. Scanning the wavelength of the laser for the ionizing step in the presence of an external electric field E, the ionization threshold is marked by a sudden increase of the ion count-rate. Determination of the ionization threshold as a function of E and extrapolation to $E = 0$ yields directly the first ionization potential.

IONIZATION THRESHOLD

According to the saddle point model the excitation energy of an atom with one highly-excited electron located in a constant external electric field E relative to its electronic ground state consists of the Coulomb energy of the electron interacting with the core, its potential energy in the external field and a constant equal to the ionization potential. In one dimension ($r = z$) the excitation energy is given by

$$W(r) = -Eer - \frac{Z_{eff} e^2}{4 \pi \epsilon_0 r} + IP \tag{1}$$

where e is the charge of the electron, Z_{eff} the effective charge number of the core, r the distance of the excited electron from the nucleus, and ϵ_0 the permittivity of the vacuum.

The ionization threshold W_{th} which is equal to the saddle point value of $W(r)$ depends on the electric field strength E in the interaction region according to

$$W_{th} = IP - 2\sqrt{\frac{Z_{eff} e^3}{4 \pi \epsilon_0}} \cdot \sqrt{E} = IP - const \cdot \sqrt{E}. \tag{2}$$

Hence the ionization threshold W_{th}, where the onset of field ionization is observed is proportional to the square root of the electric field strength E. Extrapolation to $E = 0$ yields the first ionization potential of the investigated element.

EXPERIMENTAL PROCEDURE AND RESULTS

An atomic beam of the element under investigation is produced by evaporating the sample from a sandwich filament. For preparation of this filament, the actinide sample is electrolytically deposited on a tantalum foil in form of the hydroxide and subsequently coated with a thin titanium layer [10]. The evaporated atoms are ionized by pulsed laser excitation in the presence of an electric field. The ions are accelerated and mass-selectively detected in a time-of-flight mass spectrometer. A laser system with a high repetition rate is required for sensitive detection. Therefore, pulsed copper vapour lasers with a repetition rate of 6.5 kHz are used for pumping three dye lasers. Generally the experiments are performed with laser light of a bandwidth of several GHz; the wavelength is determined by a Burleigh pulsed wavemeter (WA-4500). A detailed description of the experimental set-up is given by Ruster et al. [11].

The reliability of the method has been demonstrated for thorium, neptunium [12], plutonium, americium [13] and curium [14]. The values for the lighter actinides published in literature [7,15,16] are in excellent agreement with our data. With this technique the first ionization potentials of berkelium and californium have been determined for the first time. Using two different three step exitation schemes ($\lambda_1 = 566.06$ nm, $\lambda_2 = 720.70$ nm and $\lambda_3 \approx 544$ nm or $\lambda_1 = 566.06$ nm, $\lambda_2 = 664.70$ nm and $\lambda_3 \approx 581$ nm) and extrapolation to $E = 0$ (fig. 1 and 2) yields an ionization potential of $IP(\mathrm{Bk}) = 49989(2)$ cm$^{-1} \equiv 6.1979(3)$ eV. In the case of californium one three step exitation ($\lambda_1 = 572.77$ nm, $\lambda_2 = 625.23$ nm and $\lambda_3 \approx 581$ nm) leads to $IP(\mathrm{Cf}) = 50665(2)$ cm$^{-1} = 6.2817(2)$ eV.

Both values are in good agreement with the normalized first ionization poten-

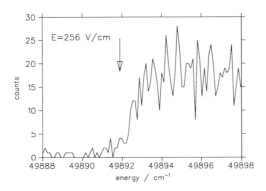

Figure 1: Ionization threshold (arrow) of ^{249}Bk when applying an electric field of $E = 256$ V·cm^{-1} to the zone of interaction between atomic beam and laser beams.

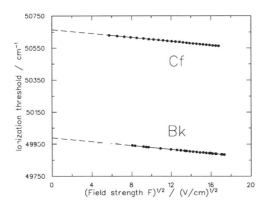

Figure 2: Plots of the ionization threshold versus the square root of the electric field strength of berkelium and californium. Extrapolation to $E = 0$ gives the first ionization potential.

269

tials, according to the ionization process $f^N s^2 \longrightarrow f^N s$ of the actinides [17], as shown in fig. 3. Ab-initio Hartree-Fock (HFC) calculations predict that the normalized ionization potentials increase linearly with N and that the slope changes at half filled shell which is in accordance with our measurements.

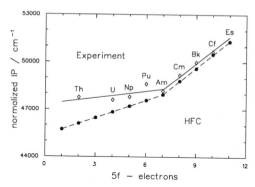

Figure 3: Normalized first ionization potentials of the ionization process $f^N s^2 \longrightarrow f^N s$: Experimental ◇ and theoretical • values.

ACKNOWLEDGMENTS

This work has been funded by the "Deutsche Forschungsgemeinschaft".

REFERENCES

1. B. Fricke et al., *Radiochim. Acta* **62**, 17 (1993).
2. P. Pyykkö, *Chem. Rev.* **88**, 563 (1988).
3. D. H. Smith and G. R. Hertel, *J. Chem. Phys.* **51**, 3105 (1969).
4. D. L. Hildenbrand and E. Murad, *J. Chem. Phys.* **61**, 5466 (1974).
5. L. R. Morss, *J. Phys. Chem.* **75**, 392 (1971).
6. E. F. Worden et al., *J. Opt. Soc. Am.* **68**, 52 (1978).
7. E. F. Worden and J. G. Conway, *J. Opt. Soc. Am.* **69**, 733 (1979).
8. J. A. Paisner et al., *Springer Series in Opt. Sci.* **7**, Laser Spectr. III, 160 (1977).
9. R. W. Solarz et al., *Phys. Rev. A* **14**, 1129 (1976).
10. K. Eberhardt et al., *AIP Conf. Ser.* **329**, 503 (1994).
11. W. Ruster et al., *Nucl. Instr. Meth. Phys. Res.* **A281**, 547 (1989).
12. J. Riegel et al., *Appl. Phys.* **B56**, 275 (1993).
13. N. Trautmann, *J. Alloys and Compounds* **28**, 213 (1994).
14. R. Deissenberger et al., *Angew. Chem. Int. Ed. Engl.* **34**, 1814 (1995).
15. S. G. Johnson et al., *Spectrochim. Acta* **47B**, 633 (1992).
16. E. F. Worden et al., *J. Opt. Soc. Am.* **B10**, 1998 (1993).
17. K. Rajnak and B. W. Shore, *J. Opt. Soc. Am.* **68**, 360 (1978).

SESSION XI: ATOMIC RIS II

Autoionizing Process of Double Rydberg States in Atom

X. Y. Xu, W. Huang, C. B. Xu, P. Xue, and D. Y. Chen

Department of Modern Applied Physics, Tsinghua University, Beijing 100084, China

Abstract. We have studied the autoionization distribution of penetrating double Rydberg (DR) states $NLnl(N < n; L, l < 4)$ experimentally in calcium by using five-laser resonance excitation and sequential ionization with a pulsed and a strong constant electric field, as well as theoretically in helium by using the hyperspherical close-coupling method. We have found the DR states autoionize with the ejected electron having its average kinetic energy nearly independent of n but apparently related to the binding energy of the ionic Rydberg orbit NL. We have also discussed the dynamics in DR states and described two types of autoionizing processes, i.e., "penetration autoionization" and "polarization autoionization" in DR states.

INTRODUCTION

In recent years double Rydberg (DR) states play an increasingly important role in the investigation of atomic three-body problem. The correlation effects in DR states have been investigated in a great part on the structure of the excitation spectrum and much less done on the dynamic processes. An atom in a doubly excited (DE) state above the first ionization threshold will generally autoionize and eject electron(s). The autoionization paths and rates reflect the dynamic dielectronic correlation in the state. The probability of double escape which is only possible for those DE states above the double-ionization threshold has been extensively studied since Wannier's effort. In the case of single escape most of the investigations were made on the DE states in which one electron is highly excited (in a Rydberg orbit) and another electron is weakly excited (in an orbit near the valence shell). In this case the possible autoionization paths are limited. When both electrons are highly excited (DR state) this limitation becomes unimportant.

There are two basic types for DR states. One is the planetary state in which the mutual penetration of the wave functions of the two excited electrons is negligible. The other is the penetrating state where there exists significant mutual penetration for the wave functions of the two excited electrons. To our knowledge, there were some investigations on the total autoionization rates of DR states whereas no branching ratio (distribution) was reported. (Recently a group of partial branching ratios for the autoionization of a planetary state was measured (1).)

Here we report our study on the autoionization distribution of the DR states $NLnl(N < n; L, l < 4)$ (penetrating states) by means of the recently developed experimental and theoretical methods. Experimentally alkali-earth atoms have been proved to be good candidates for atomic three-body problem due to the relatively low excited energy of their DR states which can be achieved through sequential laser excitation. Theoretically, in the other hand, helium atom is play the most important role due to its simplicity. (These are the cases in our work). Whereas there is no substantial difference between helium and alkali-earth atoms in DR states because the excitation of the ionic core (2+) of the later can always be neglected due to its high excited energy. We will not distinguish these two types in our discussions unless specified.

EXPERIMENTAL METHOD AND RESULTS

Our experiment is performed by combining sequential laser resonance excitation and sequential field-ionization techniques (2). Ca atoms in a thermal beam are excited with two dye lasers from the ground state to a bound Rydberg state $4snd$. After a delay of 30ns, the inner $4s$ electron is excited with three dye lasers to a Rydberg orbit NL via two intermediate autoionizing states $4p_{1/2}nd$ and $5s_{1/2}nd$. We label the resulted DR state $NLnl$ (2).

An Ca atom in the DR state autoionizes promptly to a Rydberg state of Ca^+ and ejects an electron. The energy of the ejected electron can be deduced from the excited energy of the autoionized Ca^+ Rydberg state $N'L'$. In our experiment the energy (N'^*) is obtained from the threshold strength of the field-ionization for the state. In principle the autoionization may produce Ca^+ Rydberg state with high L', and then may have a wide distribution of $M_{L'}$. Because low $|M_{L'}|$ states are more favorable or as favorable as high $|M_{L'}|$ states for the autoionization of the DR states, we take $M_{L'}$ as 0 and ignore the influence of others, which may cause a small deviation for the measurement but will not affect our discussion. As a result the threshold field strength can approximately be written as

$$E_{\text{th}} = Z^3/(2N'^*)^4. \tag{1}$$

The Stark effect during the field ionization will slightly influence the experimental results but will also not affect our discussion.

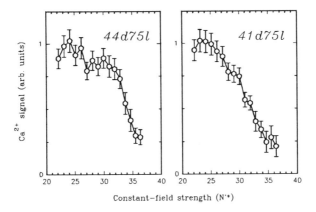

FIGURE 1. Signals of the DR states $44d75l$ and $41d75l$ in calcium versus the strength of the constant-field, obtained by fixing the frequencies of all five lasers and recording the signal of Ca^{2+} ions produced in the constant-field region. The pulsed-field strength is set at $1210V/cm(N_{high} = 38.1)$. And the influence of the pulsed-field ionization is negligible.

Shortly after the laser excitation (80ns), all ions produced with the lasers are extracted by a pulsed electric field, and then experience a strong constant electric field after travelling a field-free zone. Lastly the ions are detected with a dual-microchannel-plate detector after passing through a drift zone. The pulsed field will ionize the ionic Rydberg states with $N'^* > N_{high}$. And the constant field will ionize those with $N'^* > N_{low}$, where $N_{high} > N_{low}$. These two signals can be discriminated in our TOF (time-of-flight) spectroscopy. The influence of the decaying of the ionic Rydberg states is negligible due to their long lifetimes. The distribution of N'^* can be obtained by changing the strength of the pulsed or the constant field.

Figure 1 shows the measured autoionization signals of Ca $44d75l$ and $41d75l$ states by recording the constant-field-ionized signals for different constant-field strength (scaled to N'^* with Eq. 1). It shows that the DR states predominantly autoionize to some lower ionic Rydberg manifolds instead of the highest possible Rydberg manifold. (The autoionization to the lowest ionic states is negligible due to the poor overlap of the initial and final wave functions.)

The results exhibit that the kinetic energy of the ejected electron (on an average and the same below, e.g., about 0.017eV for $44d75l$ state) is comparable with the binding energy of the ionic Rydberg orbit NL 0.028eV for $44d$).

Measurement on other DR states $NLnl$ exhibits the similar results as in Fig. 1. To understand such specific phenomena caused by the electronic correlation, we approach the autoionization of penetrating states theoretically based on the hyperspherical close-coupling (HSCC) method (3).

THEORETICAL METHOD AND RESULTS

The Schrödinger equation for He can be written as

$$\left(-\frac{1}{2}\frac{\partial^2}{\partial R^2} + \frac{H_{\text{ad}}}{2R^2} - E\right)\Psi = 0, \tag{2}$$

where H_{ad} is the adiabatic Hamiltonian,

$$H_{\text{ad}} = -\frac{\partial^2}{\partial\alpha^2} + \frac{l_1^2}{\cos^2\alpha} + \frac{l_2^2}{\sin^2\alpha} - R\left(\frac{2Z}{\cos\alpha} + \frac{2Z}{\sin\alpha} - \frac{2}{\sqrt{1 - \sin 2\alpha\cos\theta_{12}}}\right) - \frac{1}{4}. \tag{3}$$

Here the hyperspherical coordinates R, α and θ_{12} is defined as in (3).

To solve Eq. (2) with the HSCC method (3), the whole space is divided into two parts with a certain point R_{HS}. In the inner region $(R \leq R_{\text{HS}})$, the electron-electron correlation is strong, the wave function Ψ is expanded in terms of orthogonal diabatic basis functions and the Schrödinger equation is cast into close-coupling equations. In the outer region $(R \geq R_{\text{HS}})$, the wave function is expressed in terms of the independent-particle coordinates,

$$\Psi_j^{\text{out}} = r_>^{-1}\sum_i \varphi_i(r_<, \hat{r}_>)[f_i(r_>)\delta_{ij} - g_i(r_>)K_{ij}]. \tag{4}$$

Here $r_< = \min(r_1, r_2)$ and $r_> = \max(r_1, r_2)$, and φ_i denotes a wave function of the He$^+$ ion coupled with the angular part of the wave function of the outer electron. The functions f_i and g_i are energy-normalized regular and irregular Coulomb functions. The K matrix can be calculated by matching the solutions obtained in the inner region to the above outer-region solutions at $R = R_{\text{HS}}$. From the K matrix we can get the χ matrix,

$$\chi = (1 + iK)(1 - iK)^{-1}, \tag{5}$$

and the autoionization probability from a closed channel to an open channel,

$$\Gamma_A \propto |\chi_{pk}|^2, \tag{6}$$

where p is the closed and k is the open channel index (4).

As an illustrative example, we have calculated the autoionization probability from $Npns(^1P^o)$ states to $N'L'\epsilon l'(^1P^o)(N' = 1 \sim N - 1; L', l' < 4)$.

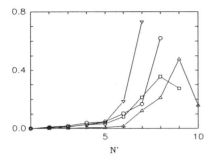

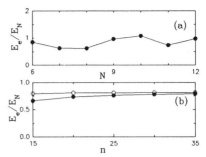

FIGURE 2. Autoionization probability from $Np20s(^1P^o)$ state to $N'L'\epsilon l'(^1P^o)$ ($L',l' < 4$) states of He: $N = 8$, ($\triangledown$); $N = 9$, ($\circ$); $N = 10$($\square$); $N = 11$($\triangle$).

FIGURE 3. The ratios of the average kinetic energy of the ejected electron (E_e) for the autoionization of He $Npns$ state to the binding energy of the ionic Rydberg orbit Np (E_N). (a) $n = 20$, $N = 6 \sim 12$; (b) $N = 11$, $n = 15 \sim 35$ (the transferred energy E_T (in $1/E_N$ scale) for the Auger decay is given with $\circ$).

For fixed $n(20)$, we obtain different curves of autoionization probability for $N = 8 \sim 11$ as shown in Fig. 2. In Fig. 2, the DR state autoionizes prior to the permitted highest ionic Rydberg manifold for small N and to lower manifolds as N increases($N \geq 10$). And in all the cases, the autoionization probability to the lowest states ($N' \ll N$) can be neglected.

In further study, we have compared the average kinetic energy of the ejected electron (E_e) with the binding energy of ionic Rydberg orbit Np (E_N). We have found for different N, E_e is always comparable with E_N (Fig. 3 (a)) as in our experiment. It is shown in Fig. 3 (b) that E_e/E_N varies slowly on n. And if the Auger decay process of the inner electron is assumed for the autoionization, the transferred energy ($E_T = E_e + 1/(2n^{*2})$) is almost independent of n.

DISCUSSIONS

The autoionization of a DR state is owing to the energy re-distribution between the two excited electrons semi-classically, or to the channel mixing in the framework of MQDT (multichannel quantum defect theory). With MQDT for a penetrating state $NLnl(N < n; L, l < 4)$, the outer electron (nl) is assumed to move in a potential due to the core (2+) and the inner electron (NL). In the outer region ($r_{nl} > R$, R is related to N, L) where the core is screened by the NL electron, the nl electron experiences a pure Coulomb potential and no channel mixing occurs. In the inner region ($r_{nl} < R$) where the penetration occurs, the nl electron is close-coupled with the NL electron, which induces the channel mixing and the autoionization therefor

(Such a correlation is called penetration correlation). For $n \gg N$ the close-coupling interaction of the two electrons is nearly independent of n whereas is dependent on N. As a result the transferred energy (E_T) between the electrons and thus the energy of the ejected electron (E_e) is dependent on N and nearly independent of n as shown in our results. The autoionization therefor is called "penetration autoionization".

For a planetary state $(N < n, L \ll l)$ the large centrifugal potential prevents the outer electron from penetrating the inner electronic wave function. The excited electrons perturb each other via a long range correlation–the polarization correlation which induces the channel mixing. The autoionization therefor is called "polarization autoionization".

It has been known a penetrating state generally has a much faster autoionizing rate than the planetary state with the same N^*, n^*. It implies that the penetration correlation is much stronger than the polarization correlation. The planetary state was shown to autoionize prior to the possible highest states (1), which is quite different from our experimental results of the penetrating states. It implies the penetration correlation and the polarization correlation may have different effects on the autoionization distribution. Whereas comparison can be performed only after new experimental results (as well as new theoretical results) of the planetary states with higher N than that in (1) are obtained, because the energy differences between the adjacent Rydberg manifolds of the possible autoionized states in (1) may be too large to observe the effects as in our case.

In conclusion, we have studied the autoionization distribution of the penetrating DR states. We have found there exist specific statistic characters in the dynamic process, which offers us an opportunity to understand the three-body interaction from a new aspect.

ACKNOWLEDGMENTS

The authors appreciate the helpful discussions with Dr. J. Z. Tang and the consistent support from Prof. J. M. Li. This work is supported by Chinese National Natural Science Foundation.

REFERENCES

1. M. Seng, M. Halka, K.-D. Heber, and W. Sandner, Phys. Rev. Lett. **74**, 3344 (1995).
2. W. Huang, X. Y. Xu, C. B. Xu, M. Xue, L. Q. Li, and D. Y. Chen, Phys. Rev. A **49**, R653 (1994).
3. J. -Z. Tang, S. Watanabe, and M. Matsuzawa, Phys. Rev. A **46**, 2437 (1992).
4. M. J. Seaton, Rep. Prog. Phys. **46**, 167 (1983).

Resonant Laser Ionization of Mercury via Core-excited Autoionizing States.

E. Bente, R. van Leeuwen, E. Buurman and W. Hogervorst

Laser Centre, Faculteit Natuurkunde en Sterrenkunde, Vrije Universiteit
De Boelelaan 1081, 1081 HV Amsterdam, The Netherlands

Abstract. Even-parity autoionizing states of atomic mercury have been excited using three Nd:YAG pumped, pulsed tunable lasers and one intermediate decay. Most of the observed resonances have linewidths of several wavenumbers down to a few GHz. They are identified as members of the $5d^96s^2$ns,nd configurations. A tentative assignment of J values of multiplet states is given.

INTRODUCTION

The bound state spectrum of mercury has been studied in great detail over the years. The use of a mercury gas discharge in fluorescent lamps explains the availability of accurate data on dipole moments, branching ratios, isotope shifts and hyperfine structures. Much less information is available on its autoionizing states. Obviously for an efficient resonant ionization scheme these states are equally relevant. The reason for the limited set of data on autoionizing states is their high energy, the first ionization limit lying at 84184 cm^{-1} (~10 eV). In addition the second and third ionization limit at 119692 cm^{-1} and 134732 cm^{-1} respectively lie much higher. Thus, even using UV photons, several steps are required to reach the autoionizing states.

An interesting aspect of Hg is its electronic structure. The bound states nearly all have a single 6s electron excited from the $5d^{10}6s^2$ $^1S_0$ ground state. The bound state Rydberg series converge to the $5d^{10}6s$ $^2S_{1/2}$ ground state of Hg$^+$. This is similar to the two electron alkaline-earth atoms. The next ionization limit, however, is $5d^96s^2(^2D_{5/2})$ and thus the autoionizing states converging to this limit have an excited 5d core electron. This is a hole-electron system. Since a $5d^96s^2(^2D_{5/2})$nl state can only decay into $5d^{10}6s\epsilon l$ autoionization must occur via electric quadrupole interaction between the hole and the electron. Since this interaction is sensitive to radial overlap of the wavefunctions of the hole and the electron and is generally considerably smaller than the dipole interaction, relatively narrow lines may be expected. Indeed Ding et al. (1) observed narrow resonances in the $5d^9(^2D_{5/2})6s^2$nf series. Due to the large energy separation between the ionization limits only a limited number of states belonging to series converging to the second limit are bound and only a few states from series converging to higher limits can affect series converging to the second limit.

Our approach to investigate autoionizing states involved a three-step excitation experiment with the $5d^{10}6s6p$ $^3P_1$ and $5d^{10}6sns,nd$ states as intermediate bound states. We discovered that (with one exception) only autoionizing states were reached, excited from bound intermediate states that were populated in decay from the $5d^{10}6sns$ $^3S_1$ states. These intermediate states have a significant admixture of $5d^96s^26p$ character. Thus even-parity autoionizing states were observed.

EXPERIMENTAL

The experiment was performed inside a vacuum chamber with a background pressure of mercury of about 10^{-7} mbar. In this chamber ions could be detected with a time-of-flight mass spectrometer with a 10 cm drift zone and a dual multichannel plate. In addition an interaction region was available where fluorescent light from the mercury vapour could be observed using a photomultiplier. This was used for tuning lasers to bound state transitions. Signals from the ion detector and/ or the photomultiplier were collected using a gated integrator and read into a computer for display, storage, analysis and control of the lasers.

With the laser system three-step excitations in the ultraviolet were possible. For the first step only the transition to $5d^{10}6s6p$ $^3P_1$ at 253 nm is practical. To determine J values from polarization dependent ionization rates the first step has to be selective for an even isotope. Only these isotopes have zero angular momentum. Therefore for this step we used a source producing single-mode output when required. It involved an injection-seeded Ti:sapphire ring laser setup, frequency tripled using a KDP crystal for doubling and a BBO crystal for mixing (both type I). The ring cavity consisted of a 15 mm long, Brewster-cut Ti:sapphire crystal in a 40 cm roundtrip length triangular and flat mirror cavity with a 70% output coupler, collinearly pumped by the second harmonic of a GCR 16 Quanta-Ray Nd:YAG laser. For injection-seeding purposes a multimode grazing-incidence Ti:sapphire laser, when single-mode operation was not required, or a single-mode Pyridine 1 ring dye laser (pumped by an argon-ion laser) were used. One of the mirrors of the Ti:sapphire ring was piezo mounted to lock the ring cavity to the frequency of the seed laser.

The other two colors were produced by frequency doubling the output of two dye lasers (Quanta-Ray PDL2) using BBO and/or KDP crystals. These dye lasers were pumped by a Quanta-Ray GCR2A pump laser. The relative delay between the output pulses from the two Nd:YAG lasers could be varied using a computer-controlled delay generator.

RESULTS AND DISCUSSION

One series of experiments involved three-step excitation of autoionizing states via the $5d^{10}6s6d$ and $7d$ $^{1,3}D_2$ states. We studied the energy range 85000 - 100000 cm^{-1} where several $5d^9(^2D_{3/2})6s^26p$ states are expected (2) as well as the energy range just below the second ionization limit in order to observe Rydberg series. Only a single 110 cm^{-1} wide Fano profile was observed near the predicted energy of $5d^9(^2D_{3/2})6s^26p$ J=3. No resonances were observed in the range where Rydberg states must be located.

Interesting results, however, were obtained using $5d^{10}6s8s$ $^3S_1$ as an intermediate state. In this case again no direct excitation from the 8s state could be observed, but strong excitations from the $5d^9(^2D_{5/2})6s^26p$ $^3P_2$ and the $5d^{10}6s7p$ $^3P_2$ states appeared. It is known that the 8s $^3S_1$ state partly decays to the two $^3P_2$ states (3). This cascading was confirmed in a pump-probe measurement. The first two dye lasers were set to populate 8s $^3S_1$ and the third, ionizing laser was scanned in time.

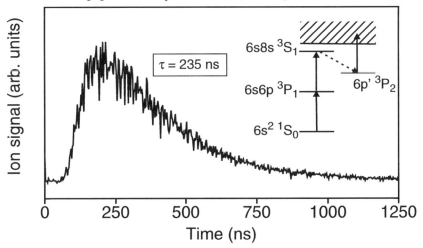

Figure 1. Resonant ionization signal as a function of delay between the excitation of $5d^{10}6s8s$ $^3S_1$ and the ionizing laser.

In Fig.1 a 22 ns growth of the signal, determined by the 8s $^3S_1$ lifetime (3), and a subsequent much longer decay time of $5d^96s^26p$ $^3P_2$ is shown. In this way the lifetimes of $5d^96s^26p$ $^3P_2$ and $5d^{10}6s7p$ $^3P_2$ were determined to be 235(10) ns and 71.5(3) ns respectively. Since a pure $5d^96s^26p$ $^3P_2$ state would be metastable these lifetimes reflect the mixing of both $^3P_2$ states. This confirms an analysis performed by Martin et al. (2).

Several multiplets excited from both intermediate $^3P_2$ states were observed. They were identified as $5d^9(^2D_{5/2})6s^2nd$ and ns states (n = 6, 7 and 8) and a $5d^9(^2D_{3/2})6s^27s$ state. Again in agreement with the analysis of Martin et al. (2) the signals in excitations from the $5d^96s^26p$ $^3P_2$ state are strongest. Spectra of a particular multiplet, obtained from both $^3P_2$ states are, apart from the overall signal strength, almost identical. This indicates that the autoionizing states have pure $5d^96s^2nl$ character. An example of a recorded spectrum is shown in Fig. 2. Indicated is the assignment in a jK-coupling scheme giving the best results of a tentative Slater-Condon analysis. Also quantum defects of the states with respect to the $5d^9(^2D_{5/2})6s^2$ ionization limit are given. The ordering of the lines is in agreement with calculations by Forest et al. (4).

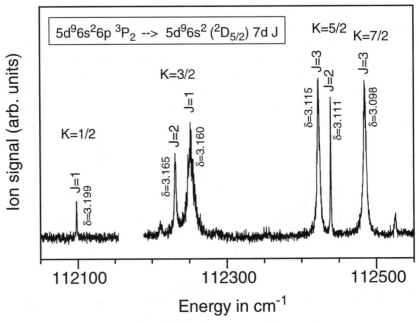

Figure 2. Spectrum showing the $5d^96s^26p^3P_2$ to the $5d^9(^2D_{5/2})6s^27d$ multiplet excitation.

ACKNOWLEDGEMENTS

The authors are indebted to Jacques Bouma for his technical assistance. Financial support from the Stichting Technische Wetenschappen and Urenco Nederland BV are gratefully acknowledged.

REFERENCES

1. Ding, R., Kaenders, W.G., Marangos, J.P., Shen, N., Connerade, J.P. and Hutchinson, M.H.R., J. Phys. B: At. Mol. Opt. Phys. **22**, L251-L256 (1989).
2. Martin, W.C., Sugar, J. and Tech, J.L., Phys. Rev A **6**, 2022-2035 (1972).
3. Benck, E.C., Lawler, J.E. and Dakin, J.T., J. Opt. Soc. Am. B **6**, 11-22 (1989).
4. Forrest, L.F., Sokhi, R., Pejcev, V., Ross, K.J. and Wilson, M., J. Phys. B: At. Mol. Phys. **18**, 4519-4527 (1985).

RIS of Mercury for Analytical Applications

Peter Bisling, Claus Weitkamp, and Harald Zobel

GKSS-Forschungszentrum Geesthacht GmbH, Postfach 11 60, D-21494 Geesthacht, Germany

Abstract. New excitation and ionization schemes of the Hg atom are investigated.

INTRODUCTION

A most promising laser-spectroscopic technique for environmental analysis is REMPI by laser excitation combined with TOF mass spectrometry in order to determine a low-concentration species in a complex matrix of unknown composition. Of the RIS studies on mercury so far, only three are dedicated to analytical applications. They report on the three-photon ionization of Hg with a two-photon resonance $Hg[\tilde{\nu}_1 \tilde{\nu}_1, \tilde{\nu}_1 e^-]Hg^+$ (1) and with a double-resonant transition $Hg[\tilde{\nu}_1, \tilde{\nu}_2, \tilde{\nu}_2 e^-]Hg^+$ (2). An alternative is the three-step excitation to Rydberg states with collisional ionization in a buffer gas B, $Hg[\tilde{\nu}_1, \tilde{\nu}_2, \tilde{\nu}_3, Be^-]Hg^+$ (3). New results on efficient multiphoton ionization schemes of mercury are presented in this work.

EXPERIMENTAL

Two dye lasers in conjunction with second harmonic and sum frequency generation provide the radiation for selective excitation and ionization of the analyte in a gaseous sample mixture which is prepared in a pulsed supersonic jet. The nozzle source is housed in a differentially-pumped expansion chamber with a skimmer as a pressure stage to the ionization chamber where the photoions are produced in a field-free Wiley-McLaren ion source. A fast dc pulse with a delay of 1.6 µs after the laser ionization ejects the particles into an ion optical lens system and further into the field-free drift region of a flight tube which is equipped with a gridless ion reflector. It redirects the ions back into the drift path for a second pass using a ten degree angle of deflection. The TOF spectrum is detected by a 3-stage microchannel plate assembly and recorded by a fast transient recorder. When low species concentrations are to be detected, then the ion source can be replaced by a quadrupole ion trap in order to increase the S/N by accumulation of the ion density before ejection into the TOF (4).

TABLE 1. Two-Color REMPI Schemes (A to D) for the Mercury Atom.

	$\lambda_{air,\,1}$ in nm	State 1	$\lambda_{air,\,2}$ in nm	State 2	$\lambda_{air,\,2}$ in nm	State 3
A	253.652	6s6p $^3P_1$	407.783	6s7s $^1S_0$	407.783	6s $^2S_{1/2}$ + e$^-$
B	253.652	6s6p $^3P_1$	435.831	6s7s $^3S_1$	435.831	6s $^2S_{1/2}$ + e$^-$
C	253.652	6s6p $^3P_1$	313.155	6s7d $^3D_1$	313.155	6s $^2S_{1/2}$ + e$^-$
D	253.652	6s6p $^3P_1$	202.58	$^2D_{3/2}$ 6p $^1P_1$	-----	6s $^2S_{1/2}$ + e$^-$

RESULTS

Different excitation schemes as summarized in Table 1 have been investigated in order to optimize the selectivity and sensitivity of Hg detection. A new Hg[$\tilde{\nu}_1$, $\tilde{\nu}_2$, $\tilde{\nu}_2$e$^-$]Hg$^+$ scheme (A) is proposed (see Fig. 1a) which is about a factor of twenty more efficient than that in the most sensitive previously reported MPI experiment (2) according to scheme B. The ionization is greatly enhanced by resonance with the $^2D_{3/2}$ 6p $^1P_1$ autoionization (AI) level when the transitions with wavenumbers $\tilde{\nu}_1$ = 39412.3 cm^{-1} ($\lambda_{air,\,1}$ = 253.652 nm) and $\tilde{\nu}_2$ = 24515.9 cm^{-1} ($\lambda_{air,\,2}$ = 407.783 nm) are used as indicated in Eqs. (1) - (4). The total energy in Eqs. (1) - (3) sums up to a difference of only -315.9 cm^{-1} with respect to the broad AI resonance maximum at 88760 cm^{-1}. In this energy range the shape of the single

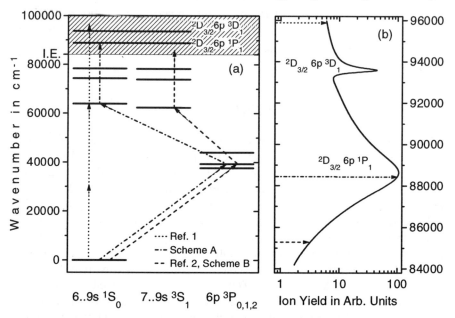

FIGURE 1. (a) Partial energy level diagram of the Hg atom, showing different multiphoton ionization schemes proposed for analytical applications. I.E. is the ionization energy. (b) Fano profiles of the autoionization resonances of Hg. Arrows indicate the amount of total energy used in different MPI experiments.

$$Hg\,(5d^{10}\,6s^2)\,{}^1S_0 \xrightarrow{\tilde{\nu}_1} Hg\,(5d^{10}\,6s\,6p)\,{}^3P_1 \qquad (1)$$

$$Hg\,(5d^{10}\,6s\,6p)\,{}^3P_1 \xrightarrow{\tilde{\nu}_2} Hg\,(5d^{10}\,6s\,7s)\,{}^1S_0 \qquad (2)$$

$$Hg\,(5d^{10}\,6s\,7s)\,{}^1S_0 \xrightarrow{\tilde{\nu}_2} Hg^*\,[(5d^9\,6s^2)\,{}^2D_{3/2}]6p\,{}^1P_1 \qquad (3)$$

$$Hg^*\,[(5d^9\,6s^2)\,{}^2D_{3/2}]6p\,{}^1P_1 \xrightarrow{V_{ac}} Hg^+\,(5d^{10}\,6s)\,{}^2S_{1/2} + e^- \qquad (4)$$

photon ionization continuum of Hg is almost completely determined by AI states and can be reproduced by the Fano parameters (5) as shown in Fig. 1b. A huge ionization cross section is thus obtained via configuration interaction V_{ac} in Eq. (4). The ion yield of another $Hg[\tilde{\nu}_1,\ \tilde{\nu}_2,\ \tilde{\nu}_2\,e^-]Hg^+$ transition (scheme C in Table 1) via 6s6p $^3P_1$ and 6s7d $^3D_1$ is comparable to that of scheme A, but the excess energy above the ionization energy is 19075.9 cm^{-1}.

In scheme D, the $^2D_{3/2}$ 6p $^1P_1$ level is reached exactly by the resonant two-photon ionization $Hg[\tilde{\nu}_1,\ \tilde{\nu}_2\,e^-]Hg^+$ with $\tilde{\nu}_1 = 39412.3$ cm^{-1} (6s6p $^3P_1$) for the first step and $\tilde{\nu}_2 = 49347.7$ cm^{-1} ($\lambda_{air,\ 2} = 202.58$ nm) for the second step to the AI maximum. In order to compare qualitatively the ionization efficiencies of schemes A to D, the ion yields are plotted versus laser pulse energy in Fig. 2.

Finally the wavelength of the second step for the resonant two-photon ionization of Hg 6s6p $^3P_1$ is scanned in the region of the ionization threshold ($\lambda_{air,\ 2} = $ 223.205 to 223.895 nm). The wavelength is calibrated with the well-known transition 6s6p $^3P_1 \rightarrow$ 6s7s $^3S_1$ by use of the fundamental wavelength of the laser at

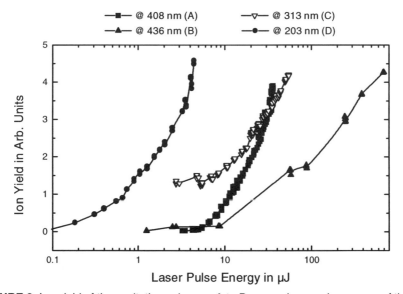

FIGURE 2. Ion yield of the excitation schemes A to D versus laser pulse energy of the laser for the ionization.

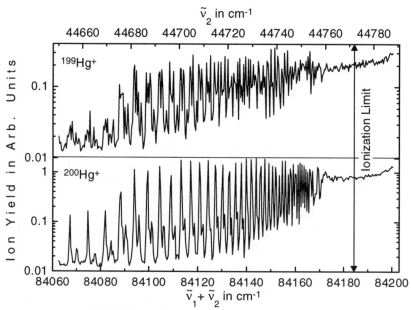

FIGURE 3. Rydberg series of mercury isotopes.

$\lambda_{air, 2}$ = 435.831 nm. In this way, the typical Rydberg spectral behavior of mercury isotopes (3) is obtained which is shown in Fig. 3. A comparison of the peaks belonging to isotopes with odd and even nucleon numbers reveals effects due to the hyperfine structure of the odd isotopes, although the resolution of the measurement is only about 0.2 cm^{-1}.

The energy values of the atomic levels used in this work are taken from the National Bureau of Standards tables (6).

REFERENCES

1. Miziolek, A. W., *Anal. Chem.* **53**, 118–120 (1981).
2. Bushaw, B. A., *Anal. Chem.* **57**, 2397–2399 (1985).
3. Matveev, O. I., Cavalli, P., and Omenetto, N., "Three-step Laser Induced Ionization of Ir and Hg Atoms in an Air-Acetylene Flame and a Gas Cell," in *AIP Conference Proceedings No. 329 — Resonance Ionization Spectroscopy 1994*, New York: AIP Press, 1995, pp. 269–272.
4. Bisling, P., Heger, H. J., Michaelis, W., Weitkamp, C., and Zobel H., "New Perspectives in Laser Analytics: Resonance-Enhanced Multiphoton Ionization in a Paul Ion Trap Combined with a Time-of-Flight Mass Spectrometer," in *AIP Conference Proceedings No. 329 — Resonance Ionization Spectroscopy 1994*, New York: AIP Press, 1995, pp. 511–514.
5. Brehm, B., *Z. Naturforschg.* **21 a**, 196–209 (1966).
6. Moore, C. E, *Atomic Energy Levels, Volume III*, Nat. Stand. Ref. Data Ser. - Nat. Bur. Stand. 35, Washington, D.C.: U.S. Government Printing Office, 1971, pp. 191–195.

SESSION XII: BIOLOGICAL AND MEDICAL APPLICATIONS OF RIS

The Feasibility of RIMS for the Analysis of Potentially Toxic Element Accumulation in Neural Tissue

O. Rhodri Jones[*], Christopher J. Abraham[*], Helmut H. Telle[*]
and Arthur E. Oakley[†]

[*]Department of Physics, University of Wales Swansea, Singleton Park, Swansea, SA2 8PP, UK.
[†]MRC Neurochemical Pathology Unit, Newcastle General Hospital,
Newcastle-upon-Tyne, NE4 6BE, UK.

Abstract. A feasibility study was conducted into the possibility of using resonance ionisation mass spectrometry for the detection of focal accumulation of neuro-toxic elements in neural tissue. Experiments were performed using a ToFMS system in conjunction with an Ar^+ source for target sputtering and a pulsed tuneable dye laser system for resonance ionisation. Detection limits of ~3ppm for Al in brain tissue homogenates were achieved, with a spatial resolution of less than 100µm.

INTRODUCTION

The potential neurotoxicity of element accumulations in neural tissue has been recognised for several decades, and has been linked with several disease states. Two of the main elements of interest are aluminium in connection with Alzheimer's disease and dialysis encephalopathy, and iron in connection with Parkinsonian Dementia. The most common techniques for the determination of trace element concentration involve measurement of bulk tissue samples by atomic emission or absorption spectroscopy or neutron activation analysis. Studies of gross concentrations can however be misleading, since they do not adequately reflect the presence of focal accumulations of the element of interest. The microanalytical techniques available for such spatially resolved analysis include secondary ion mass spectrometry (SIMS), laser microprobe mass spectrometry, electron and proton microprobe X-ray microanalysis, and now resonance ionisation mass spectrometry (RIMS).

Although SIMS has been used with some success to study focal accumulation of such elements in biological tissue (1), accurate quantification is difficult to achieve due to the problem of mass interference and the inherently large matrix effects associated with the technique. In the present work we demonstrate the feasibility of RIMS for the analysis of aluminium at the ppm level in biological tissue. It will also be shown how the problems associated with SIMS are largely eliminated using RIMS.

EXPERIMENTAL SET-UP

The experimental work was conducted using a UHV, sputter initiated , time-of-flight, reflectron mass spectrometer capable of acquiring both SIMS and RIMS spectra (2). A duoplasmatron Ar^+ primary ion source provided sputter currents of up to 1.5μA, which could be focused to give a minimum spot size of ~80μm. Samples could be accurately positioned using micrometer drives in combination with a CCD camera and secondary electron imaging system. The tuneable laser radiation required for the RIMS experiments was focused into the vacuum chamber by a Nd:YAG laser pumped dye laser system, pulsed at 20Hz. The output from the dye laser could be frequency doubled to allow the use of both visible and uv photons for resonance ionisation processes.

The resonant transitions attempted for aluminium were from the $3s^23p$ $^2P^0$ ground state configuration to the $3s^23d$ ^2D levels using radiation at 308.2nm and 309.3nm (3). Ionisation was enhanced by allowing the higher intensity visible dye laser radiation to enter the interaction region, which also made use of the larger photo-ionisation cross sections of the excited levels at such wavelengths (4).

Several samples were prepared to determine the viability of biological analysis using RIMS. The initial experiments were carried out on Al-loaded gelatin samples with an Al concentration of 100ppm and 500ppm. These were replaced for the later experimental work by uniformly Al-loaded brain tissue homogenate samples. All the samples were sputter-coated with a gold layer of less than 1μm thickness, to prevent contamination and reduce space charge effects during the analysis.

RESULTS AND DISCUSSION

Before analysis of the biological samples could be carried out it was necessary to remove the gold layer covering the area to be analysed. This was achieved by leaving the ion gun on continuously at a current of 0.85μA over an area of 1mm^2 to 'etch' away the gold layer. Multi-photon ionisation spectra, taken to include both non-resonant multi-photon ionisation (NRMPI) and resonant multi-photon ionisation (REMPI) were acquired at frequent intervals during this etching process to produce a depth profile of the gold layer to tissue interface. Each MPI spectrum was averaged over 1000 laser pulses, intersecting the sputter plume produced by a sputter ion pulse of 5μs duration. The result of such a depth profile is shown in

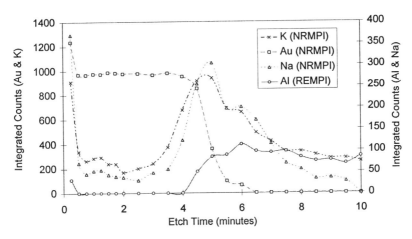

FIGURE 1. ToF-MPI depth profile of a gold-coated 500ppm Al-doped brain tissue homogenate.

Figure 1. The gold-brain tissue boundary is clearly visible, with the decay of the gold count rate and the appearance of the Al signal. Potassium (K) and Sodium (Na) concentrations are seen to be high both on the surface-gold and the gold-brain tissue boundary. A similar profile acquired using SIMS was subject to large point-to-point variations thought to be caused by variations in the secondary ion yield due to slight changes in the sample matrix with depth.

The highly selective nature of RIMS is illustrated in Figure 2. When the laser is tuned to an Al resonance, a strong Al peak is seen. Once the laser is tuned off-

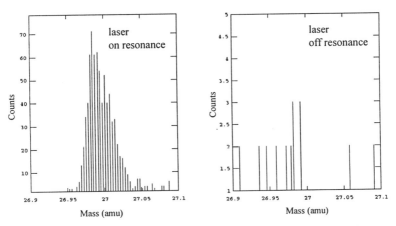

FIGURE 2. Al RIMS signals from a 100ppm Al-doped brain tissue sample.

resonance, the Al peak disappears, with the signal seen at 27amu becoming comparable in height to that of the background noise level. Hence, when analysing biological species using RIMS it can be conclusively demonstrated that all the signal seen at 27amu is due solely to aluminium when on an aluminium resonance, with any contributions from organic fragments of similar mass being comparable to the background noise level. This is not what was observed with SIMS spectra taken from the same brain tissue, where contributions from CNH and C_2H_3 could not be resolved using our time-of-flight system, leading to a broadening of the mass peak (5)

Analysis of the uniformly Al-loaded 500ppm and 100ppm brain tissue homogenate samples at various locations generated reproducible results. When averaged, these gave an Al ratio of 1:5.1±0.5 for the two samples, indicating the feasibility of using such samples as 'standards' for the future analysis of tissue samples containing unknown concentrations of aluminium. The ultimate detection limit using the current set-up was measured to be ~3ppm. Several improvements are currently being implemented to lower this figure by a factor of 10-100.

The spatial resolution achieved using the duoplasmatron Ar^+ ion source for sputtering was about 80-100μm at currents of up to 1.5μA. This is just sufficient to probe individual neurones in neural tissue, which should allow focal accumulations of Al (or other metals) to be distinguished from the background concentration in the bulk of the sample.

CONCLUSIONS

The dual selectivity of RIMS has been shown to make it an ideal tool for performing quantitative trace element analysis on complex matrices where traditional analytical techniques, such as SIMS, require very high resolution to distinguish between isobaric interferences. It has also been demonstrated that trace elemental concentrations of the order of ppm can be routinely detected. This is of importance in the investigation of focal element accumulations in neural tissue, where isobaric-free analysis is required at ppm levels.

REFERENCES

1. Candy, J.M., Oakley, A.E., Mountfort, S.A., Taylor, G.A., Morris, C.M., Bishop, H.E. and Edwardson, J.A., *Biol. Cell.* **74**, 109-118 (1992).
2. Jones, O.R., *Ph.D. Thesis*, University of Wales Swansea (1996).
3. Saloman, E.B., *Spectrochimica Acta.* **46B**, 319-378 (1991).
4. Kimock, F.M., Baxter, J.P, Pappas, D.L., Kobrin, P.H. and Winograd, N., in *Analytical Spectroscopy*, Ed. W.S. Lyon, Amsterdam: Elsevier, 1984, pp179-184.
5. Jones, O.R., Perks, R.M., Abraham, C.J. and Telle, H.H., to be published in *Rapid Commun. Mass Spectrom.* (1996).

Resonant Photoionization and Fragmentation of Dipeptides by Nanosecond and Femtosecond Lasers

N. P. Lockyer and J. C. Vickerman

Surface Analysis Research Centre, Department of Chemistry,
UMIST, P.O. Box 88, Manchester M60 1QD, U.K

Abstract. This paper reports on the multiphoton ionization and dissociation (MPID) processes which occur in gas phase biomolecules excited with femtosecond (*fs*) and nanosecond (*ns*) laser pulses. The MPID products of the dipeptides tyr-tyr, tyr-leu and val-val resulting from excitation with 5 ns and 250 fs pulses of 266 nm photons have been analysed in a time-of-flight mass spectrometer. There are two important conclusions from this study. Firstly, the efficiency of the ionization with *fs* pulses seems relatively insensitive to the presence or absence of a chromophore whereas this is not true for *ns* pulses. This suggests that while coherent processes clearly play little part in ionization with *ns* pulses, these effects become significantly more important as the time scale of the ionization reaches the subpicosecond regime. Secondly, photofragmentation is generally less extensive and possibly tuneable with *fs* pulses but much less so with *ns* pulses.

INTRODUCTION

Secondary Ion Mass Spectrometry (SIMS) has proved to be a chemically sensitive method for the detailed surface characterisation of complex organic material. However, ion yields are low which limits its application at high spatial resolution. Ion yields can be increased by post-ionizing the high yield of neutrals emitted in the sputtering process. This work forms part of an ongoing study aimed at determining the optimum conditions for post-ionizing neutral organic molecules desorbed from a solid surface under particle bombardment.

Control of the photon wavelength, pulse width and energy density provide an opportunity for optimising the ionization efficiency and tuning the degree of accompanying photofragmentation. Recent work suggests ultrahigh intensity subpicosecond pulses can more efficiently ionize systems which have fast relaxation processes in competition with optical pumping on the ns time scale (1,2,3).

EXPERIMENTAL

All experiments were carried out on a reflectron-type time-of-flight mass spectrometer. The dipeptides tyrosyl-tyrosine, tyrosyl-leucine and valyl-valine were thermally desorbed into the gas phase at about 430 K. For *ns* excitation we used a Nd:YAG (Spectra-Physics DCR-11) frequency-quadrupled to 266 nm with an output energy of up to 30 mJ in a 5 ns pulse. A regeneratively amplified, frequency-tripled, Ti:sapphire system (Clark-MXR) gave 250 fs pulses of up to 150 μJ at 266 nm.

RESULTS & DISCUSSION

Nanosecond vs. Femtosecond Ionization

Two photons at 266 nm are energetically required to ionize the chosen dipeptides, which are representative of the delicate biomaterials we wish to study. The content of aromatic chromophores in the sample molecules was varied intentionally. The presence of tyrosine (Tyr) should enable us to study the effect of resonant enhancement by a $(\pi-\pi^*)$ transition at the 1-photon level.

Ionization of Tyr-Tyr and Tyr-Leu was facile with 5 ns pulses (10^8-10^9 W cm^{-2}) but no significant ion current observed from Val-Val (Fig. 1b). The *ns* MPID spectra shown in Fig. 1b were acquired under identical laser conditions (5 J cm^{-2}). The available power density from the *ns* laser is insufficient to bring about a 2-photon nonresonant ionization of Val-Val. At these power densities coherent processes are very inefficient and the presence of a tyrosyl chromophore seems to be important.

Excitation with 250 fs pulses (10^{10}-10^{12} W cm^{-2}) produced ion currents of comparable magnitude for all three dipeptides, independent of their chromophore content (Fig. 1a). This implies that the role of nonresonant, coherent absorption has increased at these ultrahigh laser intensities to a level whereby it may dominate the ionization process, at the expense of resonant, sequential absorption. This may in part be due to a rapid AC Stark shift of resonant levels in chromophores. The *fs* MPID spectra were acquired under identical laser conditions (12 mJ cm^{-2}) and the ion intensities have been normalised to the number of neutrals in the focal volume during the acquisition. The kHz repetition rate of the Ti:sapphire system allowed the acquisition time to be reduced by a factor of 20 relative to the Nd:YAG laser. The ion signal displays a quadratic dependence on the laser intensity at the focus, suggesting we are neither optically saturating the intermediate level or hole-burning in the neutral cloud.

Thermal desorption or sputtering is expected to result in a broad vibrational and rotational state distribution in the neutral ground state of the molecules so that the absorption band is significantly broadened compared to cold molecules. This may shorten the intermediate state lifetimes to the subnanosecond time scale and hence reduce the ionization efficiency of *ns* excitation compared to cold molecules. A direct comparison of the effect of pulse duration on ionization efficiency is not possible with the present instrumentation because the bandwidths of the two lasers are different. The Ti:sapphire laser has a much wider bandwidth (>500 cm^{-1}) than the Nd:YAG laser (<1.0 cm^{-1}) and therefore the possibility exists of resonantly exciting a greater proportion of the vibrational distribution of the neutral ensemble.

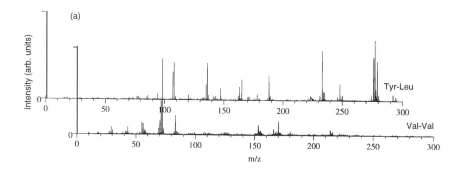

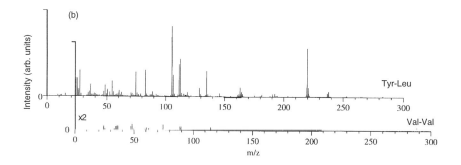

FIGURE 1. MPID spectra of Tyr-Leu and Val-Val with (a) 250 fs and (b) 5 ns, 266 nm pulses.

Nanosecond vs. Femtosecond Fragmentation

There are a number of fragment ions which are observed only in either the *ns* or *fs* MPID spectra, which suggests that the fragmentation mechanism may be dependent on the time scale of the ionization.

A power study revealed that no molecular ions could be produced from any of the dipeptides using *ns* pulses. The largest fragment ions resulting from simple bond cleavage were due to loss of a complete aromatic *or* aliphatic amino acid side group. For example, the low power *ns* MPID spectra of Tyr-Leu show a fragment at m/z 237 due to the loss of the leucine side chain (C_4H_9). A further loss of OH gives a very high relative intensity peak at m/z 220 (Figs. 1b & 2). An amide elimination reaction involving hydrogen transfer in the parent molecule results in a peak at m/z 114. This suggests that energy migration can compete with ionization on the nanosecond time scale. None of these fragments are observed in the *fs* MPID spectra at any laser power.

The degree of fragmentation observed in the *fs* MPID spectra is much less at the lower end of the power range compared to the *ns* spectra. In general the most intense fragments in the *fs* spectra arise from losses of small groups such as H_2O, OH and NH_2, and simple bond cleavages to form N-terminal acyliminium (immonium) and (by breaking the peptide bond) acylium ions (Figs. 1a & 2). Molecular ions are observed even at the highest laser intensity. The *fs* MPID spectra of Tyr-Leu contain a peak at m/z 188 due to loss of the tyrosyl side chain. This fragment is not observed with *ns* excitation.

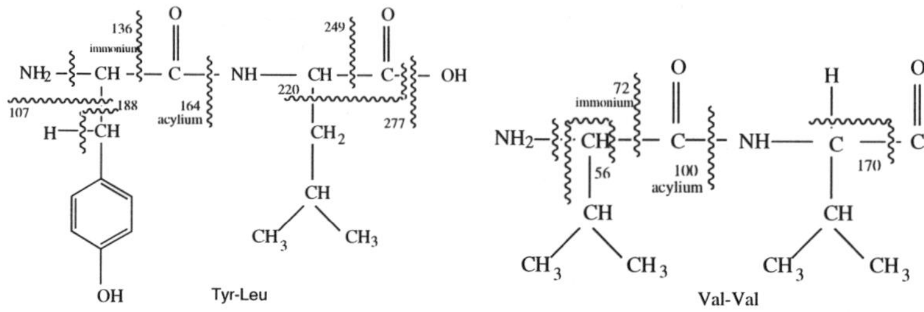

FIGURE 2. Simple bond cleavages leading to observed fragments in the MPID mass spectra of Tyr-Leu and Val-Val.

These observations, and similar ones in Tyr-Tyr (4) may be rationalised by considering the competition between intramolecular vibrational energy redistribution and ionization. Assuming the preferred site for photon absorption is the chromophoric tyrosyl ring, vibrational energy migrates through the molecule and provided the ionization rate is not too large, fragmentation can result at sites far removed from the chromophore. As the laser pulse intensity and the ionization rate increase, slow fragmentation pathways involving atomic transfer or energy migration over significant distances will no longer be able to compete and fragmentation local to the chromophore is more likely.

ACKNOWLEDGMENTS

The authors would like to thank Prof. Nick Winograd of the Pennsylvania State University for use of the Ti:sapphire system and Drs. Chris Brummel and Ken Willey for assistance in acquiring the fs data. The financial support of the EPSRC is gratefully acknowledged.

REFERENCES

1. Brummel, C. L., Willey, K. F., Vickerman, J. C., and Winograd, N, *Int. J. Mass Spec. Ion Proc.* **143**, 257 (1995).
2. Weinkauf, R., Aicher, P., Wesley, G., Grotemeyer, J., and Schlag, E. W., *J. Phys. Chem.*, **98**, 8381 (1994).
3. Mollers, R., Terhorst, M., Niehuis, E., Benninghoven, A., *Org. Mass. Spectrom.* **27**, 1393 (1992).
4. Lockyer, N. P., Vickerman, J. C., *Proc. 10th Int. Conference on SIMS*, 1996 (in press).

POSTER SESSION I:
ATOMIC RIS

Observation of the highly excited states of Lanthanum

P.Xue, X.Y.Xu, W.Huang, C.B.Xu, R.C.Zhao, X.P.Xie

Department of Modern Applied Physics, Tsinghua University
Beijing,P.R.China. 100084

Abstract. The highly excited states of Lanthanum are studied by means of laser resonance ionization time-of -flight spectrometer. Based on the two-step laser resonance excitation with intermediate state $5d^2(^3F)6p$ $^2D^0_{5/2}$, three new Rydberg state (RS) series ($5d^2(a^3F_2)ns$, $5d^2(a^3F_3)nd$ and $5d^2(a^1D_2)ns$) and a number of autoionizing states (AIS) are obtained. Theoretical calculation leads the quantum defects of ns and nd series to the value δs=4.35 and δ_d=2.80 respectively, which are very close to the experimental results. The Rydberg state series $5d^2(a^3F_2)ns$ gives the first ionization limit to be 44979.8 ± 0.3cm^{-1}, which is an order more accurate than ever.

INTRODUCTION

Much investigation has been carried out on the highly excited states of simple atoms but less done on those of complex atoms, especially the rare-earth-metal atoms, due to their complicated spectrum structures. Recently, it is believed that the rare-earth metal plays an important role to influence the ecology and environment. The study of the highly excited states of the metal is strongly needed for trace detection and other applications.

EXPERIMENT DESCRIPTION

The present paper reports our observation of the highly excited states of La. The experimental setup is similar to the former one (1). As we know, the melting point of La is only 920° C while its boiling point as high as 3469° C. This property of La makes it very difficult to achieve the atomization of the chip La with a graphite crucible. Besides, the products of La-C compound interfere with the experimental result. A newly-designed molybdenum-graphite crucible heated by an AC current 140A to about 1400°C has proved to be feasible. At such a temperature the thermal kinetic energy is close to the gap of the first two energy levels of atom La. The level $5d6s^2$ $^2D_{5/2}$, $1053cm^{-1}$ above ground state $5d6s^2$ $^2D_{3/2}$, will thus gain a comparative population. In the experiment some transition lines between this level and higher ones are chosen as the wavelength standards to calibrate the measurement. The La atom is excited to the Rydberg state or autoionizing state with two dye lasers (linewidth ~0.2cm^{-1}, pulse duration~20ns) pumped by an excimer XeCl laser (Lambda Physik EMG 203 MSC). La atoms of Rydberg state series are ionized and then attracted by a pulse electric field (~500 V/cm) about 200ns after the laser pulse, and finally detected by two microchannel plates .The signal via fast preamplifier is linked to a computer by a charge-digit converter.

EXPERIMENTAL RESULTS AND DISCUSSIONS

In fact, the spectrum of La is so complicated that it is very difficult to obtain pure structure of its Rydberg series. We have chosen many excitation channels with the intermediate states $5d^2(^3P)6p$ $^4P^0_{3/2,1/2}$; $5d6s(^3D)6p$ $^2D^0_{1/2}$; $5d^2(^3P)6p$ $^4D^0_{5/2}$; $5d^2(^3F)6p$ $^4F^0_{5/2}$ and $5d^2(^3F)6p$ $^2D^0_{5/2}$ respectively.

Careful comparisons give the best channel as:

$$5d6s^2 \ ^2D_{3/2} \xrightarrow{515.868nm} 5d^2 \ (^3F) \ 6p \ ^2D^0_{5/2} \xrightarrow{394\rightarrow371nm} RS. \ or \ AIS.$$

Fortunately, four resonance lines of La atom itself lie in the wavelength range of the second laser 394~371nm and thus are chosen to be the standard to measure the correspond wavelengths of observed Rydberg series and autoionization states. The four lines are (2):

$$5d6s^2 \ ^2D_{3/2} \rightarrow 5d6s(^3D)6p \ ^2D^0_{1/2} : \quad 25453.95 \ cm^{-1}$$
$$5d6s^2 \ ^2D_{3/2} \rightarrow 5d^2(^3P)6p \ ^4P^0_{1/2} : \quad 25616.95 \ cm^{-1}$$
$$5d6s^2 \ ^2D_{3/2} \rightarrow 5d^2(^3P)6p \ ^4P^0_{3/2} : \quad 25643.00 \ cm^{-1}$$
$$5d6s^2 \ ^2D_{5/2} \rightarrow 5d6s(^1D)6p \ ^2F^0_{7/2} : \quad 26986.25 \ cm^{-1}$$

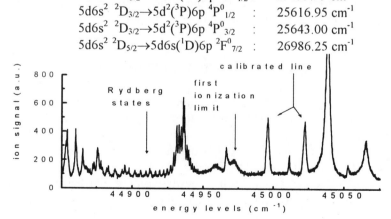

FIGURE 1. Partial spectra of Rydberg state and autoionizing state in two-step laser resonance excitation of La

300

TABLE A. The effect quantum number **n** * and energy level v_n (cm^{-1}) of the three Rydberg state series.

$5d^2(a^3F_2)ns$		$5d^2(a^1D_2)ns$		$5d^2(a^3F_3)nd$	
n*	v_n	n*	v_n	n*	v_n
34.52	44887.88	37.82	46300.84	47.28	45944.13
35.53	44893.02	36.84	46296.69	46.32	45942.10
36.58	44897.94	35.73	46291.61	45.32	45939.80
37.56	44902.18	34.99	46287.72	44.29	45937.30
38.57	44906.17	33.76	46281.27	43.29	45934.68
39.56	44909.02				
40.54	44913.18				
41.56	44916.42				
42.61	44919.50				
43.63	44922.31				
44.59	44924.76				
45.62	44927.22				
46.54	44929.29				

Figure 1 shows partial experimental spectra . Three new Rydberg state series are found and measured. Their levels are listed in Table A. We may determine the configuration of the series according their quantum defects. Our self- consistent - field theoretical calculation gives the quantum defects for ns and nd series to be the value $\delta s = 4.35$ and $\delta_d = 2.80$ respectively. The measured quantum defects for the three new Rydberg state series are 4.47 , 2.70 and 4.18, respectively. Thus the three Rydberg series are $5d^2(a^3F_2)ns$, $5d^2(a^3F_3)nd$ and $5d^2(a^1D_2)ns$. As the electron configuration of La is very complicated , very strong electron correlation exists and has a big influence on quantum defect. The small deference between the experimental result and theoretical calculation shows a success of this theory in handling complex atomic system.

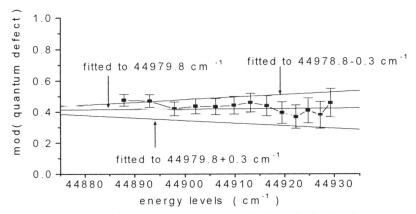

Figure 2. The experimental data are fitted to obtain the first ionization threshold. The fitted value is 44979.8±0.3 cm^{-1}.

TABLE B. The energy levels v and the widths Γ of nine new autoionizing states in two-step laser resonance for trace analysis (unit: cm^{-1} , intermediate state $5d^2\,(^3F)\,6p\ ^2D^0_{5/2}$)

v	Γ	v	Γ	v	Γ
45038.55	3.9	45272.52	5.8	45428.05	2.0
45179.92	2.7	45288.71	2.1	45493.55	1.5
45202.65	1.2	45365.62	2.8	45566.56	1.7

Although it is very hard to mark the spectrum terms of the observed Rydberg states and autoionization states , we may conduct the first ionization threshold of La based on the measured wavelengths of the Redberg states with the method indicated elsewhere (3). To date, the first ionization threshold of La (2) is known as 44981 ± 5 cm^{-1}. Within this range we fit the experimental data from the single channel quantum defect theory $v_n = v_\infty - R / n^{*2}$. The fitted result is shown in Figure 2. It's seen that our measurement gives the newest value 44979.8 ± 0.3 cm^{-1}, which is an order more accurate than ever.

Table B shows some of the obtained autoionizing state levels and their widths. These lines have rather high ion signals and thus are very useful in ultrasensitive analysis.

The present work is supported by National Natural Science Foundation of China.

REFERENCES

1. Xu,X.Y. et al, *J.Phys.* **B23**, 3315, (1990)
2. W.C.Martin et al., *Atomic Energy Levels--The Rare-Earth Element*, (1978)
3. Zhou H. J. et al, *ACTA Physica Sinica*, **Vol.1**, 19, (1992)

On-Line Hyperfine Structure and Isotope Shift Measurements with Diffuse Light Collection and Photon Burst Detection

Jens Lassen, Eric C. Benck, and Hans A. Schuessler

Department of Physics, TexasA&M University,
College Station, TX 77843-4242, USA

Abstract. An experiment is presently being set up which combines collinear-fast-beam laser spectroscopy with photon burst spectroscopy. Selectivity is provided by the large kinetic isotope shifts together with the practically Doppler free linewidth of the fluorescence from the fast atom beam. The photon burst detection, based on photon correlations in the resonance fluorescence, increases the sensitivity, so that on-line optical isotope shift and hyperfine structure measurements on low intensity radioactive beams become feasible. In order to improve photon burst detection the solid angle of detection and the observation time have to be optimized. To this end a diffuse reflecting cavity has been designed and built, which collects fluorescence over a 45 cm length of the beam and covers the full solid angle. The light collection efficiency of the cavity is calculated to be about 45%. The cavity is being tested with a 11 keV beam of krypton atoms, probing the near infrared transitions in our apparatus at Texas A&M University.

INTRODUCTION

With the projected beams at the Holifield Radioactive Ion Beam Facility at Oak Ridge National Laboratory, the efforts of our present work goes into enhancing spectroscopy and trace detection by the use of the photon burst techniques. In principle, photon burst detection is applicable whenever optical spectroscopy is done on a recyclable transition. In order to detect bursts of photons efficiently in fast beams, the observation time should be in the micro second range which necessitates observation regions of 10 cm or more along the beam. In our detector the imaging optics will be replaced with diffuse reflection light collection (1) for increased efficiency.

THEORETICAL CONSIDERATIONS

The problems inherent in low level detection lies in the achievable signal to noise ratios. In the previous collinear fast beam resonance ionization detection one of the limiting factors was the vacuum requirement of better than 10^{-10} mbar. In photon counting the limiting factors are the scattered laser light, which buries the signal, and the overall light collection efficiency of the detection system. One of the factors determining the overall detector sensitivity is the finite solid angle of detection. Even with a fiberoptic coupler that covers a solid angle of 4π there is a loss of approximately 30% in the fibers. Above that, the quantum efficiency of the photomultipliers is typically on the order of 30% down to 1%. So the n-fold photon burst statistics, as discussed by Greenlees et. al. (2,3), suffers because not all emitted photons will actually be detected. Noise and fluorescence are reduced by a common factor, therefore requiring longer integration times.

In the limit of Poissonian statistics, the photon distribution has the form of eq.1. Here R is the rate of photons scattered by the atom and T is the transit time through the detection region. For a beam of atoms with a flux of N atoms per second, dN/dt is related to the n-fold photon burst probability P_n by eq.2. Whereas the background scattered light is given by eq.3.

$$P_n = (RT)^n exp(-RT)/n! \qquad (1)$$

$$R_n = \frac{dN}{dt} P_n \qquad (2)$$

$$R_{n,back} = R_{back} exp(-R_{back}T)/[(1+R_{back}T)(n-1)!] \qquad (3)$$

An improvement in detection is achieved when, with increasing multiplicity, the background signal reduces quicker than the actual signal. As shown previously (2,4), for low level detection high multiplicities have to be detected.

EXPERIMENTAL CONDITIONS, FIRST RESULTS

Under these conditions the detection system has to be optimized to achieve significant rates of high multiplicities. Fig. 1 shows the experimental arrangement that will be implemented on-line. The detector set up used to obtain the presented spectra images a 25 mm region of the beam via a Fresnel lens onto the active area of one photomultiplier. The second spectrum in Fig. 2 shows a non optimized photon burst signal of multiplicity five below the direct photomultiplier signal. It was recorded in a single scan with 20 ms integration time in each channel in a low resolution scan of 1V/channel. The only 4% in solid angle detection of the conventional optical detection system used in the Texas A&M beamline was accounted for by having about 2000 atoms simultaneously in the detection region. The major limitation however resulted from the limited interaction region of only 25 mm. For the used 7 keV Kr beam this leads to an interaction time T of less than 0.2 µs in low level detection.

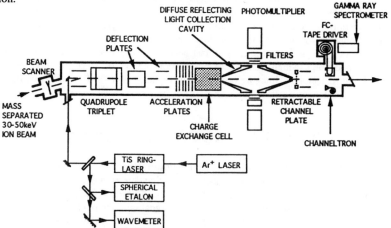

FIGURE 1. Experimental set-up for collinear-fast-beam photon-burst detection with diffuse light collection.

To improve on the fiberoptic coupler incorporated in the Oak Ridge beamline, and to eliminate some of its inherent loss factors, we have built a light collection cavity. This cavity is based on a diffuse reflection material, Spectralon ®, which is usually found in integrating spheres. The idea is to funnel the fluorescence photons emitted by the atom beam into the photomultipliers.

The photomultipliers are located next to two narrow slits on the sides of the cavity as indicated in Fig. 1. The results of Monte-Carlo simulations of photons emitted inside the cavity in random directions, taking into account the reflectivity of the Spectralon® and a random scattering, 45% of the photons emitted will be collected by the photomultipliers within 40 ns of the photon emission. The collection length of this cavity is 45 cm as compared to the 15 cm of the fiberoptic coupler that is presently installed in our Oak Ridge set up (4,5). For 105000 photons randomly released along the axis of the cavity, 966 photons will be absorbed on the cavity walls, and 49903 photons reach the photomultipliers within 35 ns. The majority of photons will however only need about 3.5 ns and 95% of all scattered photons will be collected within 15 ns.

The Spectralon® cavity will be put to the test in refining our measurements of the isotope shifts in the $5s[3/2]_2^\circ$ - $5p[1/2]_1$ transition in Kr at 892 nm. The measured hyperfine structure parameters A and B for the upper level are A = -143.2 (0.7) MHz and B = -21.3 (4.4) MHz. The previously unpublished isotope shifts from our conventional measurements are given in Table I. Also given are values in the $5s[3/2]_2^\circ$ - $5p[5/2]_3$ transition at 811 nm in Kr. The measurement in the 892 nm line was hampered by the before mentioned 4% solid angle of the TAMU collinear fast beam apparatus and the about 1% quantum efficiency of the photomultiplier (RCA 31034-2) at this wavelength. Fig. 2 shows first part of the obtained spectrum after 10 s integration per channel in a high resolution scan of 0.1V/ channel. Then a low resolution scan (1V/channel) with the direct photomultiplier signal in the upper, and the photon burst signal of multiplicity five, in the lower trace are shown. All HFS lines appear in the photon burst spectrum because the measurement was done with 2000 atoms in the detection region to compensate for the losses of the 4% of 4π solid angle detector. With the Spectralon® cavity photon bursts from single atoms will be detectable.

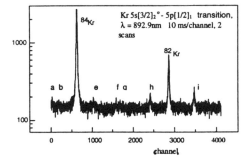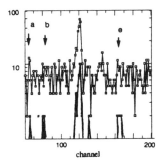

FIGURE 2. Shown are part of the $5s[3/2]_2^\circ$ - $5p[1/2]_1$ spectrum of krypton obtained in collinear fast beam laser spectroscopy, in a high and low resolution scan (upper trace). In the low resolution spectrum the photon burst spectroscopy signal (lower trace) of multiplicity five is displayed. In the low resolution two scans with 10 ms/channel are integrated. Photon burst spectroscopy will be used to recover the signal from the [78] Kr isotope that is buried within the noise without a substantial increase in integration time. In measurements with one or fewer atoms at a time in the interaction region only recyclable HFS lines will be observable.

TABLE I. Given are the measured isotope shifts of the even, naturally occurring krypton isotopes in two transitions from the metastable Kr state. The isotope shifts in the 892 nm transition are reported for the first time. The values in parenthesis are uncertainties given in MHz.

transition	wavelength [nm]	86-84 [MHz]		86-82 [MHz]		86-80 [MHz]		86-78 [MHz]	
$5s[3/2]_2^0$ -$5p[5/2]_3$	811.3	68.2	(0.7)	132.6	(0.5)	207.1	(0.5)	287.5	(0.7)
$5s[3/2]_2^0$ -$5p[1/2]_1$	892.9	63.6	(2.5)	123.7	(2.7)	195.6	(3.5)	246.3	(4.3)

DISCUSSION AND OUTLOOK

Increasing the number of photomultiplier detectors for photon burst detection is the natural, most selective and sensitive method for photon burst detection. In this work we try to push the detection limits by using only two photomultiplier tubes and a diffuse reflector to efficiently collect light over an extended detection region for photon burst detection. While the Spectralon/diffuse reflector cavity, like the fiberoptic coupler sacrifices the spatial resolution of the conventional optical imaging of the fluorescence from the beam onto the active area of the photomultiplier tube, the Spectralon® cavity covers a larger portion of the beam. The enhancement in sensitivity with respect to the fiberoptic set up will enable further on-line optical hyperfine structure and isotope shift measurements on short lived, radioactive isotopes with a flux below 10^5 atoms/s (5) in the new radioactive ion beam facility at Oak Ridge National Laboratory. Thereby continuing the successful optical hyperfine structure and isotope shift measurements on beams with intensities previously not accessible to this technique.

ACKNOWLEDGMENTS

Part of this work has been supported by the Department of Energy.
We also thank Dr. Petr V. Grigoriev, General Physics Institute, Moscow, Russia for updating all data acquisition programs and valuable discussions.

REFERENCES

1. A. I. Ludin, B. E. Lehmann, Appl. Phys. B **61**, pp 461-465 (1995)
2. D. Lewis, J. Tonn, S. Kaufman and G. Greenlees, Phys. Rev. A **19** (4), pp 1580-1588 (1979).
3. Tom J. Whitaker, Bruce A. Bushaw and Bret D. Cannon, Laser Focus / Electro Optics **2**, pp 88-101 (1988).
4. E. C. Benck, H. A. Schuessler, F. Buchinger and K. Carter, "Detection of rare isotopes by collinear-fast-beam photon-burst spectroscopy", presented at the Resonance Ionization Spectroscopy, Santa Fe, 1992, pp 329-333.
5. H. A. Schuessler, E. C. Benck, F. Buchinger, H. Imura, Y.F. Li *et al.*, Hyperf. Interact. **74**, pp 13-21 (1992).

Two-color resonance ionization spectroscopy of autoionization states of Pb

Shuichi Hasegawa, Hiroshi Yamamoto, Naoki Kawasaki,
Hidematsu Ikeda* and Atsuyuki Suzuki

Department of Quantum Engineering and Systems Science
*Research Center for Nuclear Science and Technology
The University of Tokyo
7-3-1 Hongo, Bunkyo-ku, Tokyo 113, JAPAN

abstract. Two-color resonance ionization spectroscopy was performed to observe even parity autoionization states. Following Rydberg series converging to the second ionization limit, we determined the second ionization limit $6s^2 6p\ ^2P^o_{3/2}$ as 73899.5 cm^{-1}.

1 Introduction

Resonance Ionization Spectroscopy (RIS) has revealed a lot of characteristics of high-lying states of atoms. Two-color RIS is especially good for investigating even-parity states above the first ionization limit as they cannot be accessed by one-photon UV spectroscopy. The autoionization states of two-electron atoms are interesting from the viewpoint of not only spectroscopic data but planetary atoms.

Carbon group is interesting since it has two low-lying ionic core like alkaline earth atoms. Among the group, Pb is the heaviest element and its vapor pressure is high enough to generate atomic beam by a resistively heated oven. The ground state of PbI is $6s^2 6p^2\ ^3P_0$, and the lowest ionic core PbII is $6s^2 6p$, which splits into two fine structure levels, $^2P_{1/2}$ (59819cm^{-1}) and $^2P_{3/2}$ (73900cm^{-1}). The bound states below the first ionization limit have been investigated experimentally in detail and analyzed with Multichannel Quantum Defect Theory (MQDT)(1,2). Reference (2) tabulated the states beyond the first ionization limit, but the spectroscopic method of Ref. (2) restricted the observed levels to odd parity states. Lead is also interesting from the viewpoint of the applications such as age determination, environmental pollution. To isotopically-analyze the trace sample of Pb is important for such field. Resonance ionization spectroscopy with mass spectrometer is a good candidate to solve the problem. In our experiment, we use two color resonance ionization spectroscopy to access even parity states above the first ionization limit.

2 Experiment

The experimental setup consists of laser system, vacuum chamber, and data acquisition (Fig. 1). A Nd:YAG pumped dye laser (Continuum) is used for exciting the intermediate state $6p7s\ ^3P_1^o$ (35287.24 cm^{-1}). The resonant wavelength 283.3 nm is generated by doubling 566.6 nm with BBO nonlinear crystal. The second exciting laser is an excimer pumped dye laser (Lambda Physik) with SHG crystal control option in order to access the states beyond the first ionization limit. Wavelength of the second laser is calibrated with a Ne optogalvanic spectrum. Timing of two lasers is controlled by delay generator.

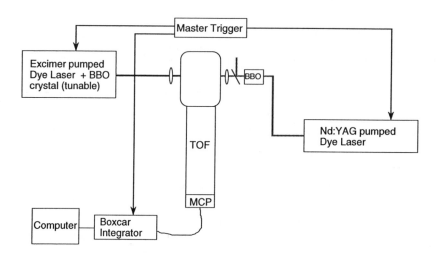

Figure 1:Experimental Setup

Atomic vapor of Pb is generated by a resistively heated oven and collimated to be atomic beam. In the interaction region, the first laser (283.3nm) excites the intermediate state 6p7s $^3P^o_1$ (35287.24 cm^{-1}) and the following scanning laser excites the states above the first ionization limit. Negative electric pulse delayed 500ns after the excitation extracts the ions from the interaction region into Time of Flight mass spectrometer. The ions are separated into each isotopes and detected microchannel plate. The ion signal is fed into boxcar integrator and the integrated output over the gate opening at Pb arrival time is digitized by AD converter.

3   Results and Discussion

Figure 2 shows an example of the resonance ionization spectrum of Pb converging to the second ionization limit. The variation in pseudo-quantum defect $(n - n^*)$ versus n (integer nearest the effective quantum number) with various ionization limit is shown in Fig. 3. The effective quantum number is determined in the following Rydberg formula.

$$n^* = \sqrt{\frac{R}{I - E}} \tag{1}$$

For high lying Rydberg series, the series limit gives smooth and constant value of quantum defect as a function of principal quantum number (3). Among various assumed values, the most suitable value is selected for the ionization limit.

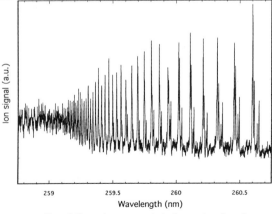

Figure 2: Two-color resonance ionization spectrum through 6p7s $^3P_1$ as a function of the second laser wavelength. Rydberg series are converging to the second ionization limit.

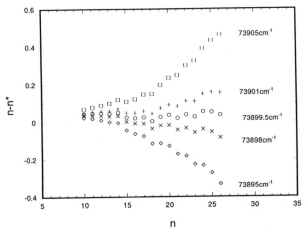

Figure 3: pseudo-quanutum defect versus n (integer nearest the effective quantum number) with assumed ionization limits.

We followed the highest peak of the several Rydberg series observed in the experiment. From Fig. 3, we determined the second ionization limit $6s^2 6p\ ^2P^o_{3/2}$ as 73899.5 cm^{-1}.

4 Summary

We have observed even parity autoionization states of Pb converging to the second ionization limit by two-color resonance ionization spectroscopy. The intermediate states used this study is $6p7s\ ^3P^o_1$. Following the Rydberg formula, the second ionization limit $6s^2 6p\ ^2P^o_{3/2}$ is determined as 73899.5 cm^{-1}. To assign their configuration, further spectroscopic data determining the J value of the states, such as one-color two photon, through other intermediate states, and using polarization, are expected.

5 References

1. Hasegawa, S. and Suzuki, A., *Phys. Rev. A* **53**, 3014-3022 (1996).

2. Brown, C. M. and Tilford, S. G., *J. Opt. Soc. Am.* **67**, 1240-1252 (1977).

3. for example, Callender, C. L., Hackett, P. A., and Rayner, D. M., *J. Opt. Soc. Am.* B **5**, 614-618 (1988).

A Study of Xenon Isotopes in a Martian Meteorite Using the RELAX Ultrasensitive Mass Spectrometer

J A Whitby, J D Gilmour, G Turner

Department of Earth Sciences, Manchester University, Oxford Road, Manchester M13 9PL, UK.

Abstract. The Refrigerator Enhanced Analyser for Xenon (RELAX), an ultrasensitive resonance ionization time-of-flight mass spectrometer, has been used with a laser microprobe to investigate the isotopic composition of xenon trapped in the martian meteorite ALH84001. The laser microprobe has a spatial resolution of the order of 100μm thus allowing the *in situ* analysis of individual mineral grains in a polished section when combined with ultrasensitive, low blank sample analysis. We present results showing that the mineral orthopyroxene in ALH84001 contains a trapped xenon component consistent with a martian origin. Additionally, a cosmic ray exposure age of 15Ma for ALH84001 is obtained from spallation derived xenon trapped within an apatite grain.

INTRODUCTION

It is now accepted that the SNC meteorites (Shergottites, Nakhlites and Chassigny) originated on the planet Mars. Oxygen isotopes suggest that all members of the group have a common parent body (1), and their recent crystallization ages indicate that their parent body has been geologically active until at least 1Ga before the present (2). Furthermore, the gases trapped in shock-induced melt glasses within some members of the group exhibit the same relative abundances as those measured in the martian atmosphere by the Viking probe (3).

Petrographic and oxygen isotope data identify Allan Hills 84001 as an SNC meteorite (4). However, it is unusual in having an Ar-Ar derived shock age of 4.0Ga (5) and this, combined with its crystallization age of 4.5Ga (6), argues for an origin in the heavily cratered Noachian-age terrains of Mars (5). Understanding the xenon isotope systematics (particularly the nature of the host phases) of this meteorite thus has the potential to constrain models of the evolution of the martian

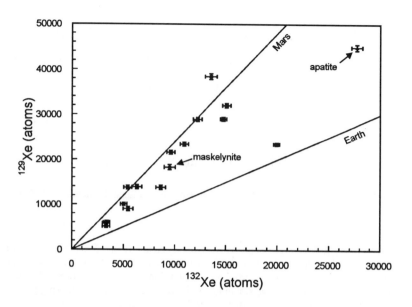

FIGURE 1. Microprobe results for orthopyroxene, maskelynite and apatite. All points are consistent with a mixture of martian and terrestrial atmospheres, which are represented by the lines of constant $^{129}Xe/^{132}Xe$ shown.

crust and atmosphere. The meteorite is almost entirely orthopyroxene, significant minor phases being maskelynite (a feldspathic shock glass), apatite and carbonate.

In this study gas was extracted from a polished section of ALH84001 by laser microprobe; scanning electron microscopy with x-ray dispersive analysis was used to determine the mineralogical composition of sampled areas before and after xenon analysis. (The CW argon ion laser formerly used (7) has been upgraded to a Nd:YAG laser with a gaussian beam profile operating at 1064nm in either CW or Q-switched modes). A CW power of 3W applied for approximately one second yielded well defined melt craters of around 100μm diameter. Smaller spots were initially used to investigate the gas content of minor phases. Active gases were removed by exposure to a hot zirconium getter before the evolved gas was admitted to the spectrometer.

The RELAX mass spectrometer has been previously described in detail (7,8). No major changes have been made to the spectrometer although refurbishment of the ion source and detector and the adoption of an improved operating protocol have enabled us to reduce the procedural blank to about 1×10^{-16} cm^3 STP ^{132}Xe (<3000 atoms). Modifications have also been made to the data reduction software resulting in improved precision and accuracy in correcting for spectrometer blank and build up when very small aliquots of gas are analysed.

RESULTS

Figure 1 shows accumulated results from laser probe analyses of the polished section. The evolved gas is consistent with mixing between a xenon component derived from the martian atmosphere ($^{129}Xe/^{132}Xe=2.4$) and a reservoir with a lower $^{129}Xe/^{132}Xe$ ratio; this latter may be terrestrial atmosphere or similar to the component observed in Chassigny, both of which have $^{129}Xe/^{132}Xe$ close to unity (9). The concentrations of gas inferred for the minor mineral phases are too low for maskelynite or apatite to be the major gas carriers, while the concentration found in spot analyses of orthopyroxene is similar to that of the bulk meteorite. Carbonate has also been eliminated as a major carrier of martian xenon in a separate series of stepped heating experiments. We therefore conclude that orthopyroxene is the major carrier of martian xenon in this meteorite.

During the meteorite's journey between Mars and the Earth, the light isotopes of xenon ^{124}Xe and ^{126}Xe were produced by the interaction of cosmic rays with suitable target elements, chiefly barium and the light rare earth elements (LREEs). LREEs are known to be highly concentrated in apatite in ALH84001 (10) and this is reflected in the anomalously high ^{124}Xe content (figure 2).

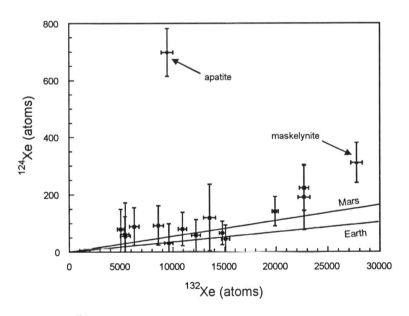

FIGURE 2. Excess ^{124}Xe in the apatite is due to spallation effects on rare earth elements, and may be used to derive a cosmic ray exposure age of 15Ma.

Combining the measured excesses of ^{124}Xe and ^{126}Xe with the known production rates (11) and the estimated mass of apatite melted yields a cosmic ray exposure age consistent with the accepted value of 15Ma (12). This illustrates the utility of the combination of ultrasensitive xenon isotope determinations with the low-blank laser microprobe sample extraction technique.

ACKNOWLEDGEMENTS

We wish to thank D J Blagburn and B Clementson of the University of Manchester for their technical support, and Dr R Ash and the Antarctic Meteorite Working Group for supplying the specimen of ALH84001. This work was supported by the PPARC, and Dr Gilmour is supported by the Royal Society under their University Research Fellowship scheme.

REFERENCES

1. Clayton R N, Mayeda T K, *Earth and Planetary Science Letters,* **62** p1–6 (1983)
2. McSween H Y Jr, *Reviews of Geophysics* **23** p391–416 (1985)
3. Ott U, *Geochimica et Cosmochimica Acta* **52** p1937–1948 (1988)
4. Mittlefehld D W, *Meteoritics* **29** p214–221 (1994)
5. Ash R, Knott S F, Turner G *Nature* **380** p57–59 (1996)
6. Harper C L, Nyquist L E, Bansal B, Wiesman H, Shih C Y, *Science* **267** p213–217 (1995)
7. Gilmour J D, Ash R, Lyon I C, Turner G in AIP Conf. Proc **329** p233–236 (1994)
8. Gilmour J D, Lyon I C, Johnston W A, Turner G *Review of Scientific Instruments* **65** p617–625 (1994)
9. Bogard D D, Nyquist L E, Johnson P *Geochimica et Cosmochimica.. Acta* **48** p1723–1739 (1984)
10. Dreibus G, Burghele A, Jochum K P, Spettel B, Wlotzka F, Wanke H *Meteoritics* **29** p461 (1994)
11. Hohenberg C M, Kennedy B M, Podosek F A, *Geochimica et Cosmochimica Acta* **45** p1909–1915 (1981)
12. Swindle T D, Grier J A, Burkland M K, *Geochimica et Cosmochimica Acta* **59** p793–801 (1995)

Temporal Behavior of Mercury LEI Signal in a Buffer Gas

W.L. Clevenger, L.S. Mordoh, O.I. Matveev, N. Omenetto, B.W. Smith and J.D. Winefordner

Department of Chemistry, University of Florida, Gainesville, FL 32611-7200

Abstract. The temporal behavior of the laser enhanced ionization signal of mercury was studied in a quartz cell under low buffer gas pressure. Using fast electronics and a short (34 ns) laser pulse, it was possible to distinguish between the non-selective photoionization component of the signal and that which was due to collisional ionization from selected levels in one time-resolved ionization waveform.

In laser enhanced ionization (LEI) spectroscopy, the process of ion formation due to collisional ionization of analyte atoms can practically always be accompanied by their resonance photoionization and by non-selective ionization of molecules and/or atoms of buffer gas. Often, in order to improve the limit of detection, one needs to acquire separate information about the contribution of the LEI signal bearing useful analytical information and the signal due to non-selective photoionization. It has been suggested[1] that when using flames as atomizers, this information can be obtained by observing a single time-resolved ionization waveform under conditions when the faster photoionization component and the slower collisional ionization component of the signal can be distinguished from one another. This idea can be extended to other atom reservoirs, such as a low pressure buffer gas. The deconvolution becomes possible when (1) the duration of the laser pulse is shorter than the average time in the collisional ionization process and (2) the distance between the collection electrode and the laser beam is less than the average distance traveled by the electron as a result of its drift velocity during the process of the collisional ionization; that is, the duration of the ionization signal from a single electron should be less than the reciprocal of the rate of collisional ionization.

In this work, the temporal behavior of the ionization signal was studied in a quartz cell containing mercury vapor and a buffer gas at low pressure in order to elucidate the processes of ion formation. A three-step LEI scheme of mercury[2] was utilized, in which the third step excited the analyte to a Rydberg level. Ionization waveforms of the signal were obtained by tuning the third step of laser radiation to levels with different principal quantum number (n = 10-26) under different buffer gas pressures (0.05-600 Torr) with variable voltage and beam position between electrodes. In this way, two clearly distinguishable components of the signal were observed: collisional ionization and photoionization.

EXPERIMENTAL

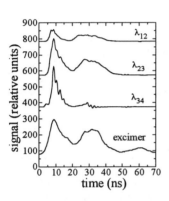

Figure 1. Laser pulse shapes from dye lasers 1, 2, and 3.

An excimer laser was used to pump three dye lasers. The frequencies of the first and second lasers remained constant, while the wavelength of the third laser was tuned to levels with different principal quantum number. Pulse energies were typically 150 mJ for the excimer, 100-150 μJ for λ_{12}, 200-300 μJ for λ_{23}, and 0.5-1 mJ for λ_{34}. Laser pulse durations were found to be 34 ns(FWHM), rather than the LPX 240i nominal duration of 14 ns. The temporal behavior of these unusual laser pulses are presented in Figure 1. After observing the pulse from the excimer itself, it was confirmed that this behavior is characteristic of the pump beam and not the dye output. Rather than hindering our work, this unusual behavior was very helpful in identifying the different sources of ionization, because any photoionization resembles the pulse which created it. In this work, the signal due only to photoionization was observed when only the first two laser connected steps were excited, λ_{12} and λ_{23}; so the photoionization resembles the shape of the λ_{23} because its power exceeds that of λ_{12} and is enough to saturate the third level. Therefore, photoionization was very easy to identify and distinguish from other sources of ionization.

The beams were directed with mirrors and prisms into a quartz cell which contained a small amount of mercury and varying pressures of buffer gas (air or argon). The cell also housed two planar Ni electrodes and was connected to a vacuum pump. One electrode was connected to a high voltage source which applied varying potentials (0-1.5 kV) to the cell, while the other was connected to a low-noise high-frequency preamplifier. The amplified LEI waveform was displayed on a fast digitizing oscilloscope. These waveforms were generated using different pressures, voltages, and beam positions for different Rydberg levels with the goal of distinguishing between different ionization mechanisms.

RESULTS AND DISCUSSION

The first Rydberg levels studied experimentally were n=10-14. The temporal differences in behavior for these levels are shown in Figure 2(a) (along with the signal from only $\lambda_{12} + \lambda_{23}$, a signal due only to photoionization). A trend emerges as one observes levels n=10 to n=14. The waveform corresponding to n=10 resembles the signal with no λ_{34}, with the addition of a small, longer temporal signal. This second component steadily increased from n=11 to n=12 until at n=13, it clearly began to dominate. In fact, at n=14, the photoionization component became almost hidden by the second component. The reasons for this are as follows: at n=10, the fourth excited level was still too far from the ionization continuum (0.39 eV) for significant collisional ionization to occur at 60 mTorr buffer gas; for this reason, the signal was clearly dominated by photoionization. From n=11-14, however, the second, longer signal component that began to emerge was a result of collisional ionization. For n=14, the energy gap to the ionization continuum is only 0.14 eV where collisional ionization is very significant. Most importantly, the rate of collisional ionization increased with increasing principle quantum number. This trend can be seen even more dramatically from the higher-lying levels.

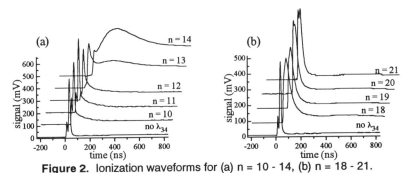

Figure 2. Ionization waveforms for (a) n = 10 - 14, (b) n = 18 - 21.

The three-step ionization waveforms from Rydberg levels n=18-21 were also observed. For these levels, the energy gap between the excited level and the ionization continuum became even smaller (0.071 eV for n=18, 0.049 eV for n=21), resulting in high rates of collisional ionization. This can be clearly observed in Figure 2(b). As λ_{34} populated principal quantum numbers from n=18 to n=21, it is clear that the time duration of the signal shortened due to the increasing rate. What was most dramatic here was the re-emergence at n=20 of the laser pulse shape. Although this shape was observed for the lower quantum energy levels (n=10-12) because of the low efficiency of collisional ionization, its reappearance at these high-lying levels was for an entirely different reason: the rate of collisional ionization was so large that its reciprocal value was much less than the duration of the laser pulse. At n=21, collisional ionization was so fast that the waveform resembled that of the laser beam used to excite the atoms; at these higher levels, deconvolution of the collisional and photoionization components again became difficult.

Some studies were carried out to further prove that what is being observed is indeed collisional ionization. Figure 3 shows the FWHM of the collisional component (from n=13) as a function of pressure. As the pressure increased to about 10 Torr, a decrease in signal duration occurred. Above about 60 Torr, an increase in signal duration occurred with increase in pressure. At lower pressures ($\leq$ 10 Torr), the FWHM depended only on the rate of collisional ionization. On the descending part of the curve, photoionization and collisional ionization could be independently observed as they occur and identified as a function of their rates. From ~10 Torr to ~60 Torr, the FWHM was no longer a function of the collisional rate, but instead a function of electron mobility; that is, the time it takes the electrons to travel from the point of ionization to the electrode. At pressures >60 Torr, the FWHM began to increase because of a lower electron velocity (drift) caused by the increase of buffer gas concentration in the cell. In this case, the two components could not be identified or distinguished from each other without difficulty. Therefore, the signal observed was temporally dependent primarily on the electron drift velocity and was not an actual observation of ion formation like that seen at $\leq$10 Torr.

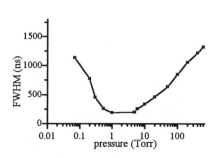

Figure 3. FWHM of collisional ionization signal vs. buffer gas pressure.

ACKNOWLEDGMENTS

The authors would like to thank K. Niemax for helpful discussions concerning this work and the National Institutes of Health which funded this research (5-R01-GM49638-03).

REFERENCES

[1] Matveev, O.I., Omenetto, N., *Spectrochim. Acta* **49B**, 691-702 (1994).
[2] Matveev, O.I., Omenetto, N, Cavalli, P., "Three-step Laser Induced Ionization of Ir and Hg atoms in an Air-Acetylene Flame and a Gas Cell," in *AIP Procedings No. 329 (RIS 94)*, 1994, pp. 269-272.

Resonance ionization spectroscopy of Gadolinium

Masabumi Miyabe and Ikuo Wakaida

Department of Chemistry and Fuel Research,
Japan Atomic Energy Research Institute, Tokai-mura, Ibaraki-ken, 319-11, Japan

Abstract. A spectroscopic study of 3-step resonance ionization was performed for atomic gadolinium. Many high-lying odd states and autoionizing states were identified. The ionization potential of gadolinium was determined to be 49601.44 ± 0.30 cm^{-1}. Photo-absorption cross-sections were measured for many transitions, enabling several efficient photoionization schemes to be determined.

INTRODUCTION

Highly efficient multistep photoionization schemes of Gd are of great importance in various applications such as laser isotope separation, laser-based ultrasensitive trace analysis, etc. In particular, for the laser isotope separation using the difference in the polarization selection rules between odd and even isotopes, an efficient 3-step scheme, including the autoionizing state of J=0 is required (1,2). The atomic structure of Gd is, however, extremely complicated because of its electronic configuration having three outermost electrons outside a half filled f shell, and as a consequence many of the spectroscopic data required for realizing an efficient photoionization are still unknown. Thus we have been working on spectroscopic studies of Gd using a 3-step resonance ionization technique.

Our experimental setup consists of three main parts: three dye lasers pumped by three excimer lasers, a resistively heated crucible and a TOF mass spectrometer. Photoionization was accomplished in a field free region; the ions produced were introduced into the mass spectrometer by a pulsed electric field.

HIGH-LYING STATE SURVEY

Using many of the 1st states excited from the $4f^7 5d 6s^2$ (^9D_j j=2-6) ground term states, the odd-parity intermediate states and even-parity autoionizing (AI) states lying in the 31000-36000 cm^{-1} and 49600-52500 cm^{-1} regions were identified. A typical measured ionization spectrum is shown in Figure 1. The wavelength for each peak was determined to within an accuracy of about 0.3 cm^{-1} using a Fizeau wavemeter calibrated by some optogalvanic lines of Ne. At present, 90 odd states, including 40 new states, and over 200 AI states were revealed (3).

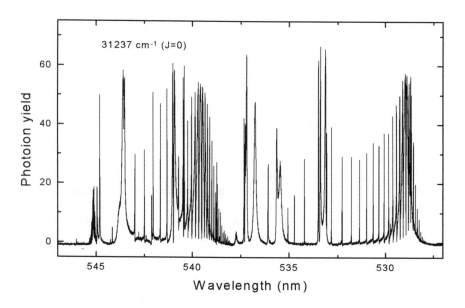

FIGURE 1. Typical ionization spectrum observed from the intermediate state of J=0

The J-value for each state was assigned according to the transition selection rules, viz., $\Delta J = 0, \pm 1$. An unambiguous assignment of J requires observation of the excitation from at least 3 lower states with different J-values. However, in the case where these 3 sets of excitations cannot be observed like the intermediate states of J=0 or J=1, two or three possible J-values remain. Accordingly, we also utilized an alternative J-assignment method for such small J states using the dependence of the ion yield of even isotopes on the polarization combinations. For example, the AI transition to the J=0 state can occur only for the $\lambda_2 \perp \lambda_3$ polarization with the J=2→1(2)→1→ 0 scheme. On the other hand, the yield of the AI transition to the J=1 state for the $\lambda_2 \perp \lambda_3$ polarization is half of that for $\lambda_2 // \lambda_3$ with this scheme. Thus, comparing the ion yield of the spectra measured with parallel and perpendicular polarization, the J-values for several AI states were first assigned. Then, some AI states appearing in the ionization spectra of intermediate states with unknown J-values were identified based on their energy and line profile, and finally the J-values of the intermediate states were determined. In this way, five previously unreported J=1 and two J=0 states were assigned.

RYDBERG SERIES AND IONIZATION LIMIT

Whereas the photoionization spectra of Gd are generally complicated, those of the

J=0 intermediate states were found to be quite simple as shown in figure 1. This is because the J=0 state has only one magnetic sublevel and only the transitions to the AI states of J=1 are allowed. Thus the ionization spectra of the J=0 states are considered to be most helpful for the study of autoionizing states of heavy elements having complicated atomic structure.

The relationship between the principal quantum number and the quantum defect for the Rydberg series converging to the 2nd excited state f^7ds ($^{10}D_{9/2}$) of Gd^+ is shown in Figure 2. The quantum defects are found to be almost constant except in the perturbation region around n=35 when the ionization potential of Gd is assumed to be 19601.44 ± 0.30 cm^{-1}. This is in good agreement with the previous value of 49603 ± 5 cm^{-1} (4) and the accuracy is improved by about one order of magnitude. In the present measured energy region, three series converging to the 1st (J=7/2), 2nd (J=9/2) and 8th (J=7/2) excited states of Gd^+ were observed. The Rydberg electron of these series is considered to be nf_j (j=5/2,7/2), since only the coupling between the odd electron for $\ell \geq 3$ and the odd ion core state for these series can produce the even-parity AI state of J=1.

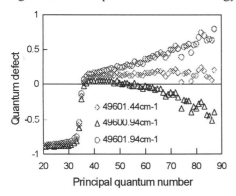

FIGURE 2. Quantum defect for the $(nf_{7/2})$J=1 series converging to the 2nd excited state of Gd ion calculated with three different limits.

CROSS-SECTION AND IONIZATION SCHEME

To determine highly efficient ionization schemes, we have been measuring photo-absorption cross-sections of many transitions. The methods utilized are the branching ratio measurement using the time-resolved pump-probe technique and the saturation cross-section measurement. A typical measured saturation curve of an ionization transition is shown in Figure 3. For this measurement, the 1st and 2nd step were sufficiently saturated, and each laser pulse was separated by 20 nsec. Several efficient schemes were selected among about 200 candidates by calculating the scheme cross-sections defined as follows (5) ;

$$\sigma_s = \frac{1}{(\sqrt{\frac{1}{\sigma_1}} + \sqrt{\frac{2}{\sigma_2}} + \sqrt{\frac{3}{\sigma_3}})^2} \qquad (1)$$

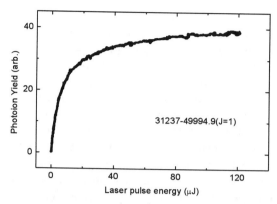

FIGURE 3. Typical saturation curve of photoion yield for an autoionization transition.

Some of the most efficient schemes are listed in Table 1. In this table, some of the efficient schemes having J=2→1(2) →1→ 0 applicable to the laser isotope separation based on the polarization selection rules are also listed (marked with asterisks). Since all of these AI states have narrow linewidths in comparison with our laser bandwidth of about 6 GHz, these cross-sections are expected to become larger with decreasing laser bandwidth. These schemes are highly efficient so that they can be used in the laser isotope separation and the trace analysis of gadolinium.

Table 1. Efficient photoionization schemes (cross-sections are in $10^{-16}(cm^2)$)

E_1	E_2	E_3	E_4	J_1	J_2	J_3	J_4	σ_1	σ_2	σ_3	σ_σ	remarks
533.0	18083.6	34811.8	50052.5	4	5	6	7	626	57	1416	13.4	
999.1	18070.3	34623.9	50052.5	5	6	6	7	282	60	835	11.0	
0	17228.0	34601.7	49994.6	2	1	0	1	90	23	85	2.9	
0	17380.8	34586.7	50624.7	2	2	1	0	73	21	119	2.9	※
0	17380.8	34586.7	52135.3	2	2	1	0	73	21	32	1.9	※
0	17380.8	34697.7	49885.4	2	2	1	0	73	35	15	1.6	※

REFERENCES

1. Haynam C. A., Comaskey B. J., Conway J., Eggert J., Glaser J., Ng E. W., Paisner J. A., Solarz R. W., Worden E. F., *Laser isotope separation*, SPIE vol.1859, 24-36 (1993).
2. Guyadec E. Le, Ravoire J., Botter R., Lambert F., Petit A., Opt. Comm. **76**, 34-41 (1990).
3. Miyabe M., Wakaida I., Arisawa T., J.Phys. B to be published
4. Worden E.F., Solarz R. W., Paisner J. A., Conway J. G., J. Opt. Soc. Am. **68**, 52-61 (1978).
5. Miyabe M., Wakaida I., Akaoka K., Ohba M., Arisawa T., *Resonance Ionization Spectroscopy 1992*, Bristol: IOP publishing, 1993, pp.139-142.

Angular Dependence of Photoelectrons from Sr $5p_{1/2}ns$ Autoionizing States

C. J. Dai and J. Lu

Department of Physics, Zhejiang University, Hangzhou 310027,
People's Republic of China

Abstract. The angular dependence of ejected electrons from the Sr $5p_{1/2}ns$ autoionizing states have been studied using three-step excitation scheme. Energy levels and widths of the $5p_{1/2}ns$ states with n ranging from 7 to 11 are measured. The electrons originated from the decays to different ionic states are resolved with the time-of-flight detector.

There have been numerous experiments devoted to the $mpns$ autoionizing states of the alkaline-earth atoms using three-step photoexcitation in recent years (1-3). Since the initial state of the excitation to the $mpns$ state is the spherically symmetric $msns$ Rydberg state, which represents the simplest target, angular distributions of electrons ejected from the $mpns$ states have attracted much attention. However, due to experimental difficulties, the lowest members of the $mpns$ series remain untouched. In this report, we have carried out an experimental study on the angular dependence of photoelectrons ejected from lowest members of Sr $5p_{1/2}ns$ series with n ranging from 7 to 11.

The details of the apparatus and approach have been described previously (4), so we only recount the major features here. The collimated atomic beam of Sr is crossed by three pulsed dye-laser beams perpendicularly between a plate and a grid 1cm apart inside a vacuum chamber with an environment of 1×10^{-3} Pa. The dye lasers are pumped by harmonics of the same Nd:YAG laser, and have the pulse energies of 100μJ and linewidths of 0.5cm^{-1}. To assure that laser polarization is linear and can be rotated freely, they pass through a double-Fresnel rhomb and a polarizer before entering the vacuum chamber. Sr atoms are excited to a $5sns$ $^1S_0$ state via the $5s5p$ $^1P_1$ intermediate state by the first two lasers. Then the third laser is scanned in the vicinity of the $5s_{1/2} \rightarrow 5p_{1/2}$ transition of the Sr$^+$ ion core, producing the $5p_{1/2}ns$ autoionizing state.

In order to achieve the angular dependence of ejected electrons from an autoionizing $5p_{1/2}ns$ state, a time-of flight (TOF) detecting system, which consists of a 10 cm-long aluminum drift tube mounted directly above the interaction region and a microchannel-plate (MCP) detector, is required to provide TOF energy analysis. A $5p_{1/2}ns$ state may decay into either the Sr$^+$ $4d$ or the Sr$^+$ $5s$ continua. The electrons ejected from the two decays have different kinetic energies, which can be resolved by the TOF detecting system. To reduce the possible effects of the residual magnetic field on the electrons, the whole TOF detecting system was enclosed by a μ-metal magnetic shield. The acceptance angle of the detector from the interaction region is defined by the diameter of the entrance aperture of the detector, 1.8 cm, and the free drift length, 10 cm. In order to assure a field-free flight region, the flight tube and magnetic shielding are grounded.

Either of the ions and the electrons resulting from the excitations and subsequent decay of the autoionizing states may be detected by the MCP detector. To observe the transition profile, it is convenient to detect ions by applying a 100V electric pulse to the plate about 0.5μs subsequent to the laser pulses, driving any ions to the detector. The output of the detector is fed into a boxcar integrator, from which a computer connected with the setup displays and stores the data through an analog-to-digit converter (ADC) for further analysis.

As the TOF detector is fixed in position, the angular dependence of ejected electron is obtained by rotating the rhomb, which changes the angle of linear polarization of the incident third laser beam relative to the detection axis of the TOF detector. Note that in this particular case the polarization of light used to prepared the $msns$ Rydberg atoms remains unchanged during the experiment due to the simplest symmetry of the $msns$ $^1S_0$ state. The dependence is measured in a plane perpendicular to the direction of the propagation of the laser beams and containing the mutually perpendicular atomic beam and detection axis.

Now let us examine the results of the angular dependence of ejected electrons. It is well known that angular distribution of ejected electrons corresponds to the

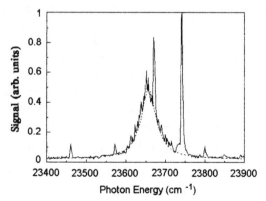

FIGURE 1. The transition profile of the $5p_{1/2}11s$ state. The profile from the best fit (dashed line) is superimposed on the experimental one.

TABLE 1. Energies, Widths and Quantum Defects of the Sr $5p_{1/2}ns$ States

n	Energy (cm^{-1})	Quantum defect	Width (cm^{-1})
7	61077.7	3.42	439.3
8	64443.0	3.41	260.0
9	66146.5	3.40	125.9
10	67130.7	3.40	81.6
11	67738.6	3.42	47.2

differential cross section for the photoionization to a given ionic state i and can be expressed as

$$\frac{d\sigma_i}{d\Omega} = I(\theta) = \frac{\sigma_i}{4\pi}\left[1 + \beta_i P_2(\cos\theta)\right] \quad , \tag{1}$$

where σ_i is the partial cross section in channel i integrated over Ω, β_i the asymmetry parameter and θ the polar angle between the laser polarization and the momentum of the ejected electron (axis of the detector). $P_2(\cos\theta)$ is the second order of Legendre polynomial.

The ion signal, which does not depend on θ, gives the total cross section, i.e., the transition profile of the autoionizing state. We have measured the transition profiles of the $5p_{1/2}ns$ states with n ranging from 7 to 11. The observed profiles are approximately Lorentzian and exhibit only one dominant feature. We have fit the observed spectra to the Lorentzian profiles, an example of which is shown in Figure 1 for the spectrum of the $5p_{1/2}11s$ state. Good agreement between them implies that the Fano interference effect (5) in our experiment is indeed negligible. The energies and widths of the states, which can be extracted from the observed profiles by fitting them to the Lorentzian profiles, are tabulated in Table 1.

To demonstrate the angular dependence of ejected electrons, we have measured several $I(\theta)$ for each $5p_{1/2}ns$ state. An example of these measurements is shown in Figure 2 for the $5p_{1/2}9s$ state. In the figure, the spectra are uniquely distinguished by the angle θ. Namely, from the top to the bottom panel, the spectra correspond to ion signal, electron signals with $\theta = 0°$, 30°, 60° and 90°, respectively. Obviously, the figure provides information in two aspects: first, for each energy, the angular dependence of ejected electrons is apparent; second, the angular distribution, or β parameter, varies strongly with energy. However, the value of β parameter can be determined at each energy. For instance, at the peak of the $5p_{1/2}9s$ resonance, $\beta \approx 1.5$. Examination of the spectra reveals that angular distribution of ejected electrons may provide more information than autoionization spectrum itself. This is expected since the spectrum itself depend only on excitation amplitudes while angular distribution depends not only excitation amplitudes but also on the continuum phase. Furthermore, autoionization spectrum provides no information

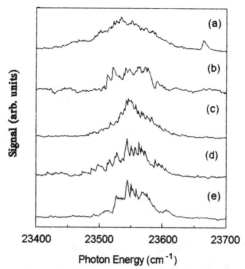

Figure 2. Angular dependence of ejected electrons from the $5p_{1/2}9s$ state. The spectra are (a) ion signal, (b) $I(0°)$, (c) $I(30°)$, (d) $I(60°)$ and (e) $I(90°)$, respectively.

on the final states of decays, while kinetic energies of ejected electrons may yield such information. On the other hand, a theoretical analysis of the experiment will be helpful. In recent year, multichannel quantum defect theory (MQDT) has been combined with R-matrix calculations to predict or analyze the experimental data on both spectrum (6) and angular distribution(7), and reaches satisfactory agreement with experiments. Such kind of comparison certainly will enhance our understanding on autoionization process.

ACKNOWLEDGMENTS

The authors thank S. F. Hu for her help in the experiment. This work has been supported by the National Natural Science Foundation of China, the State Commission of Education of China, and the Cao Kwang-Piao Foundation of advanced Science & Technology.

REFERENCES

1. Zhu, Y., Xu E. Y., and Gallagher T. F., *Phys. Rev.* **A36** 3751-3767(1987).
2. Hieronymus H., et al, *Phys. Rev.* **A41** 1477-1488 (1990).
3. Lindsay M., et al, *Phys. Rev.* **A45** 231-241 (1992).
4. Dai C. J., *Phys. Rev.* **A52** 4416-4424 (1995).
5. Fano U., *Phys. Rev.* **124** 1866-1874 (1961).
6. Dai C. J., *Phys. Rev.* **A51** 2951-2956 (1995).
7. Lindsay M., Dai C. J., Cai L. T., and Gallagher T. F., *Phys. Rev.* **A46** 3789-3806(1992).

Isotope Shift Measurement of Autoionization Rydberg States of Sm

Yi-qun Lu,Ma Jun,J.S.Zhang,Y.C.Zhang,M.G.Gao,and Y.H.Liu

Anhui Institute of Optics&Fine Mechanics,Academia Sinica
P.O.Box 1125,Hefei,Anhui,230031,PRC

Abstract. The observation and measurement of isotope shift of autoionization Rydberg states $^8G_{1/2}$ of Sm was reported.The isotope shift of the first excited state of Sm was also measured. Results were compared with that in Ref.[5].

INTRODUCTION

Autoionization states,which are advantageous to applications in atomic vapor laser isotope separation(AVLIS),rare element detection,etc.,have been studied for a long time[1-5].The isolated core excited method (ICE) used in Ref.[4] is an effective way to study the Rydberg states of atoms.In this report,we report the measurement of isotope shift of autoionization Rydberg state of Sm by using ICE method.Results about the isotope shift of the first excited state of Sm were compared with N.V.Harlov's and similarities were found.

EXPERIMENTAL PRINCIPLE AND SETUP

The energy levels of Sm is illustrated in Fig.1

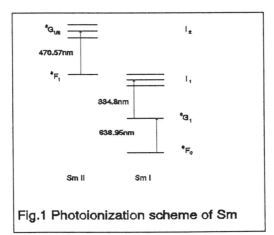

Fig.1 Photoionization scheme of Sm

The Sm atom is excited to its Rydberg states I_1 ($J=0,1,2...$) by a two-step process, one dye laser with the wavelength of 638.95nm is used for $6s^2$ to 6s6p process, while another dye laser of 334.8nm for 6s6p to I_1. The core of Sm can be considered to be isolated since the Rydberg electron is far from it. Then the third laser of 470.5nm can be used to excite the core from its $^8F_{1/2}$ to $^8G_{1/2}$. When the third laser is tuned, the Sm isotope ions can be detected by a quadrupole mass filter.

The experimental setup is shown in Fig.2,

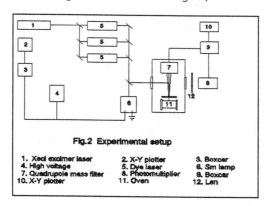

Fig.2 Experimental setup

1. Xecl excimer laser	2. X-Y plotter	3. Boxcar
4. High voltage	5. Dye laser	6. Sm lamp
7. Quadrupole mass filter	8. Photomultiplier	9. Boxcar
10. X-Y plotter	11. Oven	12. Len

three dye lasers, say, DCM dye for 638.95nm, Ox-720 dye for 334.8nm, and LD-473 dye for 470.57nm, are pumped by a XeCl excimer laser (Lambda Physik EMG 100). The linewidth of each laser is about 0.02 cm^{-1}. A hollow cathode lamp of Sm is used to calibrate laser frequencies. The fluorescence of the first excited state of Sm is detected by a photomultiplier, and the resonance ionization signal is detected by a quadrupole mass filter. The Sm sample is heated to 1100k electrically in a graphite oven ($\phi 6 \times 20$mm), vaporized there and ejected from a nozzle (d=1.5mm).

EXPERIMENTAL RESULTS

(1). Isotopic shift of the first excited state of Sm

Two dye lasers, say, λ_1=638.95nm and λ_2=334.8nm, were used in the experimental measurement. λ_1 was scanned in order to observe the isotopic shift of the first excited state of Sm. The fluorescence, optogalvanic and quadrupole mass spetroscopy results are shown in Fig.3.

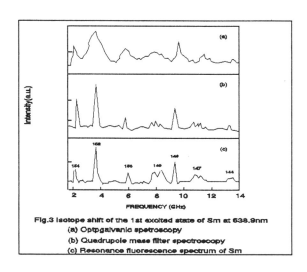

Fig.3 Isotope shift of the 1st excited state of Sm at 638.9nm
(a) Optogalvanic spectroscopy
(b) Quadrupole mass filter spectroscopy
(c) Resonance fluorescence spectrum of Sm

Our results here are similar to Karlov's[Ref.[5]],and are listed in table 1.

Table 1. Isotopic shift of Sm I

Excitation Wavelength	Isotopic shift					
	144-147	147-148	148-149	149-150	150-152	152-154
638.95nm	0.013Å	0.016Å	0.015Å	0.013Å	0.028Å	0.020Å

(2). Isotopic shift of the autoionization Rydberg state of Sm

Three dye lasers,638.95nm with a linewidth of 0.8Å,334.8nm,and 470.57nm with a linewidth of $0.02cm^{-1}$,were used in the experiment. The third laser was scanned,and the Sm isotope ions were detected by the quadrupole mass filter.Results are shown in Fig.4

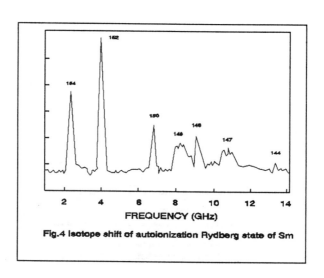

FREQUENCY (GHz)

Fig.4 Isotope shift of autoionization Rydberg state of Sm

Table 2 is the isotopic shift of the autoionization Rydberg state of Sm.

Table 2. Isotopic shift of autoionization Rydberg state of Sm

Excitation Wavelength	Isotopic shift					
	144-147	147-148	148-149	149-150	150-152	152-154
470.5nm	0.015Å	0.018Å	0.011Å	0.016Å	0.019Å	0.017Å

CONCLUSION

The isotopic shift of autoionization Rydberg state of Sm was measured and reported in this paper.Our results may find their applications in AVLIS and other areas related.

REFERENCE

1. P.Sattelberger,et.al.,Resonance Ionization Spectroscopy
   1990,in <u>Proceedings of the Fifth International Symposium on</u>
   <u>Resonance Ionization Spectroscopy and Its</u>
   <u>Application</u>,Varese,Italy,Vol.16-21,1990,pp.239-242.
2. V.S.Letokhov,and V.I.Mishin,in <u>Laser Spectroscopy</u>
   <u>VIII,Proceedings of the 8th International</u>
   <u>Conference</u>,Sweden,1987,pp.167.
3. W.E.Cooke,et.al.,Phys. Rev. Lett.,40,178,(1978).
4. P.Camus,et.al.,Resonance Ionization Spectroscopy
   1990,in <u>Proceedings of the Fifth International Symposium on</u>
   <u>Resonance Ionization Spectroscopy and Its</u>
   <u>Application</u>,Varese,Italy,Vol.16-21,1990,pp.215-218.
5. N.V.Karlov,et.al.,Appl. Opt. 17,856(1978).

Observation of autoionization states by time-resolved optogalvanic spectroscopy

Jongmin Lee, Do-Young Jeong and Yongjoo Rhee

Laboratory for Quantum Optics, Korea Atomic Energy Research Institute, P. O. Box 105, Yusong, Taejon, 305-600, Korea

Abstract. An effective method for the observation of the atomic autoionization states in a hollow cathode discharge is realized by laser optogalvanic spectroscopy. The temporal profile of the optogalvanic signal induced by three-color three-step photoionization is divided by two different components. One is a fast arising signal which corresponds to the pure photoionization, and the other is a slow signal induced by the conventional optogalvanic mechanism. By resolving the fast and slow signal in the temporal evolution, the autoionization signals are clearly extracted from the any other optogalvanic signal including bound-bound state transitions.

Introduction

The laser optogalvanic spectroscopy has known to be a very powerful tool for the study of atomic structures and interactions between light and atoms[1]. The hollow cathode discharges are also effectively used as an atomic vapor source as well as a spectroscopic detector, especially for the refractory materials such as gadolinium, zirconium etc[2,3]. Its application, however, has been restricted in the study of bound-to-bound state transitions below the ionization potential. This is because it is not easy to analyze the complex optogalvanic spectrum which contains many different types of transitions such as bound-bound transitions of a target element, buffer gas transitions, photoionizations etc.

In this work, the technique of laser optogalvanic spectroscopy has been effectively applied to the study of the photoionization via autoionization states in atomic gadolinium. The autoionization spectrum has been clearly extracted from the any other optogalvanic signals with no relation to photoionization processes. It is based on the fact that the autoionization signals are much faster than the conventional optogalvanic signals due to bound-bound state transitions below the ionization potentials[4,5].

Experiment

Three dye lasers (Lumonics Inc. HD-500) pumped by a excimer laser (Lambda Physik LPX300) are utilized to induce three-color three-step photoionization of gadolinium atoms. The first and second dye lasers are tuned to produce the two-step excitation from ground state to $34586.7\ cm^{-1}$ via $17380.8\ cm^{-1}$. And the third ionizing laser is scanned to get the autoionization spectrum of gadolinium atoms.

The dye lasers have beam diameter of $1\ mm$ and spectral linewidth of $1\ GHz$. The two dye lasers for the two-step transition from ground state are combined by a partial mirror and arranged to counter-propagate colinearly with the ionizing laser in the center of a see-through hollow cathode lamp(Cathodeon Inc. 3QQAY/Gd, clear aperture 2 mm). Optogalvanic signals are detected across the dc cut-off capacitor (0.1 microfarad) followed by a 50 ohm resistor.

Results and discussion

The typical temporal evolutions of optogalvanic signals are shown in Fig. 1. When the first dye laser is tuned to excite ground state atoms of Gd to $17380.8\ cm^{-1}$, the optogalvanic signal has only a slow component. When the second dye laser is employed to produce the transition from $17380.8\ cm^{-1}$ to $34586.7\ cm^{-1}$, the optogalvanic signal has two components, the fast and slow ones (Fig. 1-(a)). The latter is identified as a signal induced by bound-to-bound transitions followed by ionization through collisions with electrons, ions etc. The former is identified as a direct photoionization signal by two-color three-photon ionization.

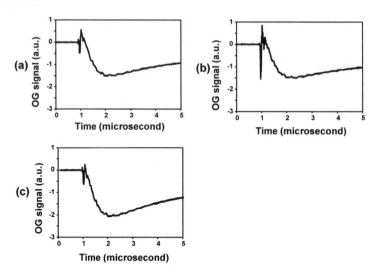

Fig. 1. Time profiles of three-color three-step optogalvanic signals : (a) two-color OG signal : $0 \to 17380.8 \to 34586.7\ cm^{-1}$ (b) OG signal when the ionizing laser is tuned to a autoionization state : $0 \to 17380.8 \to 34586.7 \to 50627.4\ cm^{-1}$ (c) OG signal when the ionizing laser is tuned to a transition line of argon atoms (buffer gas)

If the ionizing laser is tuned to excite atoms resonantly, two possible phenomena can be observed, due to the fact that the discharge cell contains only Gd and Ar. The first case is when the laser is tuned to the transition between the excited state $(34586.7 \ cm^{-1})$ and a certain autoionization state of Gd. In this case, the fast signal is greatly enhanced while the latter slow signal is not, as shown in Fig. 1-(b). On the contrary, if tuned to the other transition lines, the fast signal does not change, while only the slow signal increases (Fig. 1-(c)). The fast signal has about 15 ns temporal width which is comparable to the laser pulse width of 10 ns.

Fig. 2 shows the two different types of two-color optogalvanic spectra obtained at the same scanning of the second laser. Fig. 2-(a) comes from the fast signal, while Fig. 2-(b) from the slow signal. Comparing the relative strengths of two transition lines in slow and fast signals, it can be noticed that the 581.18 nm transition line (denoted by ** in the figure) is attributed to the near resonant two-photon transition from $17380.8 \ cm^{-1}$ to a autoionization state while 581.03 line (denoted by * in the figure) to single photon resonance two-photon ionization to the continuum state.

Three-color optogalvanic spectra with the excitation scheme of $0 \rightarrow 17380.8 \rightarrow 32660.8 cm^{-1} \rightarrow A.I. \ state$ are shown in Fig. 3. Fig. 3-(a) is evidently attributed to the autoionization states of Gd, by comparing to the earlier work by Guyadec et al[6]. Fig. 3-(b) comes from the slow signal and it is composed of the bound-to-bound transitions and partially contributed by the autoionization transitions because autoionization signals also have the small slow signal in its tail.

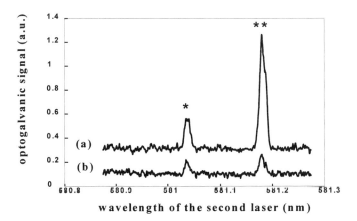

Fig. 2. Two-color optogalvanic spectra obtained at the same scanning of the second laser : (a) spectrum obtained from the fast signal (phoionization spectrum) (b) spectrum obtained from the slow signal (bound-bound transition)

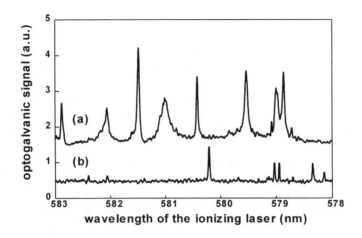

Fig. 3. Optogalvanic spectra obtained at the same scanning of the ionizing laser : (a) spectrum obtained from the fast signal (autoionization spectrum) (b) spectrum obtained from the slow signal

Conclusions

By resolving the fast and slow signals in the temporal evolution of the optogalvanic signal, the photoionization signals are effectively extracted from the conventional optogalvanic signals. This technique is a very convenient and powerful method for the study of photoionization processes, especially via autoionization states of atoms.

References

1. G. Chevalier, J. M. Gagne, and P. Pianarosa, J. Opt. Soc. Am. B 5, 1492 (1988)
2. G. Chevalier, J. M. Gagne, and P. Pianarosa, Opt. Comm. 64, 127 (1987)
3. H. -D. Kronfeldt, G. Klemz, and D. Ashkenasi, Opt. Comm. 110, 549 (1994)
4. M. Broglia, F. Catoni, A. Montone, and P. Zampetti, Phys. Rev. A. 36, 705-714 (1987)
5. F. Babin and J. -M. Gagne, Appl. Phys. B 54, 35-45 (1992)
6. E. Le Guyadec, J. Ravoire, R. Botter, F. Lambert, and A. Petit, Opt. Comm. 76, 34-41 (1990)

Atomic Lifetimes in High-lying Strontium 5snf ^3F_J States by Collinear Resonance Ionization Spectroscopy

P. Müller, G. Bhowmick, W. Nörtershäuser, K. Wendt

Institut für Physik, Johannes Gutenberg-Universität Mainz, D-55099 Mainz, Germany

Abstract: Measurements of the atomic lifetime of high-lying 5snf ^3F_J states with n=20, 23 and 32 in the spectrum of Sr I have been performed in the 5s4d ^3D_J - 5snf ^3F_J transitions using collinear resonance ionization spectroscopy. By rapidly chopping the laser light the population of the excited Rydberg state and hence the ionization signal of the fast moving atoms becomes a function of the radiative decay of both the initial and the excited states in the transition under study. A detailed analysis of the resulting time-dependence of the count rate yields experimental results the radiative lifetimes for the atomic levels involved.

INTRODUCTION

Apart from the primary application in isotope-selective ultra-trace analysis the technique of collinear resonance ionization spectroscopy is known to be an efficient tool for various atomic physics studies in high-lying Rydberg states [1,2,3]. Already the collinear excitation in a fast atomic beam renders a number of rather unique experimental advantages, which clearly favour this technique for high resolution studies on atomic physics parameters [4]:

(i) linewidths in the order of the natural linewidth are produced by the Doppler-reduction of optical resonances in collinear geometry,

(ii) the excitation of all atoms in the beam results in high sensitivity and hence allows to study weak transitions or weak hfs components and

(iii) possibility of optical excitation from metastable states (populated in the charge exchange process) gives direct access to various transitions into high-lying excited states with different angular momenta L and J without using a multitude of lasers.

Furthermore the combination of collinear excitation with field ionization and particle detection is especially well suited for the study of high-lying Rydberg states, guaranteeing high detection efficiency and low background.

At Mainz university we have carried out spectroscopic measurements in the atomic spectrum of the alkaline earth element strontium, which partly serve as preparatory work for our ultra-trace analysis applications [5]. In addition the two valence electrons alkaline earth atoms have been subject of a large number of theoretical and experimental studies on their atomic structure up to high Rydberg series [6,7,8 and references therein]. Strontium is thus an ideal test candidate for theoretical descriptions of the atomic structure. While theoretical estimates for oscillator strengths, atomic lifetimes and excitation cross sections have been determined for practically all low and medium excited states of Sr I, experimental data on members of the 5snf ^3F_J configurations are mainly limited to the lowest 5s5f configuration. We have already published an extensive study of the highly perturbed hfs in different 5s4d ^3D_J - 5snf ^3F_J transitions in ^{87}Sr, rendering insight into the configuration mixing of these excited states with high L [3]. In this paper we report on the extension of this work to the direct measurement of lifetimes, which can be used to extract transition probabilities. Measurements have been performed for 5snf ^3F_J states with n = 20, 23 and 32 and different J.

EXPERIMENTAL

The study was carried out at the 60 keV mass separator and collinear resonance ionization facility at the university of Mainz, which has already been described in detail elsewhere [9]. After charge exchange the resulting fast Sr atomic beam is superimposed in anti-collinear geometry with the 363.8 nm radiation from a single mode argon ion laser. The measurements involve pulsed excitation by rapidly switching the laser light on and off with Pockel cells. Light pulses with rectangular profile of 10 µs duration and 3 kHz repetition rate were used to excite the atomic beam via the 5s4d ^3D_J --5snf ^3F_J transition for n=20, 23 and 32. Precise Doppler frequency tuning onto the desired transition is carried out through variation of the particle velocity of the fast atomic beam in the range of 12 to 33 keV. A simplified sketch of the experimental arrangement is given in Fig. 1.

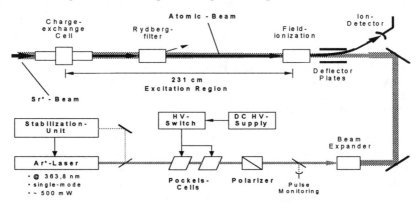

FIGURE 1.   Experimental Set-up for lifetime measurements by collinear resonance ionization spectroscopy

There is about 2 m of overlap between atomic beam and laser beam, starting at the exit of the charge exchange cell up to the location of the field ionization electrodes, which act as detector for the population of the excited state. The pulsed laser excitation results in a corresponding distribution of excited atoms along this beam path, which are travelling with a speed of about 30cm/µs after shut down of the laser pulse to the detector. Apart from the desired parameters of excitation cross section and excited state lifetime, this distribution is depending on the laser power and divergence as well as the finite lifetime of the individual initial state 5s4d ^3D_J with J = 1, 2 or 3. In addition it is altered by a number of electrostatic fields along the beam path, which are needed for background reduction especially in the trace analysis applications. Nevertheless the excited state population is an exact finger print of the light-to-atom interaction, monitoring the interplay between excitation probability and radiative decay as a function of time.

ANALYSIS AND RESULTS

Typical experimental results, i.e. resonance ion count rates as a function of delay time to the leading edge of the laser light pulse, are compiled for all five transitions under study in Fig. 2a. The curves show a pulse structure with a length of about 13 μs, which is the sum of the laser pulse length and the atomic flight time through the interaction region. The delay of about 7 μs relative to the laser pulse is caused by the flight distance of about 1.5 m between the ionization electrode and the final ion detector. The leading edge of the signal pulse is produced by atoms at the end of the excitation region interacting with the light directly after switching-on, while the right shoulder and falling edge is caused by interactions shortly before the switching-off of the light with those atoms just leaving the charge exchange cell.

Thus the rise gives primarily information on the initial state lifetime while the fall encounters dominantly the excited state lifetime. The more or less pronounced hump on the left side of the peak represents increasing transition probability and gives information on the excitation cross section. A complete numerical description of the pulse structure has been set up in a computer routine based on an incoherent rate equation theory, which describes induced and spontaneaus transitions between initial and final state as well as the different decay channels of both states. Introducing the machine dependent parameters like laser power and divergence, length and location of interaction regions and ion current density, a perfect ab-initio description of the pulse structure is achieved. This can be used to yield results for the lifetimes by incorporating a numerical fitting routine. As an example of the quality of the description a comparison between a typical fitted curve and the experimental peak shape of the 5s32f $^3F_3$ state is given in Fig. 2b. A precise completely coherent approach to the light-atom interaction using density matrix equations has been tested in addition. It does not yield a better description, which is due to the short coherence time of the gas laser of much less than 1μs.

DISCUSSION

Results for the initial and final state lifetimes are compiled in Table 1 together with theoretical expectations. For the members of the low lying 5s4d ^3D_J configurations

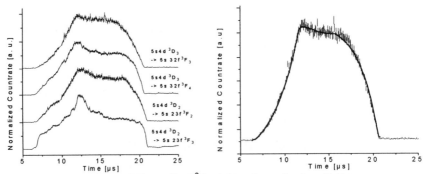

Figure 2. a: Population of different 5snf ^3F_J states after pulsed excitation,
b: Comparison between experiment and theoretical fit for the 5s32f $^3F_3$ state

we have also included experimental values from [10], which are about a factor of two larger than theory and our results. The lifetimes of the high lying 5snf ^3F_J states have been determined experimentally for the first time. Their are compared with rough estimates using the simple n^{*3} extrapolation, which are strongly depending on the uncertainty in the 5s4d state lifetime from [11]. For a detailed description, new ab-initio calculations for 5snf states, which would yield theoretical lifetime expectations, are required.

TABLE 1: Experimental results for lifetimes of 5s4d- and 5snf states

State	τ [µs] this work	τ [µs] Theory [Ref.]	τ [µs] Exp [Ref.]
5s4d $^3D_2$	1.9(4)	2.1 [10]	4.4(8) [10]
5s4d $^3D_3$	2.0(3)	2.5 [10]	3.8(9) [10]
5s23f $^3F_2$	4.6(+2.4/-1.5)	7.5(3.8) [11]	
5s23f $^3F_3$	4.2(+1.8/-1.7)	7.5(3.8) [11]	
5s32f $^3F_3$	11(+9/-4)	20(10) [11]	
5s32f $^3F_4$	15(+13/-7)	20(10) [11]	

REFERENCES

[1] K. Wendt, RIS92, Inst. Phys. Conf. Ser., No. 128, 87 (1992)
[2] S. Kunze et al., Z. Phys. D27, 111 (1993)
[3] B.A. Bushaw et al., Z. Phys. D28, 275 (1993)
[4] R. Neugart, in „Progress in Atomic Spctroscopy", part D (eds. H.J. Beyer and H. Kleinpoppen), p. 75, Plenum Press, New York (1987)
[5] K. Wendt et al., Contribution to this issue and references therein
[6] M. Aymar, Physics Rep. 110, no. 3, 163 (1984)
[7] R. Beigang, J. Opt. Soc. Am. B5, No. 12, 2423 (1988)
[8] T. F. Gallagher, *Rep. Prog. Phys.* **51**, 143 (1988).
[9] L. Monz et al., Spectrochimica Acta, 48B, 1655 (1993)
[10] E. N. Borisov, N. P. Penkin, T. P. Redko, *Opt. Spektr.* **63**, 673 (1987).
[11] H. G. C. Werij et al., *Phys. Rev. A* **46**, 1248 (1992).

An Investigation of Saturation and Power Broadening Effects in Resonance Ionization Spectroscopy

O. Rhodri Jones, Christopher J. Abraham and Helmut H. Telle

Department of Physics, University of Wales Swansea, Singleton Park, Swansea, SA2 8PP, UK.

Abstract. Experiments were carried out to study saturation and power broadening effects in the resonance ionisation spectroscopy of Al using a two-step one-colour scheme. Both saturation and broadening can be modelled with good accuracy using a reduced density matrix approach. This confirmed that at low laser fluences the variation of the RIS signal with detuning from resonance is governed by the laser linewidth, while at high fluences power broadening dominates.

INTRODUCTION

In RIS/RIMS it is frequently argued that for absolute quantification of results the transitions involved in the laser ionisation schemes should be driven into saturation. In principle, this allows one to relate ion signals to analyte particle densities in a relatively straight-forward way. However, in many cases the laser fluence required for saturation leads to power broadening of the resonant transition which may cause a reduction in the selectivity of the RIS process. In this work we compare experimental RIMS results to theoretical predictions calculated using a reduced density matrix approach which includes all the relevant experimental parameters. The effects of saturation and power broadening are discussed for the element aluminium, and in particular for RIS schemes utilising the $3s^2 3p\ ^2P_{1/2}{}^0 \rightarrow 3s^2 3d\ ^2D_{3/2}$ transition at $\lambda = 308.215$nm.

THE THEORETICAL MODEL

We consider the general case of single photon resonance excitation followed by ionisation. The density matrix approach used for the numerical calculations is presented here in matrix form with the inclusion of magnetic sub-levels:

$$\frac{d}{dt} \mathbf{Y}(M) = \hat{A}(M) \cdot \mathbf{Y}(M) \tag{1}$$

where matrix $\hat{A}$ is given by

$$\hat{A}(M) = \begin{pmatrix} 0 & A_{21} & 0 & 2B_p & 0 \\ 0 & -w_{22} & 0 & -2B_p & 0 \\ 0 & 0 & -w_{12} & -x & 0 \\ -B_p & B_p & x & -w_{12} & 0 \\ 0 & \gamma_2 & 0 & 0 & 0 \end{pmatrix} \qquad \mathbf{Y}(M) = \begin{pmatrix} P_{11} \\ P_{22} \\ R_{12} \\ I_{12} \\ P_{ion} \end{pmatrix} \tag{2}$$

with R_{12} and I_{12} representing the real and imaginary parts of the off-diagonal density matrix elements, P_{ion} representing the population ionised and P_{ii} representing the level populations. A_{21} is the radiative decay rate from sub-state $|2,M\rangle$ to $|1,M\rangle$ and x represents the detuning of the laser frequency from resonance. The term denoted by B_p is equal to half the Rabi frequency of the excitation step and is given by $E_P \mu_{12} / 2\hbar$, where E_P is the amplitude of the electric field and μ_{12} is the electric dipole moment of the transition. w_{22} is the total loss rate from the resonant level given by $1/\tau_2 + \gamma_2$, with τ_2 being the lifetime of the resonant level and γ_2 the rate of ionisation from this level. The off diagonal density matrix relaxation term is denoted as $w_{12} = w_{22} / 2 + \gamma_{LP}$, where γ_{LP} is the laser linewidth. A full derivation of this set of equations is given by Jones (1).

EXPERIMENTAL SET-UP

Experimental RIMS signals were obtained using a UHV, sputter initiated, time-of-flight, reflectron mass spectrometer in conjunction with a 20Hz Nd:YAG laser pumped dye laser system. The output pulse from the dye laser was measured to be 8ns in duration, with a linewidth of ~10GHz, and was frequency doubled before being focused into the analysis chamber. By attenuating the pump beam, the output energy of the frequency doubled dye laser radiation could be varied between 10^{-4} mJ and 0.15 mJ, while preserving the position of the focal spot above the sample. This gave UV laser fluences in the interaction region in the range $0.7\text{-}10^3$ mJ$\cdot$cm^{-2}.

All the experiments were carried out on a target made from the alloy INCONEL 718®, containing 0.5% aluminium. For each experiment the detector channel plate voltage was adjusted to allow investigation of the largest possible laser energy or wavelength variations, without saturating the detector.

RESULTS AND DISCUSSION

Experimental results of resonant ionisation of Al at 308.215nm, as a function of laser fluence, are given in Figure 1, along with the corresponding theoretical curve. All the RIMS data were collected using 1000 laser pulses and a sputter ion current of 0.7μA delivered in 5μs pulses. Each mass spectrum obtained was integrated to give an accurate indication of the total number of ions detected.

The theoretical prediction is seen to fit the experimental results quite well. It can be seen that a laser fluence of nearly 1000 mJcm^{-2} is required to saturate this resonant ionisation scheme when using the UV radiation for both excitation and ionisation. The experiment was repeated for the two-step, two-colour scheme using 308.215nm for excitation and the fundamental dye laser radiation at

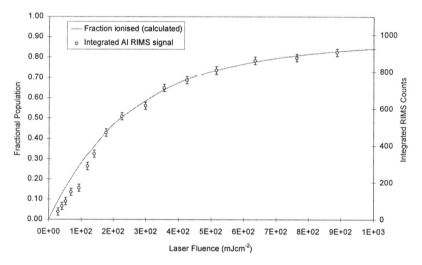

FIGURE 1. Saturation curve for Al excitation and ionisation using 308.215nm radiation

616.431nm for ionisation. With this scheme near-saturation of the ionisation step was achieved using only 100mJcm^{-2} of fundamental dye laser radiation. This enhancement in ionisation efficiency is the result of the larger photo-ionisation cross section of the excited level at such wavelengths (2).

In order to investigate the broadening of this transition with increasing laser power, a wavelength scan was carried out over the 308.215nm resonance at a UV laser fluence of both 0.8mJcm^{-2} and 16mJcm^{-2}. The corresponding results are shown in Figure 2. Both cases were modelled using the density matrix formalism outlined above. In the calculations it was assumed that the laser pulse had a

Gaussian temporal profile, while the spectral lineshape was taken to be a Lorentz function. A close agreement is once again seen between the experimental results and the theoretical predictions. It can be seen that considerable power broadening occurs even in the low fluence case, when compared to the laser linewidth of ~0.003nm.

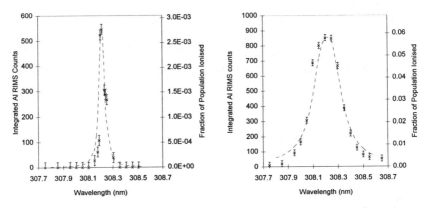

FIGURE 2. Power broadening of the 308.215nm resonant step in Al. (a) 0.8mJcm^{-2} ; (b) 16mJcm^{-2}

CONCLUSIONS

The saturation and power broadening of a one-colour, two-photon process has been successfully modelled using a density matrix approach. For such schemes it is demonstrated that the high laser fluences required to achieve efficient ionisation lead to a large broadening of the resonant transition. This will lead to a considerable decrease in the selectivity of the RIS process which may have implications when isobaric interferences are present. Such considerations are also of particular importance for isotope enrichment processes and isotope ratio measurements (3,4).

REFERENCES

1. Jones, O.R., *Ph.D. Thesis*, University of Wales Swansea (1996).
2. Saloman, E.B., *Spectrochimica Acta.* **46B**, 319-378 (1991).
3. Jones, O.R., Perks, R.M. and Telle, H.H., to be published in *Rapid Commun. Mass Spectrom.*
4. Perks, R.M., Jones, O.R. and Telle, H.H., to be published in *Rapid Commun. Mass Spectrom.*

Photodetachment Resonance in Aluminum*

B. J. Davies, C. W. Ingram, and D. J. Larson

University of Virginia

ABSTRACT

Photodetachment from Al⁻ (3s²3p² ³P) near the threshold to the first excited state of neutral aluminum has been studied. A 19 keV mass resolved Al⁻ beam was intersected by a frequency-doubled Nd:YAG-pumped dye-laser beam, and the fast atoms created by detachment processes were detected with a channeltron-based detection system. Just below the threshold of the first excited state ($3s^24s$ ²P), a large resonance peak was observed in the detachment signal which is believed to be due to the presence of a doubly-excited, autodetaching negative ion state best described as $3s^24snp$. A two-electron R-matrix calculation by Liu and Starace generates resonance structure which is qualitatively similar to the data.

*Supported in part by the NSF

INTRODUCTION

Observation of the properties of doubly-excited autoionizing states in multi-electron systems can provide detailed information on electron correlation and stringent experimental checks for multi-electron theory. The intrinsic nature of negative ions allows an excellent opportunity to observe electron-electron interactions which are not masked by stronger electron-nucleus forces. Several studies of doubly-excited negative ion states have been undertaken by observing the resonance structure in the total detachment cross section which is produced by these states. The most extensive work has been done on the closed s-subshell ions of hydrogen [1] and various alkalis [2]. Studies of p-subshell ions have thus far been limited to Si⁻ [3] and B⁻ [4].

THE ALUMINUM ION/ATOM SYSTEM

The configuration of the ground state of the negative ion is best described as $3s^23p^2$ ³P. The electron affinity of aluminum has recently been measured to be 3556.4 (+5.3/-3.9) cm⁻¹ [5]. This small electron affinity combined with the low energy of the first several excited states allows single photon detachment in the vicinity of the first four excited states using readily available dye-laser technology.

In this experiment we have explored the relative cross section near the threshold for the first excited atomic state. This state, at an energy of 25347.7 cm^{-1}, has a configuration of $3s^24s$ ^2S [6]. Since no direct one-photon, one-electron transition is possible between the ground state negative ion and this excited atomic state, any increase in the cross section at this threshold, if present at all, should be rather small. This feature allows us to observe resonance structure without the added complication of a threshold rise.

RESULTS

Photodetachment data near the threshold for the lowest excited state of the neutral atom are shown in Figure 1. The location of the threshold is indicated by the arrow at 28904 cm^{-1}. The large peak in the data about 175 cm^{-1} below the threshold is strong evidence for a doubly-excited, autodetaching negative ion state associated with the $3s^24s$ atomic level. The dipole selection rule combined with the configuration of the parent state makes $3s^24snp$ its most likely configuration. The peak is slightly asymmetric, which is consistent with the usual characteristics of Feshbach resonances.

The peak in the data is about twice the level of the signal away from the resonance structure. However, some fraction of the baseline observed in the data results from the non-resonant detachment of those aluminum ions which are initially in the ^1D state. In addition, the observation of ion current at all the masses near that of aluminum suggests the possible presence of 27 amu molecular ions as well. The signal from photodetachment or photodissociation of these ions could be significant. Since the contribution from such detachment is difficult to determine, the ratio of the peak height to the baseline signal must be assumed to underestimate the value for Al$^-$ ($3s^23p^2$) detachment.

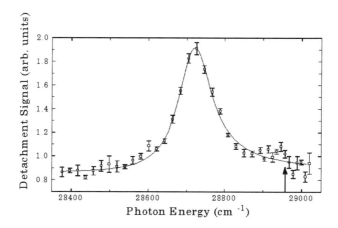

FIGURE 1. Data near the $3s^24s$ Al* state with a fit to a Beutler-Fano profile shown for comparison.

The structure observed can be reasonably well parameterized by a simple Beutler-Fano profile with an added baseline, written as

$$S = A\left[(q+\varepsilon)^2/1+\varepsilon^2\right] + B$$

where S is the neutral atom signal, A is the value of the resonant component of the signal far away from the resonance, and B is the non-resonant background signal. Here q is the lineshape parameter and ε is a scaled energy parameter,

$$\varepsilon = \left(E_p - E_0\right)/\tfrac{1}{2}\Gamma$$

where E_p is the photon energy, E_0 is the location of the resonance, and Γ is its energy width. The result of a Beutler-Fano profile fit to the data is shown in Figure 5, along with the data. The best fit parameters are found to be q = 12.4 ± 2.7, Γ = 108 ± 6 cm^{-1}, and E_0=28717 ± 3 cm^{-1}.

The need for the fit parameter B is not only a result of signal from any non-aluminum detachment present. The fine structure splitting of the ground state aluminum atom can itself provide a baseline signal for the Beutler-Fano profile by introducing multiple continuum channels into the detachment process. Fano showed that such multiple continua simply add a baseline to the resonance profile [7]. Describing the structure observed with a single profile ignores this and other complications. Thus a single Beutler-Fano profile is used here simply as a parameterization of the data.

347

CONCLUSIONS

We have observed resonance structure in the photodetachment cross section of Al$^-$ near the threshold to the first excited atomic state. The doubly-excited negative ion state responsible for the resonance structure has been tentatively identified as having a $3s^24snp$ configuration. A two electron R-matrix calculation by Liu and Starace produces structure which is qualitatively similar to the data, and supports the identification of the doubly-excited state.

Work is underway to calibrate the data presented here on an absolute scale by comparison with the known cross section of O$^-$. Preliminary observations near the next three excited states (^4P, ^2D, ^2P) have been made. There are some indications of resonance structure near the ^2D and ^2P thresholds, and work continues near these states.

- The authors would like to thank J. E. Thoma for his technical assistance.

REFERENCES

[1] see, for example, P. G. Harris, H. C. Bryant, A. H. Mohagheghi, R. A. Reeder, H. Sharifian, C. Y. Tang, H. Tootoonchi, J. B. Donahue, C. R. Quick, D. C. Rislove, W. W. Smith, and J. E. Stewart, *Phys. Rev. Lett.*, **65**, 309 (1990).

[2] see, for example, U. Berzinsh, G. Haeffler, D. Hanstorp, A. Klinkmuller, E. Lindroth, U. Ljungblad, and D. J. Pegg, *Phys. Rev. Lett.*, **74**, 4795 (1995).

[3] P. Balling, P. Kristensen, H. Stapelfeldt, T. Andersen, and H. K. Haugen, *J. Phys. B*, **26**, 3531 (1993).

[4] P. Kristensen, H. H. Andersen, P. Balling, L. D. Steele, and T. Andersen, *Phys. Rev. A*, **52**, 2847 (1995).

[5] A. M. Covington, R. W. Marawar, D. Calabrese, J. S. Thompson, and J. W. Farley, to be published.

[6] C. E. Moore, Atomic Energy Levels, Vol. 1, Circular of the National Bureau of Standards 467, 1958.

[7] U. Fano, *Phys. Rev.*, **124**, 1866 (1961).

Abnormal branching ratios in laser-excited Rydberg series of Yb$^+$

X. Y. Xu, R. C. Zhao, W. Huang, C. B. Xu, P. Xue

Department of Modern Applied Physics, Tsinghua University, Beijing 100084, China

Abstract. Using five dye lasers excitation, the Yb$^+$ Rydberg series of $np_{1/2}$, $np_{3/2}$ have been achieved. And abnormal branching ratios of $np_{1/2}$:$np_{3/2}$ excited from Yb$^+$ $7s_{1/2}$ are found. In the relativistic central-field independent-particle approximation, oscillator-strength distributions of Yb$^+$ are calculated. The calculated results are in good agreement with our experiment on Yb$^+$. We have found that the abnormal branching ratios are owing to the appearance of Cooper minima.

INTRODUCTION

The Rydberg series of an ion are more difficult to be realized than those of a neutral atom because of the high ionization threshold. Thus the systematic experimental study was scarce on ionic Rydberg series [1]. Theoretically there is also short of calculation on ionic Rydberg states to compare with existent experimental results.

On the other hand, the ionic Rydberg state is a special double Rydberg state where one electron is at $n = \infty$ and therefore the correlation of two electrons vanishes. It can be studied as the spectator of double Rydberg states where the correlation effects of the electrons are strong.

In our experiment, we have studied the Rydberg series of Yb$^+$ $np_{1/2}$, $np_{3/2}$ and found the branching ratios of $np_{1/2}$:$np_{3/2}$ excited from $7s_{1/2}$ are abnormal. Based on relativistic central-field independent-particle approximation [2], we have studied the oscillator-strength distributions and branching ratios of Yb$^+$ Rydberg states.

EXPERIMENT

In our experiment the excitation of $np_{1/2}$, $np_{3/2}$ of Yb$^+$ Rydberg series are achieved by using five dye lasers (all linear polarized). In the first step Yb atoms in ground state are excited to an intermediate state $6s6p(^1P_1^\circ)$.

Then the Yb^+ ions in ground state is produced via an autoionization state which is excited by the second laser. The third and forth lasers excite Yb^+ from ground state to Yb^+ $7s_{1/2}$ state through $6p_{1/2}$. Finally, the fifth laser (linewidth is $0.3cm^{-1}$) is scanned over an energy range from $41638cm^{-1}$ to $43863cm^{-1}$ to excite Yb^+ $np_{1/2}$, $np_{3/2}$ series. After five lasers excitation, the Yb^+ ions in the Rydberg states are field-ionized to Yb^{2+} ions, then the ions travel a distance of about one meter before striking a dual-microchannel plate (MCP) detector.

From the spectra of $np_{1/2}$, $np_{3/2}$ series, the ionization threshold of Yb^+ can be deduced [1]. And the principal quantum number n and the quantum defect for each energy level can be obtained.

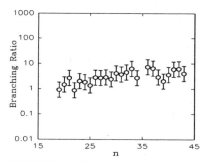

FIGURE 1. The branching ratios of $np_{1/2} : np_{3/2}$ excited from Yb^+ $7s_{1/2}$.

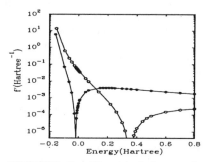

FIGURE 2. The calculating results of the differential oscillator strengths of Yb^+ $\epsilon'p_{1/2}$(• dots), $\epsilon'p_{3/2}$(○ dots) series excited from $7s_{1/2}$ vs. ϵ'. Notice that near $\epsilon_{3/2}^{CM}$, the branching ratio of $\epsilon'p_{1/2} : \epsilon'p_{3/2}$ is far larger than 1:2.

In experiment we have measured the line intensities of the spectra. In Fig. 1, the branching ratios of $np_{1/2}$:$np_{3/2}$ versus n from 19 to 43 are given (lack of n=35). The results are obtained by integrating the area of the resonance peaks. Because the linewidths of the peaks are basically that of the last dye laser, we neglect the influence of the saturation effects. It is obvious that the branching ratios of $np_{1/2}$:$np_{3/2}$ apparently deviate from the ratios of the angular part integrals in their oscillator strengths (which is the same as the ratios of their statistic weights, i.e., 0.5 for the excitation of $s \rightarrow p$ with linear polarized laser). As can be seen in Fig. 1, the ratios are about $1 \sim 10$.

THEORETICAL ANALYSIS

The relativistic Dirac-Slater self-consistent field [2] is a kind of independent-electron average field. All the electrons in an atom can be assumed to move in the same central field independently. The relativistic one-electron wavefunction can be written as,

$$\Psi_{n\kappa m} = \frac{1}{r} \begin{pmatrix} G_{n\kappa}(r) & \Omega_{\kappa jm}(\theta, \varphi) \\ iF_{n\kappa}(r) & \Omega_{-\kappa jm}(\theta, \varphi) \end{pmatrix}, \tag{1}$$

where $G_{n\kappa}(r)$ and $F_{n\kappa}(r)$ are the large and small components of radial wave function respectively. The $\Omega_{\kappa jm}$, $\Omega_{-\kappa jm}, \kappa$, j and l are the same as [2]. Ψ satisfies Dirac self-consistent equation,

$$\left(-\frac{d}{dr} + \frac{\kappa}{r}\right) F_{n\kappa}(r) - [E_{n\kappa} - 1 \quad V(r)]G_{n\kappa}(r) \tag{2}$$

$$\left(\frac{d}{dr} + \frac{\kappa}{r}\right) G_{n\kappa}(r) = [E_{n\kappa} + 1 - V(r)]F_{n\kappa}(r) \tag{3}$$

where the potential $V(r)$ is the function of the electronic wave functions, therefore it must be solved on the self-consistent condition.

With considering the core to be unrelaxed, the dipole matrix can be reduced to the integral of only the wave functions of the outer electron in the initial and final state [3], so the differential oscillator strength (orientation averaged for initial state) can be written as

$$f' = \frac{df_{n\kappa,\epsilon'\kappa'}}{d\epsilon'} = \frac{2}{\Delta E(2j+1)} \sum_{mm'} |\langle n\kappa m \mid \alpha \cdot A_p^1 \mid \epsilon'\kappa'm' \rangle|^2, \tag{4}$$

where $\alpha \cdot A_p^1$ is the electric dipole operator [4] and ϵ' is the orbital energy of the outer electron on the final state (a continuum state for $\epsilon' > 0$ and a bound state for $\epsilon' < 0$).

Figure 2 shows the calculated results of the differential oscillator strengths f' for Yb$^+$ $\epsilon'p_{1/2}, \epsilon'p_{3/2}$ series excited from $7s_{1/2}$. Apparently, there are absorption cross section minima which is usually called Cooper minima[3]. We have calculated quantum defects μ of $\epsilon'p_{1/2}, \epsilon'p_{3/2}$ series and found $\mu_{1/2}$ is a little larger than $\mu_{3/2}$ because the the spin-orbit force is repulsive for $j = l+1/2$ and attractive for $j = l - 1/2$ [5]. Both calculated results of $\mu_{1/2}$ and $\mu_{3/2}$ at the threshold (3.41 and 3.35 respectively) are in good agreement with the values derived from the spectra (3.42 and 3.36 respectively) [1]. And quantum defect of Yb$^+$ $7s_{1/2}$ state (3.83), is also in good agreement with the experimental result (3.84) . For the transitions of $7s_{1/2} \rightarrow \epsilon'p_{1/2}$ and $\epsilon'p_{3/2}$, the differences of the quantum defects between the initial and final states (at threshold) are 0.42 and 0.48. At ϵ_{CM} of $\epsilon'p_{1/2}$ series the nodes of the wave function of the $\epsilon'p_{3/2}$ series locate a little outside those of the $\epsilon'p_{1/2}$ series. With ϵ' increasing the nodes will move in towards the nucleus, therefore the Cooper minimum of $\epsilon'p_{3/2}$ series occurs at a little higher energy above that of $\epsilon'p_{1/2}$ series (Fig. 2). Clearly despite of the accurate energy points where Cooper minima occur (ϵ^{CM}), the energy range of our experiment is located between the two Cooper minima, and more accurately at a small domain near the crossing point of two curves. In this energy range the branching ratio will apparently deviate from the statistic weight 0.5.

It should be point out that in our experimental range, for the transitions from $7s_{1/2}$ to $np_{1/2,3/2}$, the differential oscillator strengths are very small($f'_{1/2} < 0.01$ and $f'_{3/2} \leq 0.001$ in Fig. 2) . Our experimental setup has a high excitation efficiency for ionic Rydberg states.

DISCUSSION

We have made the first quantitative measurement of non-vanishing branching ratios in laser excitation of the Yb^+ Rydberg series. Based on the relativistic Dirac-Slater self-consistent field approximation, the branching ratios of the oscillator strengths of the Yb^+ Rydberg states can be well explained. The method offers us an effective way to select adequate excitation channels in the future experiments.

Apparently the calculated $\epsilon^{CM}_{1/2,3/2}$ deviates a little from (indeed higher than) the experimental ones, which presumably arises from the neglect of the core-relaxation and the core-polarization effects. In further accurate calculation one may follow the recently developed relativistic multi-channel theoretical method to obtain accurate $\epsilon^{CM}_{1/2,3/2}$.

ACKNOWLEDEGMENTS

The authors appreciate the helpful discussions with Prof. J. M. Li and Dr. X. M. Tong. This work is supported partially by the National Natural Sciences Foundation of China, the Chinese Association of Atomic and Molecular Data, SSTCC, Science and Technology Funds of CAEP, and National High-Tech ICF Committee in China.

REFERENCES

1. W. Huang, X. Y. Xu, C. B. Xu, M. Xue, and D. Y. Chen, J. Opt. Soc. Am. B **12**, 961 (1995).

2. J. M. Li and Z. X. Zhao, Act. Phys. Sin. **31**, 97 (1982).

3. J. W. Cooper, Phys. Rev. **128**, 681 (1962).

4. I. P. Grant, Adv. Phy. **19** 1, 747 (1970).

5. S. T. Manson, C. J. Lee, R. H. Pratt,I. B. Goldberg, B. R. Tambe, and A. Ron, Phys. Rev. A **28**, 2885 (1983).

Resonance Ionization Detection of 253.7 nm Photons from Mercury Atoms

O.I. Matveev, W.L. Clevenger, B.W. Smith, N. Omenetto and J.D. Winefordner

Department of Chemistry, University of Florida, Gainesville, FL 32611-7200

Abstract. Avalanche detection of a laser enhanced ionization (LEI) signal has been studied in a resonance ionization detector (RID) cell containing mercury vapor at room temperature. Single photoelectron events were detected with an avalanche multiplication factor of more than 8000.

The detection of photons *via* resonance ionization has been the subject of many studies in recent years(1-6). Experimentally, the limit of detection (LOD) has been limited by preamplifier noise and shot noise of the background current in the atom reservoir to a level of 10^3 resonance photons during the laser pulse(7). There remains a gap of several orders of magnitude between these experimentally achieved results and the ultimate theoretically estimated LOD(1,3,7).

In this work, a novel type of Hg RID cell was developed to detect single photo-electron events after avalanche ionization of a buffer gas in the cell. Previous work on mercury (5,6,8) has shown the usefulness of several ionization schemes. Mercury was chosen as the atomic vapor for several reasons. At room temperature, the vapor pressure of mercury is high enough to absorb more than 90% of resonance photons over an optical path length of several cm. In addition, there are several important potential practical applications of a Hg RID.

EXPERIMENTAL

The experimental setup included an excimer laser (Questek. Model 2110. Billeriica. MA.) which was used to simultaneously pump two dye lasers (DL II, Molectron Corp. Santa Clara, Calif.). The first dye laser was tuned to the Hg transition $6^3P_1^0 \rightarrow 7^3S_1$ (λ_2=435.8 nm) and the second dye laser to the transition $7^3S_1 \rightarrow 13\,^3P_1^0$ (λ_3=489.0 nm). The dimensions and pulse duration of first and second laser beams were 3×4 and 2×3 mm and 8 and 11 ns respectively. The pulse energy was 0.2 mJ for the first laser and 1.5 mJ for the second laser beam. The laser beams were directed into the quartz RID cell (25 mm diameter and 55 mm length) between a plane nickel electrode and a nichrome wire of diameter 0.27 mm. The distance between the electrodes was 10 mm. The high voltage power supply (Series 205B, Bertan Associates, Inc.) was connected to the plane electrode and the nichrome wire was connected to the signal detection system, consisting of a home built charge sensitive preamplifier (capable of detecting 1500 electrons per pulse) and a digital oscilloscope (Tektronix TDS 620A). Radiation from a cw Hg discharge lamp (SCT-1, Spectronic Corp, Westubry, NY) was collected by a lens and directed with a pierced mirror into the RID cell. In order to estimate the quantum efficiency and the resulting sensitivity of the RID, it was necessary to measure the absolute power of the lamp in 4π steradian. These measurements were done using a sensitive laser power meter (OPHIR model 10A) with a Keithley 182 digital voltmeter. The total photon flux of the lamp at $\lambda=253.7$nm was estimated to be 5 mW or 6.4×10^{15} quanta/s. Spectral isolation of the resonance radiation of the lamp was accomplished using an interference filter with known transmittance.

RESULTS AND DISCUSSION

In earlier experiments (8), it was shown that 87% of narrow band laser radiation was absorbed in a 1 cm length cuvette filled with saturated mercury vapor at room temperature. Since the temperature of the Hg discharge lamp used here was much higher than the room temperature, its emission line was much broader than the halfwidth of the absorption line in the RID cell. To estimate the absorption efficiency, at the resonance transition, of the RID cell filled with saturated mercury vapor, the cell transmission was monitored as a function of the buffer gas pressure. The results of these measurements indicated that less than 65% of the lamp resonance radiation was absorbed by the cell, confirming that the halfwidth of the emission line is broader than the absorption profile of the atoms in the cell.

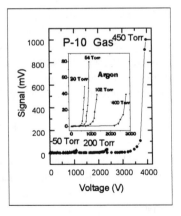

Figure 1. Signal vs. Voltage.

To find the optimal conditions for the detection of very small quantities of charged particles in the RID cell, the dependence of the signal versus pressure for argon and for P-10 gas (10% methane in argon) was studied. The results are shown in Figure 1. For these measurements, instead of the cw Hg discharge lamp, a dye laser tuned to the resonance Hg line at 253.7 nm was used. A clear avalanche effect was observed for both gases and all pressures. However, in the case of pure argon at low pressure, an acceptable signal level could not be observed due to the occurrence of electrical breakdown when the voltage was increased above a certain level. The maximum multiplication factor, M=8000, was attained only for the P-10 gas. For all other cases, M ranged from 100 to 300.

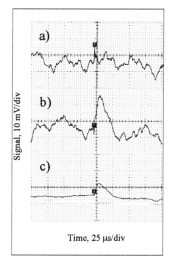

Figure 2. Oscilloscope traces.

With the lasers tuned to λ_2 and λ_3 the signal obtained with the cw Hg lamp source was measured as a function of the electrode voltage. The addition of λ_3 increased the signal by 12 times and the maximum signal to noise ratio was found to be more than 10^4. When the lamp was placed at a distance of 2.5 m from the RID cell behind the pierced mirror (in this case, there was no direct optical imaging of the lamp into the detector and only randomly diffused radiation was detected), a very small signal was observed. Typical averaged and real-time signal waveforms are shown in Figure 2. Trace 2a shows the background noise (with no averaging) with the Hg lamp turned off. With the Hg lamp turned on, a clearly distinguishable pulse (Fig. 2b) was observed on approximately every fourth laser pulse. The averaged signal (Fig. 2c, 72 pulses averaged) was three times less than without averaging. The amplitude of the real time signal exceeded the noise level by a factor of 5. Based upon the detection limit of our preamplifier (1500 electrons) and the avalanche gain of 8000, we can assume that the signal observed can be interpreted as a single-electron event. The working voltage during these measurements was 3.92 kV.

The efficiency of ionization, η, i.e., the ratio between the number of charges created and the number of excited Hg atoms in the $6^3P_1^0$ state, of the RID cell was also evaluated using two approaches(12). η was estimated to be 0.1 ± 0.05 and 0.12 ± 0.6, both of which are self consistent and validate the reasonableness of the conclusions regarding single electron event detection.

Analyzing the data from Figs. 2b and 2c about the signal and noise level for a single photo- electron event, one can easily estimate the limiting number of mercury resonance quanta which can be detected after averaging. Comparing the typical amplitude values shown in Fig. 2 and extracting the level of noise, one can estimate that in our experiment the S/N ratio was of the order of 22 for the averaged measurement. The reciprocal product of $(S/N\times\eta)^{-1} = N_{min}$ gives a minimal number of photons $N_{min} = 0.5$ which can be detected by the RID in our experiment for a S/N ratio of unity.

Thus, the experimental data obtained show that the mercury RID can potentially be an extremely low noise photon detector and that there is still possibility for further improvement. The most important question which remains to be solved is to further improve the efficiency of ionization of the excited Hg atoms in the RID cell. It is clear that this can be most easily achieved by increasing the available fluence in λ_2 and λ_3.

ACKNOWLEDGEMENTS

The authors gratefully acknowledge support from the United States Department of Energy, Grant number DOE-DE-FG05-88-ER13881.

REFERENCES

(1) Matveev, O. I., *J. Appl. Spectrosc. USSR* **46**, 359 (1987).

(2) Okada,T., Andou, H., Moriyama,Y., Maeda, M., *Opt. Lett.* **14**, 987 (1989).

(3) Smith,B. W. , Farnsworth,P. B., Winefordner, J. D. and Omenetto, N., *Opt. Lett.* **15**, 823 (1990).

(4) Smith,B. W., Omenetto, N. and Winefordner,J. D., *Spectrochim. Acta* 50th Anniversary Issue, 101 (1990).

(5) Ganeev, A. A., Matveev, O. I., Sholupov, S. E., Grigor'yan, V. N., and Slyadnev, M. N., *Optics and Spectrosc. Russ.* **76**, 769 (1994).

(6) Ganeev, A. A., Matveev, O. I., Sholupov, S. E., Grigor'yan, V. N., and Slyadnev, M. N., *Optics and Spectrosc. Russ.* **77**, 197 (1994).

(7) Petrucci, G. A. and Winefordner, J.D., *Spectrochim. Acta.* **47B**, 437 (1992).

(8) Matveev, O. I., Cavalli, P. and Omenetto, N.,"Three-step Laser Induced Ionization of Ir and Hg Atoms in an Air-Acetylene Flame and a Gas Cell," in *AIP Proceedings No. 329 (RIS 94)*, 1994, p. 269.

(9) Letokhov, V. S., *Laser Photoionization Spectroscopy*, New York: Academic Press, 1987, p. 58.

Resonant 2-Photon-Ionization of Xe

M. Meyer*, J. Lacoursière#, L. Nahon*§, M. Gisselbrecht*,
P. Morin*§ and M. Larzillière#

* L.U.R.E., Bâtiment 209D, Université Paris-Sud, 91405 Orsay, France
§ C.E.A., DRECAM, SPAM, CEN Saclay, 91191 Gif-sur-Yvette, France
# L.P.A.M., Département de physique, Université Laval, Québec, Canada, G1K7P4

Abstract. The combination of laser and synchrotron radiation has been used to investigate in a pump-probe arrangement the ionization of Xe atoms via the resonant state Xe*$5p^5 5d$ [3/2]$_1$. In a first type of experiments the synchronization between the pulses of a mode-locked Ar$^+$ laser and the synchrotron radiation has been demonstrated by measuring the lifetime of the intermediate, resonantly excited states. In addition, a tuneable dye laser has been used to excite the Xe* $5p^5 4f$ [5/2]$_2$ autoionization resonance.

INTRODUCTION

The different, partly complementary characteristics of lasers and synchrotron radiation offer new experimental possibilities and some particular advantages when the two light sources are used in combination. The main advantage of synchrotron radiation (S.R.) is the very large energy range of the emitted photons (≈ 1eV $\leq h\nu_{S.R.} \leq \approx 10$keV) and its easy tuneability. Studies of photoexcitation and photoionization in different subshells of many atoms and molecules have been performed during the last years (see for example [1,2,3]. The combination of lasers and S.R. for the study of laser excited atoms was introduced by F. Wuilleumier et al.[4]. In these experiments the high photon flux and the high wavelength resolution of a c.w. ring dye laser was used to excite the outermost electron of an alkaline atom, e.g. Li $1s^2$ 2s + $h\nu_{laser}$ --> Li* $1s^2$ 2p. Up to 30% of the atoms can be pumped in the excited states by the laser radiation. The subsequent photoionization by the S.R. allows for a direct investigation of the complex correlation effects introduced by the change of the orbital of the outer electron.

For studies of highly excited states or even resonant autoionization states the pump-probe scheme has to be inverted and S.R. has to be used for the pump process. Due to the much lower number of photons of the S.R. (about a factor of 10^5 compared to conventional laser sources) the excitation yield is less favourable. In addition, the S.R. has a time structure, i.e. in the case of the SuperACO storage ring in Orsay a short lightpulse (width of about 700ps) is emitted every 120ns. In consequence the temporal overlap (duty factor) between S.R. and a laser can be increased by a factor of 10 to 100 by using a pulsed, mode-locked laser (repetition rate of some MHz) perfectly synchronized to the synchrotron pulses. The first results on a gaseous target have been obtained by J. Lacoursière et al.[5] where a mode-locked Ar^+ laser was used to photoionize He atoms which have been excited to the He*1s3p ^1P level by the S.R.. As a continuation of this work we have presently investigated the two-photon ionization of Xe atoms via the Xe* $5s^2 5p^5$ 5d $[3/2]_1$ resonance.

EXPERIMENTAL

The experiments have been performed at the SuperACO storage ring in Orsay/France. Synchrotron light monochromatized by the 3m normal incidence monochromator (SA63) was used to excite the Xe atoms to the Xe* $5p^55d$ $[3/2]_1$ resonance at $h\nu=10.401eV$ [6]. The energy resolution of the monochromator was determined to $\Delta(h\nu)\approx7.5meV$ (entrance and exit slit were set to 600μm). A detailed description of the synchronization technique is given elsewhere [5]. Briefly, the Radio-frequency of the mode-locker crystal of the Ar^+ laser (Coherent Innova 100-20) was triggered with the master clock of the Radio-frequency of the SuperACO storage ring. In the two-bunch operation mode of SuperACO ($f_{S.R.}=8.32MHz$) precisely 9 laser pulses matched the interpulse period of the S.R. ($f_{laser}=74.9Mhz=9xf_{S.R.}$). The relative delay between the laser and S.R. pulses could be manipulated by a voltage-controlled phase shifter. The temporal width of the pulses have been measured to about 300ps and 700ps for the laser and S.R., respectively. An average power of 1W for the 514nm line of the Ar^+ laser could be achieved in this mode of operation. In second type of experiment a linear c.w. dye laser (Rhodamine 6G) providing an average power of about 400mW in the interaction region was combined with the S.R..

A conventional time-of-flight analyser has been used to detect and to analyse the Xe^+ ions which were produced in the region of intersection of the two lightbeams and an effusive beam of rare gas atoms. With the use of a LiF window (cut-off wavelength at about 1000Å) in the beamline the higher order light of the monochromator has efficiently be suppressed and the experimental spectra could be measured without significant non-resonant background signal .

RESULTS

The pump-probe excitation scheme used for the 2-photon ionization of Xe is displayed in figure 1. The synchrotron radiation ($h\nu=10.4eV$) excites the Xe* $5p^55d$ $[3/2]_1$ resonance lying below the first ionization threshold. The linewidth of this resonance ($\Delta E\approx0.01eV$) mainly determined by Doppler broadening illustrates the poor excitation efficiency obtained with the S.R. ($\Delta h\nu_{S.R.}=7.5meV$). The excited atoms are ionized in a second step by the laser pulse (514nm line of the Ar^+ laser), if the two light pulses are both present in the interaction region within a short time window determined by the lifetime of the Xe*$5p^55d$ state ($\Gamma=600ps$). By changing the delay between the S.R. and the laser pulses the decreasing population of the excited state and the corresponding lifetime can be determined.

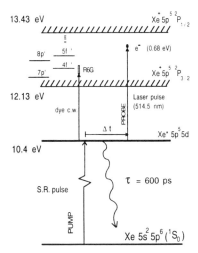

FIGURE 1. Schematic energy diagram illustrating the excitation pathway used to ionize atomic Xe via a resonant 2-photon process.

Figure 2 shows the resulting ionization signal as function of the delay between the two light pulses. The asymmetric lineshape has to be described by a convolution of an exponential function taking into account the population of the excited state and two Gaussian profiles accounting for the temporal width of the S.R. and the laser pulses, respectively. A simulation performed with the known values given above reproduces the experimental data points and demonstrates that we are able to synchronize and to control on a MHz time scale the pulses of a mode-locked Ar^+ laser with respect to the S.R. radiation pulses of the SuperACO storage ring. The weak background which is not affected by the variation of the time-delay between the pulses is probably due to the ionization of the metastable Xe^*5p^56p levels which can be populated by radiative decay of the excited Xe^*5p^55d resonance [7].

Secondly, a c.w. dye laser was used in order to change the wavelength of the ionizing photon. Within the energy range accessible with Rhodamine 6G it was possible to study the Xe^*5p^54f autoionization resonance (cf. figure 1). The observed ionization signal as function of the laser wavelength is given in figure 2 (left-hand side). The obtained asymmetric lineprofile is characteristic for an autoionizing resonance and can be described by the formalism developed by Fano [8]. The determined Fano parameter of the resonance (q=7.3) differs from the value obtained in earlier experiments by Rundel et al. [9] (q=-16.4). This difference is partly related to the fact that the two-photon excitation process used in our experiment imposes clearly the following transitions:

$$Xe5p^6\ ^1S_0 + h\nu_{SR} \rightarrow Xe^*5p^55d\ [3/2]_1 + h\nu_{las} \rightarrow Xe^*5p^54f\ [5/2]_2,$$

whereas transitions leading to j=2 and j=3 final states are possible, when starting from the metastable Xe^*5p^56s $[3/2]_2$ level [9]. A more detailed characterisation of the initial as well as the final state, which have to be described by multi-configuration interaction calculations, is necessary to get further insight in the experimentally obtained spectra.

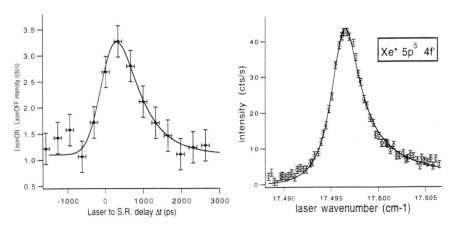

FIGURE 2. (left-hand side) Two-photon ionization signal of Xenon as a function of the delay between the laser and the S.R. pulses. (right-hand side) Two-photon ionization signal of xenon as a function of the wavelength of the ionizing laser.

CONCLUSION

In the present work we have demonstrated the synchronization between the pulses of a mode-locked Ar^+ laser and the S.R. of the SuperACO storage ring in Orsay. This new technique enables us to study short-lived states by maximising the overlap between the two light sources and to determine lifetimes of these excited states. In a more general sense the method allows to perform time resolved measurements sensitive to dynamical effects on a nanosecond timescale. The combination of a tuneable dye laser and the S.R. used for the excitation of the Xe^*5p^54f' resonance has shown that, in spite of the small number of excited species, it is possible to perform energy resolved spectroscopy on excited states which have been prepared by the S.R..

REFERENCES

1. Schmidt, V., Rep.Prog.Phys. 55, 1483 (1992)

2. Sonntag, B., and Zimmermann, P., Rep.Prog.Phys. 55, 911 (1992)

3. Hitchcock, A. P. and Mancini, D.C., J.Electr.Spectr. 67, 1-132 (1994)

4. Wuilleumier, F.J., Cubaynes, D., and Bizau, J.M. , Ann.Physique 17, C1 (1992)

5 Lacoursière, J. , Meyer, M., Nahon, L., Morin, P., and Larzillière, M., Nucl.Instrum.Meth. Phys.Res. A351, 545 (1994)

6. Radzig, A.A., and Smirnov. B.M., *Reference Data on Atoms, Molecules and Ions,* Springer Series in Chemical Physics. Vol. 31 (Springer Verlag, Germany), (1985)

7 Meyer, M., Nahon, L., Lacoursière, J., Gisselbrecht, M., Morin, P., Larzillière, M., J.Electr.Spectr.Relat.Phenom. (1996) in press

8. Fano, U., Phys.Rev. 124, 1866 (1961)

9. Rundel, R.D., Dunning, F.B., Goldwire, H.C., Stebbings, R.F.,,J.Opt.Soc.Am 65,628 (1975)

^{89,90}Sr-Determination in Various Environmental Samples by Collinear Resonance Ionization Spectroscopy

K. Wendt, B.A. Bushaw, G. Bhowmick, V.A. Bystrow[#], N. Kotovski[#],
J.V. Kratz, J. Lantzsch, P. Müller, W. Nörtershäuser, E.W. Otten,
A. Seibert[*], N. Trautmann[*], A. Waldek[*], Y. Yushkevich[#]

*Institut für Physik, *Institut für Kernchemie, Johannes Gutenberg-Universität, D-55099 Mainz,
#Joint Institute for Nuclear Research, Dubna, Russia*

Abstract: The application of Collinear Resonance Ionization Spectroscopy for the determination of ^{89,90}Sr contaminations in various environmental samples is presented, demonstrating the versatility of this highly specialized and isotope-selective ultra-trace determination technique. Additionally to these studies we have analysed the different processes, which lead to nonselective ionization and hence cause background. Optimization of the experimental conditions resulted in a significant lowering of the detection limit to a value of 3×10^6 atoms of ^{90}Sr (corresponding to an activity of 2mBq) in the presence of up to 10^{17} atoms of stable ^{88}Sr. In a parallel measurement trace determination of both isotopes ^{89}Sr and ^{90}Sr is possible with a detection limit for ^{89}Sr of 10^8 atoms. Measurements on different samples, ranging from air filter samples of 1958 up to today, soil, vegetation, milk and human urine samples as well as cooling water from nuclear power plants document the applicability and the specifications of the technique.

INTRODUCTION

The development and first application of Collinear Resonance Ionization Spectroscopy in an on-line combination to a 60 keV mass separator for the fast and sensitive detection of the radiotoxic isotopes ^{89}Sr and ^{90}Sr in the environment has already been presented earlier [1,2,3]. In this paper we report on further applications and refinements of this highly specialized RIMS technique, which enabled the isotope-selective ultra-trace analysis of ^{89,90}Sr in various environmental and technical samples. For further improvement of the technique we have additionally analysed and separated the different processes, which lead to non-selective ionization and hence cause background. This approach resulted in a significant reduction of the detection limits for both isotopes ^{89,90}Sr.

EXPERIMENTAL

A simplified sketch of the experimental set-up for collinear resonance ionization spectroscopy is given in Fig. 1. Detailed presentations of the technique and the experimental apparatus can be found in [1,2,3]. After a suitable rapid chemical extraction and preparation, usually applying ion exchange column chromatography, the sample is introduced into the ion source of the mass separator. Following efficient surface ionization of strontium in a tantalum-tungsten ion source a mass separated 33 keV strontium ion beam is formed and neutralized in cesium vapour.

The fast strontium atoms are selectively excited by the 363.8 nm light of a narrow band cw argon ion laser, which intersects the atomic beam in quasi-collinear geometry. The highly-excited Rydberg atoms are quantitatively detected after field-ionization and energy selection on a channeltron detector. Measurements of the number of atoms of e.g. ^{90}Sr relative to the known amount of stable Sr are carried out by varying the beam energy and detecting corresponding resonance signals. Background, which sets the detection limits, is produced in this technique by a number of processes:

(i) the direct population of high-lying Rydberg states during the charge exchange process,

(ii) the population of these states by collisions with residual gas molecules along the atomic beam path and

(iii) the contribution of direct collisional ionization processes.

The last contribution (iii) has already been minimized earlier by specially designing the field ionization region [4]. The contribution from (i) could be strongly suppressed by increasing the field strength in filtering capacitors along the atomic beam path to values of up to 22 kV/cm; contribution (ii) required further optimization of the ultrahigh vacuum conditions along the beam path.

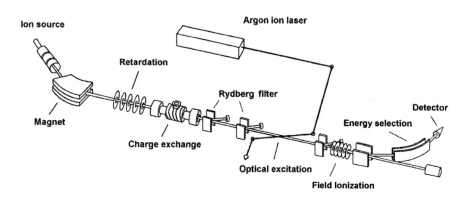

FIGURE 1. Simplified sketch of the experimental apparatus for collinear resonance ionization mass spectrometry

The study was performed on strontium and the isobaric isotopes of zirkonium and yielded relative charge exchange rates into low- and high-lying excited states in the atomic spectrum of strontium as well as precise values for cross sections of exciting collisions between Sr atoms and different residual gas molecules, e.g. H_2, He and Ar [5]. After corresponding optimization of the experimental conditions by installation of additional getter pumps along the excitation region a reduction of the detection limits down to values of 3×10^6 atoms for ^{90}Sr and 5×10^7 atoms for ^{89}Sr in the presence of up to 10^{17} atoms of ^{88}Sr were realized. The lower value for ^{89}Sr is primarily due to the complex hfs splitting of the transition under study, which has been analysed in detail for the stable odd isotope ^{87}Sr [6].

RESULTS

In combination with a refined and simplified chemical preparation procedure, which applies ion exchange column chromatography with commercial crown ethers, trace determination measurements on a variety of different samples were carried out. Aerosol filter samples, stemming from collections 1958 during the time of the above ground nuclear weapon tests, have been analysed. For present-day aerosol samples, taken 1995 in Mainz, very low upper limits for the ^{90}Sr contaminations could be obtained. The successful determination of ^{90}Sr-contents in liquids (e.g. water and human urine), soil, vegetation (grass) and food (milk) shows the wide applicability of this RIMS technique; it is used to extract transfer factors within the human food chain. In Table 1 a compilation of our RIMS results on a number of samples is collected together with comparison values, which have been measured for these samples by different radiometrical techniques.

TABLE 1: Experimental results for the ^{90}Sr content in various sample materials

Sample Material	Collection Location and Date	Sample Size	^{90}Sr Content	^{90}Sr-Comparison Value	Ref.
Aerosol Filter[†]	Heidelberg February 1958	300 m^3	8(4)x10^{-5} Bq/m^3	1.4(3)x10^{-4} Bq/m^3	[7]
Aerosol Filter	Munich April 1986	1,000 m^3	1.2(4)x10^{-3} Bq/m^3	1.18(8)x10^{-3} Bq/m^3	[8]
Aerosol Filter	Mainz February 1995	3x10^4 m^3	<1x10^{-7} Bq/m^3	-	
Soil	Clincy, 1994	1.5 g	3.6(1.1) Bq/g	3.1(5) Bq/g	[9]
Grass, spiked	unknown, 1994	5.5 g	16(3) Bq/g	16(2) Bq/g	[9]
Milk, spiked	unknown, 1994	250 ml	11.2(2.4) Bq/l	12(2) Bq/l	[9]
Human Urine	PTB-Laboratory				
Sample 1	Intercomparison.,	100 ml	191(23) Bq/l	178(37) Bq/l	[10]
Sample 2	1995	100 ml	2.66(60) Bq/l	2.49(38) Bq/l	[10]
Cooling Water	Reactor plant Phillipsburg, 95	200 ml	1.6(3) Bq/l	1 to 5 Bq/l	[11]

† Daily variations in the ^{90}Sr contamination of about a factor of 3 have been detected

The parallel measurement of ^{89}Sr and ^{90}Sr in the same sample has been demonstrated successfully on synthetic samples with detection limits of 1×10^7 atoms for ^{90}Sr and 1×10^8 atoms for ^{89}Sr. These measurements can be applied for the rapid and sensitive analysis of cooling water and low contaminated radioactive waste e.g. from nuclear power or reprocession plants.

CONCLUSION AND OUTLOOK

Collinear Resonance Ionization Spectroscopy in on-line combination with conventional mass spectrometry has shown to be an extremely versatile tool for rapid ultra trace analysis of ^{90}Sr and ^{89}Sr in environmental and technical samples. With the set up presented in this work an overall efficiency of better than 10^{-5} has been demonstrated and suppression of stable strontium with an abundance sensitivity (= isotopic selectivity) of more than 10^{10} is realized for ^{90}Sr. This specifications result in a detection limit of 3×10^6 atoms per sample for ^{90}Sr (2mBq) and 5×10^7 atoms per sample for ^{89}Sr (7 Bq) within a preparation and measurement time of about 6 hours. The applicability of this technique to a wide variety of sample materials has been demonstrated by studying numerous environmental and technical samples. A limitation of the technique is found for materials which contain large quantities of strontium salts, e.g. sea water. Here the selectivity is still not high enough to enable ultra-trace determination of ^{89,90}Sr contaminations.

ACKNOWLEDGEMENT

We thank the BfS and the GSF at Neuherberg, Munich, as well as the Institut für Umweltphysik at the university of Heidelberg for kindly submitting different environmental samples, Prof. Knöchel and his collaborators for chemical preparation of soil, milk and vegetation samples, the DFG and the Deutsch-Indische Gesellschaft for providing support to our international collaborators.

REFERENCES

[1]   K. Zimmer et al., *Appl. Phys.* B59, 1(1994)
[2]   J. Lantzsch et al, *Ang. Chem, Int. Ed.* 34, 181 (1995)
[3]   L. Monz et al., *Spectrochim. Acta*, 48B, 1655 (1993)
[4]   K. Stratmann et al., *Rev. Sci. Instr.* 65, 1847 (1994)
[5]   W. Nörtershäuser et al., *Z. Phys. D*, to be published
[6]   B.A. Bushaw, et a., *Z. Phys. D28*, 275 (1994)
[7]   G. Schumann und G. Eulitz, *Naturwissenschaften* 47, 13 (1960)
[8]   G. Rosner, GSF, private communication (1993)
[9]   J. Alfaro et al., *Angew. Chem. Int. Ed. Engl.* 34, 186 (1995) and priv. comm. (1996)
[10]   E. Günther, PTB Braunschweig, Laboratory Intercomparison Run FS-AKI-RV-Sr-95
[11]   D. Rühle, Reactorplant Phillipsburg, private communication (1996)

Two-XUV-Photon Ionization of Argon
Using a Harmonic Radiation Source

D. Xenakis[*], O. Faucher[†], D. Charalambidis[*‡] and C. Fotakis[*‡]

[*]*Foundation for Research and Technology–Hellas, Institute of Electronic Structure and Laser,*
PO Box 1527, GR-711 10 Heraklion, Greece.
[†]*Laboratoire de Physique, Université de Bourgogne, Faculté des Sciences Mirande,*
BP 138, FR-21004 Dijon, France.
[‡]*Physics Department, University of Crete, Greece.*

Abstract. The third harmonic of an intense sub-picosecond KrF excimer laser was employed in a near-resonant 1 + 1 ionization scheme in argon demonstrating two-photon ionization in the XUV wavelength regime. The implications of high-intensity applications beyond the ultraviolet region utilizing the high peak power of harmonic radiation sources are discussed.

Introduction. When an atomic system is subjected to intense electromagnetic (laser) radiation the response of the system becomes highly non-linear and can produce coherent radiation at multiple harmonics of the incident laser field. Harmonics in the XUV region can be produced with brightness exceeding that obtainable from conventional sources such as synchrotrons (1).The investigation of this phenomenon, known as higher order harmonic generation (HOHG), has been an area of very active research over the past years, both experimentally and theoretically.

Recently, the HOHG process has been utilized to develop novel table-top XUV light sources. The feasibility of such XUV sources for spectroscopic applications has been demonstrated very recently in accurate measurements of relative photoionization cross-sections and autoionization profiles (2). Two properties of such sources stand out in particular due to their unique nature, compared to other, conventional sources: their short pulse duration and the high peak power of the radiation produced. Demonstration of an application of HOHG sources utilizing their high peak power has not been forthcoming. The purpose of this work is to initiate applications of intense laser-field-induced harmonics that do exactly that: reach beyond the limits of conventional XUV sources due to the high peak power of the harmonic source. In particular, a clear signature of that property would be the detection of a *non-linear* process induced by the harmonic photons.

As it is to our knowledge the first attempt to observe a multiphoton process with XUV photons, the demonstration was designed to simplify as much as possible the conditions: i) a two-photon near-resonant ionization scheme was chosen in

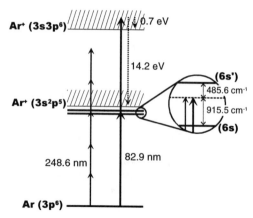

FIGURE 1. The ionization scheme of argon used in this experiment.

order to maximize the cross-section and ii) the third harmonic of the laser beam was used in order to maximize XUV intensity. It should be noted that although the third harmonic is the lowest harmonic that can be produced in an atomic field-free medium, its photon energy of ~15 eV lies in the XUV region due to the 5 eV energy of the photons of the fundamental. Thus it is comparable to higher order harmonics (between the 7th and 15th) of e.g. Ti:Sapphire, Nd:YAG or dye lasers.

Experimental Description. The experiment has been performed using Ar as the target atom. The scheme is shown in figure 1. It involves the two-photon ionization of Ar at 82.9 nm, i.e. the third harmonic of the KrF excimer. The energy of the first XUV photon lies between the 6s'[1/2]$_1$ and 6s[3/2]$_1$ Rydberg states of Ar, which are single-photon allowed, and detuned by 486 and -915 cm^{-1} respectively (3). Absorption of a second XUV photon takes the atom to the continuum 0.7 eV above the first excited state of Ar$^+$ (3s3p^6), so that the decay is to both the ionic ground state as well as to the inner subshell excited state of the ion.

The laser beam came from a hybrid dye-KrF excimer laser system (4) with wavelength 248.6 nm, bandwidth of about 170 cm^{-1}, pulse duration ~450 fs and maximum energy 14 mJ. The XUV source/gas target interaction complex consisted of three interconnected, differentially pumped vacuum chambers maintaining a background pressure better than 10^{-6} mbar. The XUV radiation was generated by focusing the KrF beam to a Xe magnetic pulsed gas jet in the first chamber. The second chamber was a home-made XUV monochromator having as dispersive element a gold-coated concave reflection grating with 600 groves/mm and 400 mm radius of curvature. The XUV radiation was incident on a piezoelectrically pulsed Ar gas jet in the third chamber. The selected geometry was such that the Xe and Ar jets lay on the Rowland circle defined by the grating, thus imaging the XUV to a line into the Ar-jet. A simple time-of-flight mass spectrometer with a microchannel-plate detector was employed for ion detection.

No absolute measurements were performed of the XUV radiation intensity. Upper and lower limits were estimated by measuring the ion signal produced when the third harmonic was interacting with different rare gases such as Xe, Kr and He. No two-photon ionization could be observed in He. Contrariwise, strong single photon ionization was produced in Xe and Kr which could be varied by more than three orders of magnitude by changing the laser intensity. Using the known single-photon ionization cross-sections of Xe and Kr (5), the two-photon ionization cross-section of He (6, 7), the estimated ion extraction volume and the gas pressure in the interaction region, the third harmonic intensity was estimated to be in the range of or 10^5-10^8 W/cm^2 within this interaction region. The scattered light intensity of the fundamental at the interaction region was measured to be of the order of kW/cm^2, which can be considered negligible compared with the third harmonic intensity and thus effectively not contributing to the ionization spectrum.

Results. The results of the Ar ionization experiment are indicated in figure 2. The three spectra in the figure are the time of flight mass-spectra recorded having the THG jet and the Ar jet respectively off and on (a), on and off (b) and both on (c). In case (a) the spectrum is a flat line as expected since XUV radiation is not present in the Ar target chamber. In spectrum (b) two small mass peaks are observable due to the ionization of contaminants (the one at 6 μs is compatible with the mass of H_2O) by the XUV photons. When both jets are operational a strong peak corresponding to the Ar$^+$ mass appears in the spectrum (figure 2c). These results demonstrate clearly the two XUV-photon ionization of Ar.

The Ar$^+$ yield was further measured as a function of the energy of the fundamental laser frequency. The results of this measurement are depicted in a log-log scale as an insert in figure 2. A linear dependence with a slope of three is found, indicating a single XUV-photon process, instead of a slope of six that should be expected for a 2-XUV photon process. This result can be attributed to the saturation of the first step of the ionization process due to the near-lying 6s resonances of Ar. Upon saturation of the first step the ionization yield dependence on the XUV intensity becomes a single XUV-photon dependence. Consequently the dependence on the intensity of the fundamental is defined by the order of the XUV production process which in the present experiment is three.

Conclusion. In conclusion we have observed a near resonant two-XUV-photon ionization in Argon using the intense third harmonic of a sub-ps laser beam generated in an atomic jet. This is to our knowledge the first observation of such a non-linear effect in the XUV wavelength region and was feasible due to the high intensity of the third harmonic employed. As such, the present experiment demonstrates high intensity effects of strong field harmonic generation and thus the complementary role that such a short wavelength light source can play to the conventional XUV radiation sources. The present experiment utilizes a low order harmonic but of reasonably short wavelength. It should be mentioned however that the XUV light focusing conditions in the present experiment have been rudimentary, because of the aberrations in the grating of the monochromator. These conditions and thus the XUV intensity at the focus can be dramatically improved by a few orders of magnitude by using more sophisticated XUV optics. In this case the use of higher order harmonics or non linear processes with smaller cross-sections will be feasible. Finally, since the observed two-photon ionization process is above the first

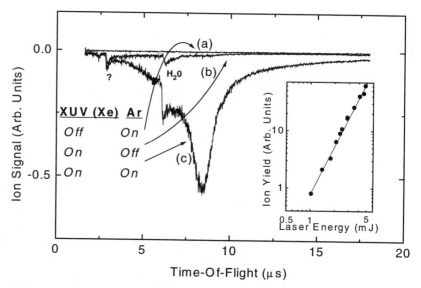

Figure 2. Time-of-flight ionization spectra recorded with the Xe /Ar jets respectively off/on (a), on/off (b) and on/on (c). The inserted picture shows the dependence of the Ar$^+$ yield on the energy of the laser beam. The slope of the line is 3.

excited state of the ion which is on the same time an inner sub-shell excited state, photoelectron- instead of ion-spectroscopy would further allow investigation of the ionization branching ratio. Further experiments demonstrating the above are currently in progress.

Acknowledgments. This work has been carried out in the Ultraviolet Laser Facility operating at FORTH-IESL (HCM contract No. ERB-CHGE-CT920007). It has further been supported by the HCM project with Contract No. CHRX-CT920028. The authors acknowledge discussions with C. J. G. J. Uiterwaal, P. Lambropoulos and E. Cormier, as well as the expert technical assistance of A. Eglezis.

REFERENCES

1. Li, X. F., L'Huillier, A., Ferray, M., Lompré, L., and Mainfray, G., *Phys. Rev.* **A 39**, 5751 (1989).
2. Balcou, Ph., Salières, P., Budil, K. S., Ditmire, T., Perry, M. D., and L'Huillier, A., *Z. Phys D* **34**, 107 (1995).
3. Moor, E., *Atomic Energy Levels* NSRDS-NBS 35, Vol. I, Washington D.C. US Govt. Printing Office (1971).
4. Szatmari, S., and Schäfer, F. P., *Opt. Commun.* **68**, 196 (1988).
5. Berkowitz, J., *Photoabsorption, photoionization and photoelectron spectroscopy,* Academic Press, 1979, p. 176.
6. Starace, A. F., and Tsin-Fu Jiang, *Phys. Rev.* A **36**, 1705 (1987).
7. Cormier, E., *private communication.*

POSTER SESSION II:
MOLECULAR RIS

Multiphoton Ionization of Ion-Beam and Laser Desorbed Molecules from Organic Surfaces

D.E. Riederer, R. Chatterjee, and N.Winograd

Department of Chemistry
The Pennsylvania State University
University Park, PA 16802-7003

Z. Postawa

Institute of Physics
Jagellonian University
ul. Reymonta 4
PL-30059 Krakow 16 Poland

Abstract. The time-of-flight (TOF) distributions of neutral molecules desorbed from a surface using a pulse of ions or photons have been measured using multiphoton ionization. Pyrenebutyric acid molecules sputtered from a thick layer of this material have kinetic energies centered around 0.1 eV. Molecules ejected from a thick layer of tryptophan undergo delayed ejection when either fast ions or photons (266 nm) initiate the desorption event.

INTRODUCTION

The low detection limits afforded by resonant multiphoton ionization provide an ideal way to study desorption processes initiated by energetic probes such as fast ions or photons. Ion beam and laser desorption are the basis behind a number of emerging techniques for organic and biological surface analysis, and understanding the details of molecular ejection has become increasingly important. Detection of sputtered neutrals by laser multiphoton ionization enables separation of the desorption and ionization processes and consequently provides information specific to the molecular ejection event. Time-of-flight (TOF) desorption profiles have been measured for molecules ejected from thick layers of pyrenebutyric acid and tryptophan using keV ions as the desorption probe. The desorption behavior of tryptophan was also investigated using 266 nm radiation.

EXPERIMENTAL

Ion-beam-induced desorption was achieved using 8 keV, 500 ns pulses of Ar^+ focused to a 3 mm diameter spot on the surface. The 266 nm radiation used for laser desorption was generated using the fourth harmonic of a Nd-YAG laser. Postionization of the desorbed neutral molecules was achieved using 280 nm (3 mJ/pulse) radiation produced from the frequency doubled output from a Nd-YAG pumped dye laser. TOF distributions were measured by systematically changing the delay between the probe pulse and ionization pulse. The distance from the surface to the ionizing laser beam was approximately 1 cm. The identity of the postionized species was determined using TOF mass spectrometry. Thick layers of pyrenebutyric acid and tryptophan were prepared as pellets.

RESULTS AND DISCUSSION

TOF distributions obtained for neutral tryptophan molecules desorbed using 8 keV Ar^+ and 266 nm photons (10^4 W/cm^2) are shown in Figure 1. The profiles were recorded for postionized fragments appearing at m/z 130, 116, and 93. No molecular ion was present in the postionization mass spectrum and m/z 130 was found to be the most intense ion regardless of the desorption probe.

Several interesting features are present in these distributions. Unique to the m/z 130 profile generated using the Ar^+ projectile is a rather well defined peak centered at approximately 35 μs. Transformation of the peak to kinetic energy coordinates indicates that these molecules are ejected from the surface with translational energies on the order of 0.1 eV which is in the range expected for a molecular collision cascade.[1] This feature, which is attributed to ejection via a ballistic mechanism is not present in any of the other distributions.

A feature which is common to all of the profiles is a high TOF tail which shows that molecules or molecular fragments continuously desorb from the surface for up to 200 μs after irradiation with ions or photons. Several important points should be noted. The remarkable similarity between the distributions obtained using ions and photons strongly suggests a common desorption mechanism. The fact that delayed desorption is observed indicates that the probe energy is not rapidly dissipated into the solid and is "stored" in a manner which facilitates ejection over an extended period of time. This is in sharp contrast to the behavior of metal surfaces under ion-beam bombardment where events leading to ejection cease approximately 200 fs after ion impact.[2]

Although 266 nm photons and fast ions produce similar TOF distributions for tryptophan, the individual fragment ions recorded have significantly different profiles. The broad profiles of m/z 130 and 116 show no significant reduction in intensity for up to 200 μs after irradiation. In contrast, the intensity of m/z 93 decreases rapidly from 50 to 150 μs. An observation which may be related to this behavior is that the ions at m/z 130 and 116 correspond to the entire amino acid side chain and the indole ring, respectively, and both may be formed by a direct bond cleavage process. However, m/z 93 does not appear to correspond to a direct cleavage and is likely formed via rearrangement. It is presently unknown whether the detected fragments are emitted directly from the surface or are formed during the ionization process.

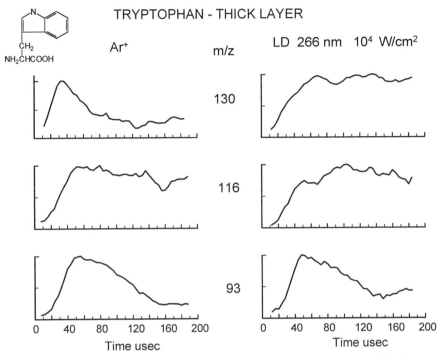

FIGURE 1. TOF profiles for fragments of tryptophan desorbed by 8 keV Ar$^+$ and 266 nm photons.

While mechanistic interpretations are preliminary, two possibilities have been considered. One explanation for delayed ejection is surface heating. Organic solids are poor conductors and localized temperature spikes may persist for some time. This explanation is reasonable in the case of laser desorption in which the 3 mm surface spot is irradiated with 10^{13} photons/pulse. However, in the case of ion bombardment, the low ion dose (10^6 ions/pulse) means that desorption is initiated by single ion events and heating under these circumstances seems less likely. Another possibility is that surface irradiation may induce electronic transitions within the surface molecules. Tryptophan is known to have a long lived triplet state. If this state is populated by transitions initiated by 266 nm photons or fast ions, the energy released upon non-radiationless decay may account for the delayed ejection behavior. Further investigations are underway to explore these possibilities.

The TOF distribution for molecules ejected from a thick layer of pyrenebutyric acid upon 8 keV Ar$^+$ bombardment is shown in Figure 2. This profile was obtained by recording the intensity of the molecular ion (m/z 288). The distribution obtained when recording a fragment at m/z 215 was nearly identical to that of that of the molecular ion which suggests that some fragmentation occurs during the ionization process. Transformation of this distribution to energy coordinates shows that molecules leave the surface with an average translational energy of 0.2 eV, implying a ballistic ejection mechanism. In contrast to tryptophan, no delayed ejection is evident for this system and no molecules are present in the laser plane 100 μs after the desorption pulse.

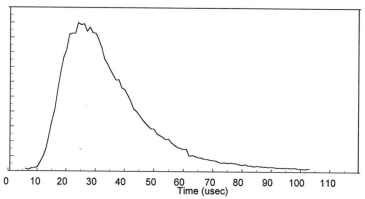

FIGURE 2. TOF distribution for pyrenebutyric acid desorbed by 8 keV Ar$^+$ ions.

ACKNOWLEDGMENTS

The authors gratefully acknowledge the financial support of the National Science Foundation and the Office of Naval Research.

[1] Hoogerbrugge, R., van der Zande, W.J., Kistemaker, P.G, *Int. J. Mass Spectrom. Ion Processes.*, **76**, 239 (1987).
[2] Garrison, B.J., *Nucl. Instrum. Methods B*, **17**, 305 (1986).

Desorption Mechanism of Benzene from C_6H_6/Ag(111) using keV Ion Bombardment and Laser Postionization

R. Chatterjee, D.E. Riederer, B.J. Garrison and N. Winograd

Department of Chemistry
The Pennsylvania State University
University Park, PA 16802

Z. Postawa

Institute of Physics
Jagellonian University
ul. Reymonta 4
PL 30-059 Krakow 16 Poland

Abstract. Laser postionization of neutral benzene sputtered from C_6H_6/Ag(111) has been performed to investigate the mechanisms that lead to desorption of molecules from surfaces. The velocity time of flight (TOF) distributions of neutral benzene were measured after 8 keV Ar^+ bombardment. Two components are present in the distributions. For sub-monolayer and monolayer coverages a low TOF , high kinetic energy (1.14 - 0.25 eV) peak dominates the distribution, whereas for multilayer coverages a high TOF, low kinetic energy (0.04 eV) peak dominates. The intensity of the high TOF peak increases with bombardment time for desorption from multilayers of benzene. The results indicate that a collision cascade in the metal initiates the benzene ejection from thin overlayers of benzene on Ag(111). An ion beam induced chemical change is proposed to occur in thick benzene overlayers.

INTRODUCTION

Ion beam induced desorption techniques have found useful applications in molecular surface characterization of organic and biological systems (1). Of fundamental interest is the interaction of ion beams with organic solids and the mechanisms that lead to molecular desorption (2,3). It is well known that a collision cascade leads to ejection of atomic species in the case of metals and semiconductors. The mechanism by which the primary ion beam dissipates energy into the substrate to lead to the ejection of molecules is still not very well understood. The present study focuses on desorption induced by ion bombardment of benzene adsorbed on Ag(111). Since the sputtering of Ag(111) is well investigated a submonolayer to multilayer coverage of benzene on Ag(111) provides a simple model system to study desorption mechanisms from molecular overlayers. To our knowledge, these are the first measurements obtained for neutral benzene molecules desorbed from monolayer films.

EXPERIMENTAL

Benzene was adsorbed on a clean Ag(111) surface cooled to 120 K. Gases dissolved in benzene were removed by several freeze-pump-thaw cycles before dosing. The benzene coverage was controlled by monitoring the chamber pressure and the dosing time.

An 8 keV, 220 ns Ar^+ pulse focused to a 3 mm spot on the sample initiated the desorption process. Ejected neutral molecules were detected after multiphoton ionization with a 6 ns laser pulse at 266 nm. The laser beam was focused to a ribbon shape approximately 1 cm above the sample. The time of flight of neutral benzene molecules from the surface to the laser plane was recorded by varying the delay between the ion pulse and the laser pulse. The ionized molecules were then detected by time of flight mass spectrometry using a gated microchannel plate detector. The experimental setup has been described in detail previously (4).

The angle of incidence of the primary ion beam was 45° and the signal was detected normal to the surface over an angular range of 0° $\pm$ 20°. The postionization mass spectrum had peaks at m/z 52 (C_4H_4 fragment), 78 (molecular C_6H_6), 108 (Ag) and 216 (Ag_2). The distributions were recorded by monitoring m/z 78 for benzene and m/z 216 for silver dimer. The silver monomer distributions were not recorded as they have contributions from dimers fragmenting in the laser plane.

RESULTS AND DISCUSSION

The velocity TOF distributions of neutral benzene desorbed after 8 keV Ar^+ bombardment of C_6H_6/Ag(111) are shown in Figure 1 for various exposures. The distributions have a striking dependence on the benzene coverage. A low TOF component (peak A) is present at low coverage and a high TOF component (peak B) becomes dominant at higher coverage. At extremely low coverage the benzene distributions look similar to that of silver. This result implies that benzene desorption at low TOF occurs due to direct collisions between the substrate and overlayer species. As the benzene overlayer is increased from sub-monolayer to monolayer coverage (1L - 5L), peak A shifts towards a higher time of flight. This corresponds to a peak shift in the kinetic energy distribution from 1.14 eV to 0.25 eV. As the benzene coverage increases collisions between the overlayer molecules become significant and a part of the primary energy is deposited into various internal degrees of freedom, thereby decreasing the kinetic energy of the molecules. Increase in benzene coverage also changes the surface binding energy which leads to the shift in the kinetic energy peak.

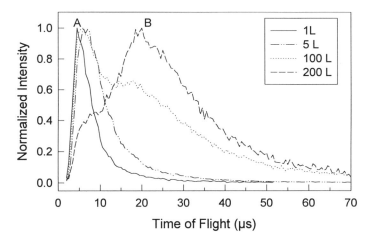

Fig.1. The velocity time of flight distribution of neutral benzene for various exposures
$(1 \text{ L} = 1 \times 10^{-6} \text{ torr} \cdot \text{sec})$.

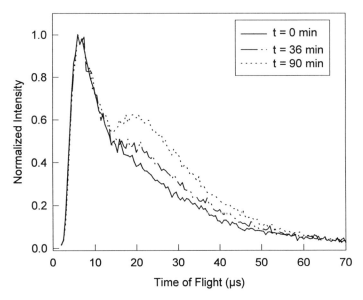

Fig. 2. The velocity time of flight distribution of neutral benzene (90 L exposure) as a function of bombardment time

As the thickness of the benzene layer is futher increased a very low kinetic energy peak (0.04 eV) emerges in the distribution (peak B in Fig. 1). The two components coexist, with an increase in peak B intensity as the benzene overlayer becomes thicker. On the other hand, the intensity of the silver as well as the low TOF peak of benzene decreases for higher exposures, and at 800 L only the high TOF peak is present. At this point no silver signal is seen in the mass spectrum. This clearly indicates that desorption at extremely low kinetic energies from multilayer coverages of benzene is not initiated by direct collisions from the substrate. Although it is tempting to ascribe the low kinetic energy component to a thermal desorption process, the Boltzmann distribution fit temperature (575 K) does not match the experimental surface temperature (120 K).

Another interesting feature, as shown in Figure 2, is that the TOF distributions depend not only on benzene coverage but also on the bombardment time. The intensity of the high TOF component increases with prolonged ion bombardment. The rate of change of the signal intensity increases as the benzene overlayer becomes thicker. After moving the ion beam to a new sample spot the signal intensity reverts to its original value. This implies that a local chemical or physical change in the sample induced by the ion beam results in desorption at these extremely low kinetic energies. Earlier studies have shown that electron beam impact of benzene overlayers on Ag(111) leads to the formation of phenyl fragments which desorb as biphenyl upon increasing the sample temperature (5). It is possible that a similar process initiated by the ion beam results in the ejection of biphenyl or larger aromatic molecules from the surface which eventually fragment into benzene in the laser plane.

ACKNOWLEDGMENTS

The authors gratefully acknowledge the financial support of the National Science Foundation, the Office of Naval Research and the Polish Committee for Scientific Research fund no. 7 T08 C053 11.

REFERENCES

1. Pachuta, S. J. and Cooks, R. G., *Chemical Reviews* **87**, 647-669 (1987).
2. Galera, R., Blais, J. C. and Bolbach, G., *International Journal of Mass Spectrometry and Ion Processes* **107**, 531-543 (1991).
3. King, B. V., Tsong, I. S. T. and Lin, S. H., *International Journal of Mass Spectrometry and Ion Processes* **78,** 341-356 (1987)
4. Kobrin, P. H., Schick, G. A., Baxter, J. P. and Winograd, N., *Review of Scientific Instruments* **57**, 1354-1362 (1986).
5. Zhou, X. L., Castro, M. E. and White, J. M., *Surface Science* **238**, 215-225 (1990)

Clusters In Intense Laser Fields:
Multiple Ionization and Coulomb Explosion

E. M. Snyder, D. A. Card, D. E. Folmer, and A. W. Castleman, Jr.

Department of Chemistry, The Pennsylvania State University, University Park, PA 16802

Abstract. Findings of the production of highly charged atomic species (e.g. Xe^{20+}, Kr^{17+}, I^{17+}, Ar^{8+}, N^{5+}, O^{5+}, and C^{4+}) resulting from the interaction of intense laser fields (up to $\sim 10^{15}$ W/cm^2) with atomic and molecular clusters, are reported herein. The processes are also investigated using ultrafast pump-probe techniques, showing distinct beating patterns for the ionization structure in the molecular systems. A comparison of our results with predictions of several different theoretical models provides strong support for the ionization ignition mechanism.

INTRODUCTION

The study of the interaction of atoms, molecules, and clusters with high intensity laser fields is a subject of extensive current interest. Investigations (1-3) have revealed that sufficiently intense interactions with clusters can lead to multiple ionization events, and many questions have been raised as to the mechanisms involved. Early theoretical models (4,5) have been unable to account for the high charge states observed in many experiments, in particular, ones involving the interactions of intense fields with van der Waals and hydrogen-bonded clusters (1-3). To account for the high charge states observed in experiments involving clusters, a mechanism based on the coherent motions of the field ionization electrons has been proposed (3). Alternatively, an ionization "ignition" mechanism has been formulated for laser-driven clusters (6).

The apparatus used in this study is a reflectron time-of-flight (TOF) mass spectrometer coupled with a femtosecond laser system which has been described in detail previously (7). The amplified femtosecond laser output is typically 2.5 mJ/pulse with a pulse duration of about 350 fs, yielding a focused power density of $\approx 1 \times 10^{15}$ W/cm^2.

RESULTS AND DISCUSSION

Xenon, krypton, argon, hydrogen iodide, ammonia, or acetone, seeded in ≈2000 torr of helium, was expanded into a vacuum chamber, and the resultant monomer and cluster species were ionized by the femtosecond laser. A time-of-flight spectrum of the multicharged xenon atoms is shown in Figure 1. The acetone was studied using a pump-probe arrangement. The pump-probe transients of the oxygen fragments are shown in Figure 2. The significant feature that should be noted is the asymmetric and irregular "beating" pattern observed. Other multicharged species that have been observed are Kr^{18+}, Ar^{8+}, I^{17+}, N^{7+}, and C^{+4}.

The two recent models proposed to explain the high charge states observed are: the coherent electron motion model (CEMM) (3) and the ionization ignition model (IIM) (6). In the CEMM, the multiple electron ejection arises from the coherent motion of the field ionized electrons; subsequent removal of electrons occurs in a fashion similar to electron impact ionization. In the IIM, after the initial ionization events, the parent ion cores are inertially confined to the cluster leaving the ion field unscreened, resulting in a very large (>10^{12} V/m) and inhomogeneous electric field. This large field lowers the ionization barrier and enables subsequent ionization events to occur, which in turn further increases the field and lowers the ionization barrier.

In their CEMM, Rhodes and co-workers have proposed (3) an expression to approximate the number of ionization events:

$$N_x \cong n^{4/3} Z \frac{\sigma_{ei}}{r_0^2} \quad (n \geq 3)$$

where n is the number of atoms in the cluster, Z is the resultant ionic charge, σ_{ei} is the inelastic electron impact ionization crossection, and r_0 is the interatomic spacing. Furthermore, the coherent electrons behave as a quasi-particle of mass Zm_e and charge Ze, requiring modification of the electron impact cross section to the form $\sigma_{ei} \rightarrow Z\sigma_{ei}$.

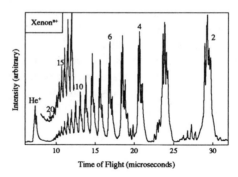

FIGURE 1. Time-of-flight mass spectrum of multicharged xenon atoms, ionized at 624 nm. Atoms with charge states up to Xe^{20+} are clearly observed.

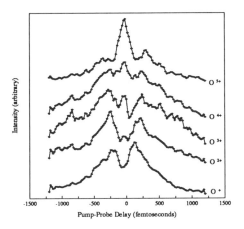

Intensity (arbitrary)

Pump-Probe Delay (femtoseconds)

O^{5+}

O^{4+}

O^{3+}

O^{2+}

O^{+}

-1500 -1000 -500 0 500 1000 1500

FIGURE 2. Pump-probe transients of multicharged oxygen ions, O^{n+} ($1 \leq n \leq 5$), formed through the Coulomb explosion of acetone monomer and acetone clusters.

In the IIM (6), the ion cores in a cluster are responsible for the high charge states obtained, rather than the large electron density. It was concluded that removal of the electrons is not expected to be very sensitive to pulse width, cluster size, or atomic weight. A related two atom version of this general model has been proposed (8) based on time-dependent quantum mechanical studies of diatomic molecules. The conclusion of this study is that the rate of ionization is highly dependent upon the internuclear distance. Additionally, it has been shown (9) that the dependence on internuclear separation is due to the role of electron localization in intense field ionization. Such effects would be expected to give rise to an irregular beating pattern, consistent with present observations.

As the various models predict, we do not observe any multicharged species unless clusters are present in our molecular beam. The charge states obtained are reasonable in considering both the CEMM and the IIM. The cluster distribution was varied by sampling different portions of the molecular beam. The multicharged species abruptly appeared in the spectra as a certain minimum critical cluster size was obtained, but the charge distribution did not vary with further variation in the cluster distribution. This is evidenced by the lack of charge state dependence on the molecular beam expansion conditions or whether the laser was positioned at the front, center, or rear of the expansion. Under our ionization conditions, multicharged krypton and argon atoms are only observed when a trace amount of HI is present (2), although pure xenon, ammonia, and acetone clusters are multiply ionized. The fact that neat Kr (IP = 14.00 eV) and Ar (IP = 15.76 eV) clusters do not exhibit multiple ionization, whereas neat Xe (IP = 12.13 eV), HI (IP = 10.39 eV), NH_3 (IP = 10.18 eV), and $(CH_3)_2CO$ (IP = 9.71 eV) clusters do, suggests that the ability to multiply ionize atoms is less sensitive to atomic number and depends more upon the threshold for single ionization.

The "beating" structure shown in Figure 2, from the pump-probe studies provide the following: The pump beam creates multicharged carbon and oxygen species within the acetone cluster. After the cluster undergoes electron loss, the probe beam arrives at the cluster at some later time and the interatomic spacings have increased to a particular distance as a result of the nuclear motion arising from the Coulomb explosion process. The structure seen for the transients in Figure 2 is due to the varying ionization rates as the interatomic spacing is changed, as predicted (8,9). In these models, the ionization rate is predicted to be a highly irregular function of interatomic distance, in contrast to the CEMM which predicts a monotonic decrease in the degree of ionization with an increase in interatomic distance (see r_0 in the above equation).

In conclusion, we have demonstrated that the high charge states obtained when van der Waals and hydrogen bonded clusters are irradiated with intense laser fields, are well described by the IIM (6). The lack of dependence on the degree of clustering or atomic weight, and the strong dependence on ionization potential and interatomic (or intermolecular) distances, support this conclusion. Our results cannot totally eliminate the possibility of coherent electron motions, but under our experimental conditions, it is not a major contribution to the multicharging and subsequent Coulomb explosion of clusters.

ACKNOWLEDGMENT

Financial support by the Air Force Office of Scientific Research, Grant No. F49620-94-1-0162, is gratefully acknowledged.

REFERENCES

1. Purnell, J., Snyder, E. M., Wei, S., and Castleman, A. W., Jr., *Chem. Phys. Lett.* **229**, 333 (1994).
2. Snyder, E. M., Wei, S., Purnell, J., Buzza, S. A., and Castleman, A. W., Jr., *Chem. Phys. Lett.* **248**, 1 (1996).
3. Boyer, K., Thompson, B. D., McPherson, A., and Rhodes, C. K., *J. Phys. B: At. Mol. Opt. Phys.* **27**, 4373 (1994).
4. Augst, S., Strickland, D., Meyerhofer, D., Chin, S. L., and Eberly, J. H., *Phys. Rev. Lett.* **63**, 2212 (1989).
5. Mevel, E., Breger, P., Trainham, R., Petite, G., Agostini, P., Migus, A., Chambaret, J. P., and Antonetti, A., *Phys. Rev. Lett.* **70**, 406 (1993).
6. Rose-Petruck, C., Schafer, K. J., and Barty, C. P. J., *Applications of Laser Plasma Radiation II, SPIE* **2523**, 272 (1995).
7. Purnell, J., Wei, S., Buzza, S. A., and Castleman, A. W., Jr., *J. Phys. Chem.* **97**, 12530 (1993).
8. Chelkowski, S., Zuo, T., Atabek, O., and Bandrauk, A. D., *Phys. Rev. A* **52**, 2977 (1995).
9. Seideman, T., Ivanov, M. Y., and Corkum, P. B., *Phys. Rev. Lett.* **75**, 2819 (1995).

The Metastable Decay of Alkene Clusters after Photoionization

L. Poth, Q. Zhong, J. V. Ford, Z. Shi and A. W. Castleman, Jr.

152 Davey Laboratory, Department of Chemistry
The Pennsylvania State University, University Park, Pennsylvania 16802

Abstract. Metastable fragmentation studies of alkene clusters, ionized with a femtosecond laser system following a supersonic expansion, have been performed. The investigation shows an interesting switching of decay channels from the loss of a molecular fragment to loss of a monomer unit. Taken together, the results indicate that, after ionization, a fast intracluster oligomerization takes place, which does not propagate further than pentamerization for ethene and trimerization for propene clusters. 1-butene and 2-butene exhibit oligomerization only up to the dimer. Larger cluster ions of all investigated alkenes appear to contain an oligomerized core solvated by monomer units.

INTRODUCTION AND EXPERIMENTAL

The ethene ion-molecule reaction has been the subject of many investigations during the last decades, what is not surprising since polyethene and its analogous compounds, polypropene and polybutene, are chemical products of great importance which have found innumerable applications in daily life. Early experiments on the uncatalyzed, cationic polymerization have generally concentrated on the relative rate constants of the ethene ion-molecule reaction and subsequent secondary reactions as a function of gas pressure or translational energy. In high pressure mass spectrometric investigations where ions may be expected to undergo many thousands of collisions prior to detection, Kebarle et al.[1,2] observed that an increase in the concentration of ethene in the ion source led to the disappearance of smaller ions $(C_2H_4)_n^+$ (n = 1,2 and 3) accompanied by large increases in the intensities of intermediate size molecules with n=4 and 5. This effect has been explained by considering the formation of oligomer ions which, due to steric effects, do not react further to form larger molecular ions.

In contrast to high-pressure sources where two and three body collisions are dominant, the use of cluster beam sources makes it possible to provide conditions for these ion-molecule reactions where the ion is initially surrounded by van der Waals bonded neutral molecules.

---

[1] Kebarle, P.; Hogg, A. M., *J. Chem. Phys.* **42**, 668-674 (1965).
[2] Kebarle, P.; Haynes, R. M., *J. Chem. Phys.* **47**, 1676-1683 (1967).

Investigation of the photoionization of ethene clusters by Ceyer et al.[3] has lead to the conclusion that ionic clusters with the empirical formula $(C_{2m}H_{4m})^+$ and $(C_{2m-1}H_{4m-3})^+$ are actually represented by $(C_4H_8)^+$ and $(C_3H_5)^+$ solvated by neutral ethene molecules whereas Ono et al.[4] and Tzeng et al.[5] arrived at the conclusion that the ionized trimer rearranges to $(C_6H_{12})^+$ before fragmentation. Coolbaugh et al.[6,7] performed experiments in which they found a magic number for n=4 in the distribution of $(C_2H_4)_n^+$ clusters under experimental conditions of low nozzle temperature and high stagnation pressure. These data, an IR photodissociation study by Feinberg et al.[8], and a recently published study on the CID of ethene clusters[9] led to the conclusion that ethene cluster ions up to pentamer are covalently bonded ions.

To investigate the influence of different substituents on this oligomerization reaction of unsaturated hydrocarbons in the presented study, the dissociation of several photoionized alkene cluster ions in the drift region of a time of flight MS was examined.

Since the reflectron time-of-flight mass spectrometer employed has been described in previous publications[10], only the details essential to this investigation are given here. Neutral clusters are formed by supersonic expansion of a premixed sample gas with the composition 20% alkene and 80% argon, at a stagnation pressure of 3.5 bar through a pulsed nozzle with a 150 μm orifice. The formed neutral clusters pass a 2 mm diameter conical skimmer to enter the ionization region, which is differentially pumped by a turbomolecular pump.

Ionization of the clusters is performed utilizing a 400 nm femtosecond laser with a pulse duration of 100 fs and a pulse energy of 0.7 - 1.0 mJ.

In this study, the reflectron is used as an energy analyzer to study metastable dissociation processes of the clusters during their passage through the drift-region. This is accomplished by performing cut-off experiments[11], i. e. allowing the ion signal to be eliminated or "cut-off" by traversing through, rather than being reflected by, the reflecting electric field. This procedure enables us to determine the mass of the daughter ion formed during the decay. To exactly determine the appropriate parent ion from which the daughter ion is formed, a time overlap procedure can be used in which the reflectron voltage is lowered down so that the trajectories of parent and daughter ions in the reflectron are the same according to

[3] Ceyer, S. T.; Tiedemann, P. W.; Ng, C. Y.; Mahan, B. H; Lee, Y. T., *J. Chem. Phys.* **70**, 2138-2143 (1979).

[4] Ono, Y.; Linn, S. H.; Tzeng, W.-B.; Ng, C. Y., *J. Chem. Phys.* **80**, 1482-1489 (1984).

[5] Tzeng, W.-B.; Ono, Y.; Linn, S. H.; Ng, C. Y., *J. Chem. Phys.* **83**, 2813-2817 (1985).

[6] Coolbaugh, M. T.; Peifer, W. R.; Garvey, J. F., *Chem. Phys. Lett.* **168**, 337-343 (1990).

[7] Coolbaugh, M. T.; Vaidyanathan, G.; Garvey, J. F., *Int. Rev. Phys. Chem.* **13**, 1-19 (1994).

[8] Feinberg, T. N.; Baer, T.; Duffy, L. M., *J. Phys. Chem.* **96**, 9162-9168 (1992).

[9] Lyktey, M. Y. M.; Rycroft, T.; Garvey, J. F., *J. Phys. Chem.* **100**, 6427-6433 (1996).

[10] Shi, Z.; Wei, S.; Ford, J. V.; Castleman, A. W.,Jr., *J. Chem. Phys.* **99**, 8009-8015 (1993).

[11] Wei, S.; Castleman, A. W., Jr., *Int. J. Mass. Spec. Ion. Proc.* **131**, 233-264 (1994).

$U_{tp}/U_{td} = M_p/M_d$. Where U_{tp} is the reflectron voltage setting for the parent TOF spectrum and U_{td} the reflectron voltage setting for the daughter TOF spectrum. M_p and M_d are the mass of parent and daughter ions, respectively.

During operation, the pressure inside the chamber is maintained at $8 \cdot 10^{-7}$ torr in the ionization region and $2 \cdot 10^{-6}$ torr in the drift region.

RESULTS AND DISCUSSION

The cluster spectra of all investigated gases show a dominating intact series with the stoichiometry $(C_mH_{2m})^+$, which can be observed up to m= 56 for ethene, m=51 for propene and m= 52 for butene. Additionally smaller series with the stoichiometries $(C_mH_{2m-1})^+$, $(C_mH_{2m-2})^+$, $(C_mH_{2m+1})^+$ and series formed by fast fragmentation losing a alkyl unit were observed. The intensity of the latter fragment series decreases very rapidly with cluster size. Using cut off potentials and the calculated time overlap for parent and daughter ions, it can be determined that the observed daughter series described in the next section are formed solely from parent ions of the intact cluster series.

Metastable Dissociation Studies

The metastable decay for ethene dimer and trimer shows as the dominant fragmentation, loss of a methyl group where, in the case of the trimer a second decay channel with loss of a 28 unit opens which is comparable in intensity to the CH_3 loss channel. For the tetramer and pentamer, loss of an ethyl unit can be observed which is accompanied by loss of a 43 amu unit in the tetramer case. From clusters with 6 and more ethene moieties, the main metastable fragmentation channel switches solely to loss of an ethene molecule.

Although the metastable decay shows less fragmentation channels open than Lyktey's CID study, the main channels are the same in both. This observation supports the proposed intracluster oligomerization reaction which leads to molecular ion structures for small cluster sizes. For all larger clusters beyond the hexamer the decay pathway changes to loss of a monomer unit indicating that intracluster oligomerization does not proceed further, and that in larger clusters at least one monomer unit is loosely bound to the ion core.

To investigate the influence of additional methyl groups on this oligomerization reaction, experiments with propene and 1- and 2-butene clusters have been performed. The observed metastable fragmentation channels observed for propene are summarized in Table 1.

For propene and butene clusters, a similar switching of the dominant decay channel can be observed. But in this case switching occurs at cluster sizes different from those observed for ethene.

TABLE 1. Metastable losses of propene cluster ions $(C_3H_6)_n^+$.

	n=1	n=2	n=3	n=4	n=5
main loss	-1	-15	-43	-42	-42
2nd loss	-15 (w)	-28 (s)	-29 (m)	-85 (w)	
3rd loss		-42 (w)	-57 (w)		

* (s), (m), and (w) are used here to express the intensities of the 2nd and 3rd losses comparing them with the main channel; s means relative intensity is greater than 50%, m for 25-50%, w for <25%.

For the propene trimer, loss of a propyl fragment and a slightly weaker channel corresponding to the loss of an ethyl group are the dominant processes. For clusters bigger than the trimer, only loss of a monomer unit can be observed.

Interesting is the fact that the metastable decay of the propene dimer, except for an additional weak decay channel losing a 42 mass unit, shows the same fragmentation pathway as the ethene trimer. This indicates that both ions have a similar molecular ion structure. A similar observation has been made for the 1-butene dimer. The metastable decay of this ion shows close similarities with the ethene tetramer, again indicating a similar ion structure.

As found for ethene and propene, 1- and 2-butene also show a change in decay channels from mainly alkyl fragment losses to monomer loss. For 2-butene cluster ions a strong monomer loss channel can already be observed for the dimer indicating that oligomerization is less favorable for 2-butene cluster ions than for all other investigated systems.

CONCLUSION

Taken together, the results of the presented study indicate that for all alkene clusters an oligomerization reaction takes place, presumably after ionization. The formation of molecular ions can be observed only for small cluster sizes, whereas larger clusters consist of a molecular ion core solvated by neutrals. The number of molecules taking part in this oligomerization process decreases from 5 in the case of ethene, to only two in the case of 2-butene. This can be accounted for due to an increasing steric hindrance which prevents further attack of the formed molecular cation at the π-bond of the neutral molecule.

ACKNOWLEDGMENTS

Financial support by the U.S. National Science Foundation, Grant No. CHE-9632771, is gratefully acknowledged. One of the authors (L. Poth, Feodor Lynen-Fellow) acknowledges support by the Alexander von Humboldt Foundation, Germany.

Multiphoton Ionization Studies of Laser Induced Chemistry in Clusters

C.S. Feigerle, M.Z. Martin, Liyu Liu, and J.C. Miller

Chemical and Biological Physics Section, Oak Ridge National Laboratory
Post Office Box 2008, Oak Ridge, Tennessee 37831-6125
and
Department of Chemistry, University of Tennessee
Knoxville, Tennessee 37996-1600

Abstract. Three examples are presented where multiphoton ionization mass spectrometry is used to study photochemistry in clusters. In the first, $NO^+(N_2O_3)_m$ and $NO_2^+(N_2O_3)_m$ are made by 266nm multiphoton ionization of the clusters produced in an expansion of $NO/CH_4/Ar$ with a trace of H_2O. Second, $H_3O^+(H_2O)_n$ and $CH_3OH_2^+(CH_3OH)_n$ are observed when sufficiently large clusters of $NO(H_2O)_m$ and $NO(CH_3OH)_m$ are ionized, suggesting laser initiation of intracluster charge transfer reactions in these systems. Thirdly, multiphoton ionization of mixed expansions of NO and $Fe(CO)_5$ leads to the production of $(Fe)_m^+$, $(Fe)_m^+(CO)_n$, $Fe^+(NO)(CO)$, Fe^+NO, and FeO^+. The mechanisms for formation of these species will be discussed and analogies drawn between intracluster and collisional chemistry.

INTRODUCTION

An important added feature in multiphoton ionization (MPI) of molecules is the possibility that fragmentation competes with or follows ionization. In MPI of weakly bound clusters, a variety of new phenomena associated with ionization are possible including: evaporation of cluster components to dissipate excess energy, charge transfer between species within the cluster, initiation of intracluster ion-molecule reactions, and others. In many cases, parallels can be found between the chemistry within a cluster and that which occurs in ion-molecule collisions.

In this paper, results on multiphoton ionization of clusters formed in mixed expansions with nitric oxide and other intracluster reaction partners are briefly summarized. In many cases, weaker bonds are retained within the cluster while stronger bonds are being broken. These examples demonstrate that clusters can be rich media for the study of photoinitiated chemistry.

TECHNIQUE

Clusters are produced by expanding mixtures of the appropriate precursors diluted in argon in a pulsed supersonic jet. The gas pulse is skimmed and the isentropic core of the expansion is crossed at 90° with the fourth harmonic (266 nm) of a 30 ps Nd:YAG laser beam. At this wavelength, the energy contained in two photons is sufficient to non-resonantly ionize NO or Fe in the cluster, but 3 or more photons are required to directly ionize most other components. The mass spectrum of the cluster ions produced is obtained by extracting the ions into a 1 m linear time-of-flight (TOF) mass spectrometer and by recording and averaging the TOF mass-spectrum on a digitizing oscilloscope. Further details on the apparatus and technique can be found elsewhere (1).

DINITROGEN TRIOXIDE CLUSTERS

The TOF mass spectrum produced when a gas mixture of 5% CH_4 / 5% NO / 90% Ar with a trace of H_2O is expanded and crossed with focused pulses of 266 nm light is shown in figure 1. In contrast to what might be expected from such a complex mixture, the spectrum is remarkably simple, consisting primarily of a single cluster ion series. This series of peaks, separated by 76 amu, has been assigned (2) to the cluster ions, $NO^+(N_2O_3)_n$ (n=1-16). In a preliminary report (3) of this mass spectrum, the simpler but incorrect interpretation was made that the cluster series was built directly from the precursors in units of $(NO)_2CH_4$ rather than the photochemical product N_2O_3, both of which are consistent with the observed mass increment of 76 amu. The correct assignment was confirmed by comparing spectra obtained using isotopically substituted, $^{15}N^{18}O$. Here the conversion of nitric oxide to clusters of dinitrogen trioxide appears to occur in a facile manner when excess energy is deposited in the cluster by electron or photon impact. This photochemical

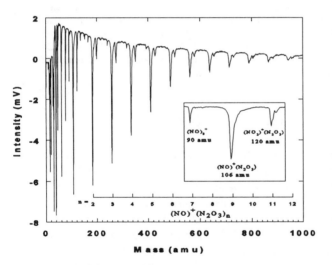

Figure 1. Multiphoton Ionization Mass Spectrum of $NO^+(N_2O_3)_n$.

conversion is not observed unless methane and water are present in the mixture, though the role that these components play in the mechanism of the photochemical production of $NO^+(N_2O_3)_n$ is not fully understood. Minor cluster series assigned to $NO_2^+(N_2O_3)_n$ and $(NO)_3^+(N_2O_3)_{n-1}$ are also observed in the MPI mass spectrum. The $(NO)_3^+$ has been shown to exhibit special stability when clustered with various partners, and the formation of $(NO)_3^+(N_2O_3)_{n-1}$ cluster appears to be another example of this. The $NO_2^+(N_2O_3)_n$ cluster ions simply involves addition of one oxygen to the major series of the spectrum. Other isotopic substitution experiments have traced the source of that additional oxygen to H_2O.

LIGAND SWITCHING REACTIONS INVOLVING NO$^+$

We have previously reported (4) that MPI of clusters produced from NO and D_2O mixtures produce predominantly $D_3O^+(D_2O)_n$ for $n \geq 4$. These water cluster ions have been attributed to the ligand switching reaction within the cluster $NO^+(D_2O)_{n+1} \rightarrow D^+(D_2O)_n + DONO$. This represents one example of a growing number of cluster reactions which exhibit size selective switching in the products formed by evaporative decay. Here similarities exist in the ion-molecule collisional chemistry between H_2O and $NO^+(H_2O)_m$ which yields the condensation product $NO^+(H_2O)_{n+1}$ for $n \leq 2$, but for $n=3$ reacts to eliminate HONO to form $H^+(H_2O)_n$.

We have recently observed another example of a size selective cluster reaction in studies of photoionization of mixed clusters of NO, H_2O and CH_3OH. Here, the mass spectra of ion products reflects differing stabilities of various cluster ion combinations. $NO^+(CH_3OH)_m$ is a significant product only for $m \leq 2$, whereas $H^+(CH_3OH)_m$ clusters are the dominant product up to $m=9$. Retention of H_2O is favored with increasing cluster size, with $H_3O^+(CH_3OH)_m$ emerging as the favored product followed by $H^+(H_2O)_2(CH_3OH)_m$. A recent study of the decay products of $NO^+(H_2O)_m(CH_3OH)_n$ has also shown cluster size dependent effects (5). These results indicate that small changes in the number or type of solvent molecules in a cluster can strongly affect the outcome of the cluster unimolecular decay reaction.

ORGANO-METALLIC PHOTOCHEMISTRY IN CLUSTERS

There have been numerous investigations into the photochemistry of organo-metallic species with labile ligands, spurred both by interest in the fundamental chemistry involved and the implications for catalysis, corrosion, and their use as chemical vapor deposition precursors. However, little is known about the photochemistry of these species in cluster environments.

We have recently studied the photoionization of clusters of $Fe(CO)_5$ produced in a dilute expansion with argon and with 1- 5% NO/Ar (6). In general, the resulting mass spectra show extensive stripping of the ligands and a large production of bare iron ions. This is not surprising since elimination of CO is typical in the photochemistry of $Fe(CO)_5$. However, several unexpected and interesting phenomena were observed in these photoionizations of clustered $Fe(CO)_5$.

When the clusters are prepared in an expansion with argon, ion series of Fe^+Ar_n and $Fe_2^+Ar_n$ are observed. As is the case in all these experiments, the clusters are ionized in a collisionless environment. Therefore, the argons in the final cluster ion must have been retained by the cluster inspite of the fact that all the ligands have been photolyzed. Curiously, in this process the strong bonds in the molecule have been severed but the weak bonds remain.

When a small amount (1%) of NO is added to the expansion mixture, Fe_m^+ and $Fe_m^+(CO)_n$ are observed to be the major product ions. A previous study has similarly reported production of Fe_m^+, but only at shorter wavelengths (7). Increasing the NO concentration to 5% leads to the production of ions such as FeO^+, $FeNO^+$, and $FeNOCO^+$. Here various ion-molecule reactions are assumed to be taking place within the cluster. With nitric oxide present, either Fe and NO could be photoionized by only 2-photons at these wavelengths, opening up the possibility of multiple pathways for the cluster ion-molecule chemistry.

ACKNOWLEDGMENTS

Research is sponsored by the Office of Health and Environmental Research, U.S. Department of Energy under contract DE-AC05-84OR21400 with Lockheed Martin Energy Systems, Inc.

REFERENCES

1. Desai, S.R., Feigerle, C.S., and Miller, J.C., *J. Chem. Phys.* **97**, 1793 (1992).
2. Martin, M.Z., Desai, S.R., Feigerle, C.S., and Miller, J.C., *J. Phys. Chem.* **100**, 8170 (1996).
3. Desai, S.R., Feigerle, C.S., and Miller, J.C., "Laser Ionization of Molecular Clusters," in *Resonance Ionization Spectroscopy 1994*, edited by Kluge, S., and Parks, J.E. (American Institute of Physics, New York, 1994), p. 179.
4. Desai, S.R., Feigerle, C.S., and Miller, J.C., *J. Chem. Phys.* **101**, 4526 (1994).
5. Winkel, J.F., and Stace, A.J., *Chem. Phys. Lett.* **221**, 431 (1994).
6. Martin, M.Z., Liu, L., Feigerle, C.S., and Miller, J.C., to be published.
7. Duncan, M.A., Dietz, T.G., and Smalley, R.E., *J. Am. Chem. Soc.* **103**, 5245 (1981).

Methods to investigate the interaction of 16 μm laser radiation with uranium hexafluoride in a supersonic expansion

Hendrik G. C. Human

Atomic Energy Corporation, Box 582, Pretoria 0001, South Africa

Abstract. Selective dissociation of UF_6 using three wavelength IR irradiation did not yield the desired results initially. Various spectroscopic methods such as UV and IR absorption of UF_6, fluorescence of UF_6 and Time-of-Flight Mass Spectrometry of the products of irradiation, were implemented to investigate the nature of the interaction. These techniques identified the source of the problem as the presence of condensates in the flow-cooled gas, and were used to select conditions to minimise this effect.

EXPERIMENTAL

A gas mixture of UF_6 , scavenger gas and argon was flow-cooled to a vibrational UF_6 temperature of approximately 100K in a 15 mm long Laval nozzle. A closed loop recirculating pumping system was used. Irradiation of the gas was done 20 mm downstream from the throat of the nozzle. A MOPA chain of CO_2 lasers generated line tuneable pulsed radiation in the 10 μm region, which was shifted to the 16 μm region by Raman conversion in para-hydrogen.

UF₆ FLUORESCENCE OF THE 375nm BAND

Fluorescence of the A-X or 375 nm band of UF_6 was excited by an excimer pumped dye laser at 395 nm and measured through an interference filter at 420 nm by a fast photomultiplier-amplifier system. With co-linear IR (single wavelength of 620 cm^{-1}) and 395 nm beams the fluorescence decay curves of Fig. 1 were obtained, for a gas mixture without scavenger gas. Without the IR pulse a normal decay of ca. 20 μs was obtained. IR irradiation directly before or after the 395 nm excitation results in severe quenching of the fluorescence. This is ascribed to strong absorption of the IR radiation in the medium, leading to vibrational excitation of UF_6 and through collisions to the release of kinetic energy and a high rate of quenching collisions.

The presence of a scavenger gas in the mixture has a dramatic influence on the

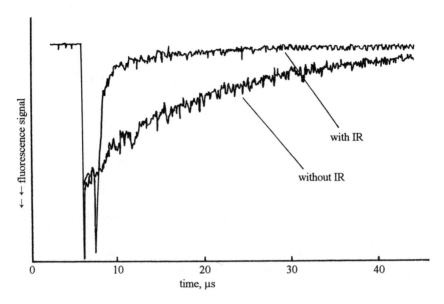

FIGURE 1. Effect of IR radiation on fluorescence. IR pulse 4 μs delayed w.r.t. 395 nm.

fluorescence of the UF_6 : it reduces the signal size and decay time significantly. With a mild IR (620 cm^{-1} , 50 mJ/cm^2) pulse before the 395 excitation, the fluorescence signal amplitude is partially restored, having the same decay time as without scavenger. These phenomena are explained as the formation of a condensate involving the scavenger and UF_6 , the absorption of such clusters being shifted in wavelength away from 395 nm. The IR radiation, however, is absorbed by the condensate and leads to breaking up of the clusters, releasing UF_6 to the gas phase capable of fluorescence.

UV ABSORPTION OF UF_6

The high absorption of UF_6 in the deep UV around 215 nm was used to monitor the change in UF_6 density as result of IR or UV dissociation. A copper hollow cathode lamp emitting strong resonance radiation at 218 nm was used as primary source in conjunction with a low resolution monochromator. Fig.2 shows the increase in transmission as a result of UF_6 dissociation by a 266 nm laser pulse. The two beams were spatially separated, the hollow cathode beam being downstream from the dissociating beam. The peak on the left is the stray light from the 266 nm laser. The increase in transmission reflects the dissociation of UF_6 in the irradiated volume as it flows past the monitoring beam.

When the UV dissociating beam is replaced by a IR pulse at 620 cm^{-1}, a similar increase in transmission is observed. However, these signals indicated a large UF_6 density decrease for low IR fluence, e.g. 14% density decrease for a 80 mJ/cm^2 IR

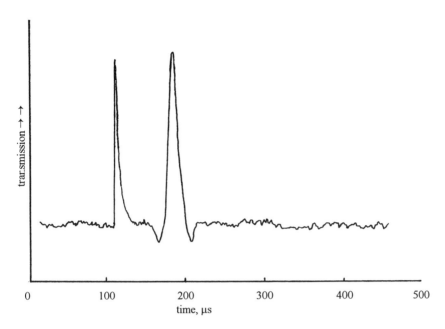

FIGURE 2. Transmission change of UF_6 at 218 nm as a result of 266 nm dissociation.

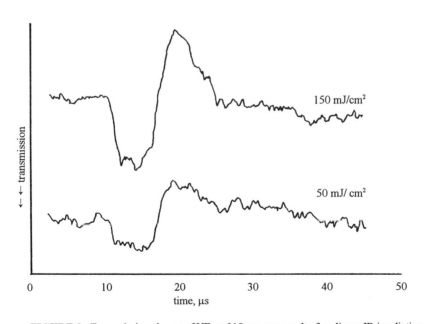

FIGURE 3. Transmission change of UF_6 at 218 nm as a result of co-linear IR irradiation.

pulse. With the 218 nm monitoring beam co-linear with the IR beam, an increase of transmission is measured, followed by a decrease, the total signal lasting about 15 microseconds (Fig. 3). These results with the IR irradiation is explained as a lower UF_6 density at the beam centre due to absorption of energy that results in molecules with high kinetic energy (temperature) expanding outwards. This expansion creates compression of the surrounding gas, therefore the peak of lower transmission.

ABSORPTION OF UF_6 AT 16 μm

Relative absorption measurements of pulsed IR radiation in the flow-cooled gas was done by positioning a piezo-detector in the gas stream near the outlet of the nozzle. The absorption signal obtained increased linearly with IR pulse energy, indicating linear absorption without threshold. With various wavelengths in the 16 μm region (Raman shifted line tunable CO_2 laser lines) between 590 and 630 cm^{-1}, a broad absorption spectrum with a peak at 614 cm^{-1} was obtained. The piezo-detector signals were calibrated by measuring the absorption for a single case (50 mJ/cm^2 of the 620 cm^{-1} line) in conventional manner. An absorption of 1.6% was measured, which is more than an order of magnitude higher than expected for cold UF_6 gas, and indicates 12 photons or 1 eV absorbed per molecule. These absorption measurements confirmed the existence of a condensate absorbing the IR energy.

TIME-OF-FLIGHT MASS SPECTROMETER MEASUREMENTS

A linear time-of-flight mass spectrometer was constructed and integrated with the gas flow system, a gas skimmer directly underneath the nozzle separating the main gas stream from the TOFMS. The product of irradiation was ionised with a focused 1064 nm laser beam, enabling 235/238 uranium isotope measurements. The results were generally in accord with the diagnostic measurements described in the previous sections. No enrichment was obtained under conditions at which the 620 cm^{-1} wavelength is absorbed by what is believed to be condensate.

CONCLUSION

Diagnostic measurements for ascertaining the conditions necessary for successful isotope enrichment in a flow-cooled supersonic expansion of UF_6 can be made employing conventional measurements of absorption and fluorescence. Applied to our system these measurements mainly indicated the detrimental effect of condensation of UF_6 in the expansion. The described techniques were implemented to delineate the conditions necessary for minimising condensation and achieving isotope separation.

The Photo-Dissociative Pathways of Nitromethane Using Femtosecond Laser Pulses at 375 nm.

H.S.Kilic[1,2], K.W.D.Ledingham[1], D.J.Smith[1], S.Wang[1], C.Kosmidis[3], T.McCanny[1], R.P.Singhal[1], A.J.Langley[4] and W. Shaikh[4]

[1]*Department of Physics & Astronomy, University of Glasgow, G12 8QQ, Scotland*
[2]*Permanent address;Department of Physics , University of Selcuk, 42079, Konya, Turkiye*
[3]*Department of Physics, University of Ioannina, Ioannina, Gr -45110 Greece*
[4]*Central Laser Facility, Rutherford Appleton Laboratory, Chilton, Didcot, OX11 OQX, UK*

ABSTRACT: The dissociative pathways of the nitromethane molecule have been studied using 10 ns, 7 ps and 90 fs laser pulses. In this paper, the differences between the dissociation channels opened using different laser pulse widths are discussed to investigate the potential for femtosecond laser mass spectrometry (FLMS). Using 90 fs laser pulses, a very large parent ion has been observed but the 10 ns and 7 ps laser pulse widths cannot defeat the lifetimes of the dissociative excited states of nitromethane.

INTRODUCTION

Nitromethane is the smallest and one of the most important members of the nitro group of molecules. The photochemistry of nitromethane has been studied by many groups and most of the dissociation channels have been investigated using several methods including multiphoton processes (1,2 and references therein) with the different wavelengths shown in Table 1.

The absorption spectrum of nitromethane is now well documented and it is known to have two absorption bands. A weak absorption band appears at 270 nm with a stronger band at 198 nm (4). The dissociative cross-section of nitromethane at 193 nm is 1.7×10^{-2} cm^{-2} (7) and molecule can reach to the excited state by absorbing two photon at 375 nm which matches a wavelength about the maximum of this stronger absorption band. The cleavage of the C-N bond in the decomposition of the nitromethane molecule is the dominant primary reaction and the present work shows a good agreement with literature by presenting large peaks at mass 15 (CH_3) and 46 (NO_2) which corresponds to a primary dissociation process via the equation (1);

Table 1: Table shows the dissociation channels opened for the neutral nitromethane. Several groups have applied different techniques for analysing the nitromethane.

D	E	M	W
CH₂O+HNO	0.68 eV	Photolysis	200-300 nm
		Photolysis	240-360 nm
CH₃O+NO	1.87 eV	MPIS	266 nm
		Flash Photolysis	199.9-206.4 nm
CH₃+NO₂	2.61 eV	Photolysis	313 nm
		Flas Photolysis	193 nm
CH₂NO+OH	2.78 eV	Pump-Probe	266 nm
CH₂+HONO	3.73 eV	Emission Spectr.	193 nm
CH₂NO₂+H	3.92 eV	Photolysis	253.7 and 313 nm
CH₃NO+O	4.07 eV	Photolysis	253.7 and 313 nm
CH₃+NO+O	5.74 eV	MPIS	193 nm

D:Dissociation channel opened; E:Energy needed to open channel; M:Method of detection; W:Wavelength used for experiment. All details of the different groups are included in reference 2.

$$CH_3NO_2 + h\nu \longrightarrow CH_3 + NO_2 \tag{1}$$

For this bond to be broken, 2.61 eV energy and/or less than one photon at 375 nm is needed. The major dissociation channel produces NO_2 in an excited state and CH_3 is produced with a little internal energy. Dissociation products continue to absorb some further photons and undergo secondary dissociation process. Dissociation of these two products contribute ion peaks of NO, O, C, CH_n and it seems that H and H_2 comes from secondary dissociation of CH_3 as well.

The main experimental arrangement consists of the femtosecond and nanosecond laser systems, the linear time-of-flight (TOF) mass spectrometer and data acquisition system. It has already been described in detail elsewhere (1,2,5). The laser beam is tightly focused down to 9.2 µm diameter in the ionization ragion to give about 10^{14} Wcm^{-2} laser intensities using 10 cm focal lenth mirror in vacuum in the case of femtosecond laser and 30 cm focal length lens about 10^7-10^9 Wcm^{-2} in the case of nanosecond laser.

Results and Discussion

The ionization potential of nitromethane is 11.28 eV and thus 4 photons of 375 nm wavelength are needed to ionize this molecule. It is energetically possible that a number of dissociation pathways are opened after the absorption of one or two photons at 375 nm (2,6). The ultra-violet absorption spectrum of nitromethane shows that the probability of absorption at 375 nm is however very small and thus the absorption of two photons is necessary to permit most of these pathways to be opened with a high probability (4).

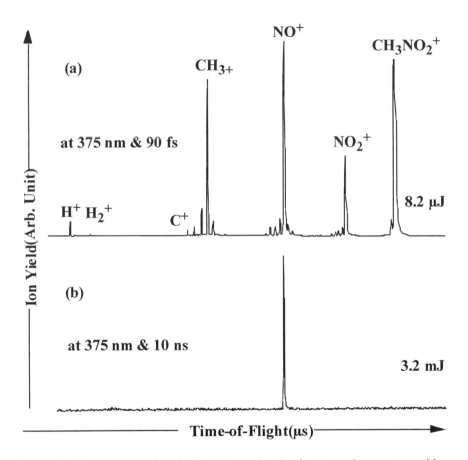

Figure 1: The mass spectra of NM taken using the femto- and nano-second laser pulses.

For the first time, the nitromethane parent ion has been detected (1,2) with a large intensity, using femtosecond multiphoton processes which have not been observed prior to this work. In the present work this molecule has been excited using 10 ns, 7 ps and 90 fs laser pulses with the same linear time-of-flight mass spectrometer. The nitromethane parent ion could not be observed with either the 7 picosecond or 10 nanosecond laser pulses which is expected since the lifetime of the dissociative excited state of the nitromethane molecule is less than 5 ps (3)

A comparison of the femtosecond and nanosecond multiphoton ionization and dissociation for nitromethane at 375 nm is shown in Fig.1. The nanosecond spectrum (b) shows only one peak at NO(m/e=30) with no parent ion and other peaks. This is indicative that with the nanosecond pulses, apart from the NO ion, neutral fragments are created which cannot be subsequently ionised with these laser fluxes. Alternatively the femtosecond spectrum (a) shows a great number of peaks with the principal peaks being m/e= $61(CH_3NO_2^+)$, $46(NO_2^+)$, $30(NO^+)$ and $15(CH_3^+)$. Many other fragment peaks were also visible shown in fig 1. This means that all the channels listed in Table 1 are opened using femtosecond laser pulses at 375 nm. At the present it is being considered in detail whether the fragmentation precedes ionization or comes after the ionization process.

Acknowledgements

H.S.K., K.W.D.L., C.K., T.McC., and R.P.S would like to thank the staff at the Rutherford Appleton Laboratory, Didcot, U.K., for all the assistance provided during the experimental period. H.S.K. would also like to acknowledge receipt of a post-graduate studentship from the Turkish Government and Selcuk University.

References

1. Ledingham, K.W.D., Kilic, H.S., Kosmidis, C., Deas, R.M., Marshall, A., McCanny, T., Singhal, R.P., Langley, A.J. and Shaikh, W., *Rapid Comm. in Mass Spectrom.* **9**, 1522, 1995.

2. Kilic, H.S., Ledingham, K.W.D., Kosmidis, C., McCanny, T., Singhal, R.P., Langley, A.J. and Shaikh, W., to be published.

3. Schoen, P.E., Marrone, M.J., Schnur, J.M. and Goldberg, L.S., *Chem.Phys.Lett.* **90**,272, 1982.

4. Nagakura, S., *Mol. Phys.* **3**, 152, 1960.

5. Singhal, R.P., Kilic, H.S., Ledingham, K.W.D., Kosmidis, C., McCanny, T., Langley, A.J. and Shaikh, W., *Chem. Phys. Lett.* **253**, 81, 1996.

6. Butler, L.J., Krajnovich, D. and Lee, Y.T., *J.Chem.Phys.* **79**, 1708, 1983.

7. Blais,N.C., *J.Chem.Phys.* **79**, 1723, 1983.

REMPI-Spectroscopy of organic molecules with unusual electronic properties: Biphenylene and [2,2]-Paracyclophane

Ralf Zimmermann

Institut für Physikalische und Theoretische Chemie, Technische Universität München,
D-85747 Garching, Germany & Lehrstuhl für Ökologische Chemie und Umweltanalytik,
Technische Universität München, D-85354 Freising, Germany

Abstract. The S_1-transitions of the jet-cooled biphenylene (BP) and [2,2]-paracyclophane (PCP) molecules are investigated, using the two-color resonance-enhanced multi-photon ionization (Jet-REMPI) technique. Both molecules show unusually large geometry changes upon $S_1 \leftarrow S_0$ excitation, inducing high Franck-Condon activity. In the case of the antiaromatic BP the large extent of π-bond fixation causes major bondlength changes within the central four membered ring. Five a_g-modes are assigned in the spectrum of BP. The BP jet-spectrum represents the first one of an antiaromatic compound. The $S_1 \leftarrow S_0$ geometry change of PCP is due to large *"through space"*-interactions between the two benzene subsystems (i.e. increased attraction of the benzene systems in the S_1). The spectrum of PCP shows an intense progression of the corresponding breathing-mode (a_g). The one-color limit of PCP is located in the well structured region of the S_1-spectrum, allowing the measurement of the ionization potential by comparison of the one- and two-color REMPI-spectra.

INTRODUCTION

The investigation of the vibronic finestructure of the UV-spectra of biphenylene (BP) as well as [2,2]-paracyclophane (PCP) is particularly interesting because both molecules exhibit unusual electronic properties, causing large $S_1 \leftarrow S_0$ geometry changes. Biphenylene (BP) depicts antiaromatic electronic character [1] and PCP has two *face to face* interacting π-systems [2]. The pattern of totally symmetric vibronic modes/progressions in the molecular UV-spectrum of the S_1-state is determined by the Franck-Condon (FC) principle, i.e. by the geometry changes upon excitation (high activity of modes which exhibit nuclear displacements similar to the geometry shift). The S_1-spectra of BP and PCP have been recorded by means of the two-color REMPI-technique, which is particularly advantageous for spectroscopic investigation of weak- or non-fluorescent molecules as e.g. biphenylene [3]. The experimental setup is described elsewhere [4]. Briefly its consists of a compact reflectron time-of-flight mass spectrometer with a homebuild, heatable (up to 230°C) supersonic jet inlet-system based on a *General Valve series 9* solenoid, an excimer-laser (XeCl, 308 nm) pumped dye-laser (excitation laser) and a Nd:YAG laser with nonlinear optics for generating 266 or 213 nm laser light (ionization laser).

RESULTS AND DISCUSSION

A) Biphenylen. Cyclic conjugated polyenes can be either stabilized or destabilized in comparison to their corresponding open chain polyenes. In the first case one speaks of aromatic- in the latter case of antiaromatic compounds. The very unstable 1,3-cyclobutadiene (CB) molecule represents the basic antiaromatic compound, just as benzene can be seen as the basic aromatic compound. Biphenylene (BP) represents the dibenzoderivative of CB. The question, to which extent the CB-type antiaromaticity is disturbed in BP due to condensation of two (aromatic) benzene rings, is discussed controversially in the literature [5]. The high thermal and chemical stability of BP is in contrast to the usually observed instability of antiaromatics. However, theoretical investigations on the geometry of BP in different electronic states suggest, that large geometrical shifts occur upon e.g. the $S_1 \leftarrow S_0$ excitation [6,7]. The first *ab initio* based (3-21G basis) comparison of S_0- and S_1-bondlengths is presented in this work (table 1). The S_1-state of BP represents a nearly pure LUMO$\leftarrow$HOMO transition [6]. Therefore an open-shell UHF-approach can be applied for the *ab initio* calculations on the S_1-

geometry. Major bondlength variations were observed in the central fourmembered ring, indicating a strong π-bond-fixation, typical for antiaromatics (aromatic π-systems are highly delocalized). This strongly support unusual spectroscopic and photophysical properties that point towards antiaromaticity in BP [1,5] (e.g. diabatic ring-current in nuclear magnetic resonance spectroscopy etc.). Although the spectroscopy of BP was investigated extensively by different research groups [1], no highly resolved vibronic spectrum of the S_1-transition is reported up to now. The first Jet-spectrum of biphenylene (i.e. the first one of an antiaromatic compound) is presented. The spectrum shows single vibronic state resolution (figure 1). The symmetry forbidden $S_1 \leftarrow S_0$ transition of BP mainly gains intensity by vibronic coupling of the $S_1(^1B_{1g})$- and the allowed $S_2(^1B_{3u})$-electronic state, involving b_{2u} vibrational modes [1a,1b]. Theoretical investigations have shown, that the v_{35} b_{2u}-vibration (S_0:1622 cm^{-1}) is by far the dominant inducing mode [7]. As depicted in table 2, high Franck-Condon (FC) activities were predicted for the v_{10}, v_9, v_7 and v_4 totally symmetric modes [7] (for the mode-notation see citation in [1a] at their table 1).

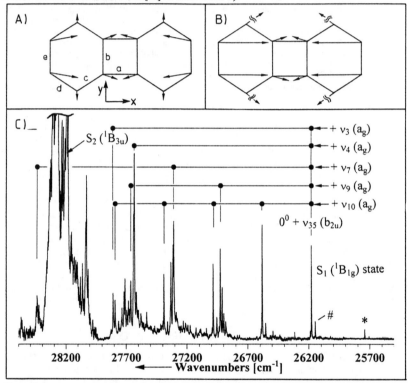

Figure 1

*a)Atomic displacements due to the $S_1 \leftarrow S_0$ geometry shift, bond-numbering and the used molecular coordinates. **b)** Atomic displacements due to the v_{10} vibrational mode. c)Two-color REMPI-spectrum of the S_1-transition of jet cooled biphenylene. Progressions and combination bands of a_g-modes build up on the main false origin ($0^0 + v_{35}$, 26180 cm^{-1}). From the main false origin of the spectrum and the (ground state) frequency of the v_{35} (1622 cm^{-1}) the forbidden electronic origin of BP can be estimated to be around 24550 cm^{-1}.The peaks marked by # and * are probably due to another weak false origin ($0^0 + v_{34}$, #) and a v_{10}-hot band transition respectively.*

Figure 1 shows the two-color REMPI-spectrum of the S_1-transition of BP as well as (exaggerated) representations of the relative atomic displacements due to the $S_1 \leftarrow S_0$ geometry shift and the v_{10}-vibrational mode (after 3-21G *ab initio* calculations). The calculated S_1 and S_0-geometries are given in table 1. The main structure of the spectrum is due to progressions and combination bands of totally symmetric modes, which mainly build up on one vibronic false origin. Five a_g-modes are assigned. Many other peaks are either due to combination bands of these a_g-modes or belong to a system based on a second weak false origin ($0^0 + v_{36}$, marked with # in figure 1). A more detailed analysis of the spectrum will be published elsewhere later [8]. The unusually high FC-activity of the a_g *in-plane* modes is obviously a symptom of strong bond-fixation in BP and the following large bond length changes (table 1). In addition to the remarkably red shifted S_1-transition [1c] (S_1-origin of BP lies at 407.3 nm, the one of the corresponding tricyclic aromatic antracene lies at 361.3 nm), the observed FC-pattern is a further, striking piece of evidence on the antiaromatic character of BP.

Table 1: Calculated C-C bondlength (ab initio,3-21G basis set) in [Å]

bond	S_0 [Å]	S_1 [Å]	Δ [Å]
a	1.526	1.411	-0.115
b	1.419	1.497	+0.078
c	1.353	1.395	+0.042
d	1.418	1.373	-0.045
e	1.371	1.429	+0.058

Table 2: Observed frequencies (S_1 and S_0,in [cm^{-1}]) and $S_1 \leftarrow S_0$ potential surface displacements (Δ_i) of a_g modes

a_g mode	S_0 [1a]	S_1 (REMPI)	$\Delta_i(S_1$-$S_0)$ [7]
v_{10}	395	403	0.94
v_9	765	749	0.53
v_7	1105	1129	0.60
v_4	1462	1455	0.79
v_3	1666	1630	-

B) [2,2]-Paracyclophane. In [2.2]-paracyclophane (PCP) two benzene subsystems are held together with a separation (3.09 Å) smaller than the van der Waals distance by two aliphatic C_2H_4-bridges. This induces strong electronic inter-ring coupling and a loss of planarity of the benzene subsystems [2b,9]. The spectroscopy of PCP has been studied frequently in the past due to its interesting electronic π-π interactions and the pronounced structural distortion (boat shape) [9a]. Theoretical studies on the electronic spectrum of PCP [9b] pointed out, that the first excited state ($^1B_{3g}$ in D_{2h}) is dominated by the HOMO→LUMO configuration. The HOMO exhibits antibonding character between the two benzene π-systems, whereas the LUMO shows attractive inter-ring interaction. As a consequence, the inter-ring spacing decreases by about 0.31 Å upon $S_1 \leftarrow S_0$ excitation [9b]. A laser induced fluorescence (LIF) spectrum of jet-cooled PCP already was assigned [2a,2b]. In this work the one/two-color REMPI-spectrum of jet-cooled PCP, performing increased spectral resolution as well as an ionization potential (IP) measurement, is presented. The Jet-REMPI-spectrum of PCP is shown in figure 2. The assigned main progression B_0^n is due to the a_g-breathing mode (237 cm^{-1}), which is efficiently FC-induced (decreased spacing between the benzene ring in the S_1-state [9b], see fig. 2a). Double excited overtones of the a_u-twist mode T_0^2 occur in combination with the B_0^n transitions ($B_0^n T_0^2$, b-peaks in figure 2). Further on, hot-bands (twist mode) of the above mentioned peaks are observed ($B_0^n T_1^1$, a-peaks in figure 2 and $B_0^n T_1^3$, c-peaks in figure 2). The assignment of the $B_0^n T_1^1$ and $B_0^n T_1^3$ hot-band progressions [2b] of the twist-mode is supported: due to the better cooling (skimmed jet) the hot-band based peaks (a- and c-peaks in fig. 2) exhibit less relative intensity in comparison to the LIF-spectrum [2b]. The one-color limit (~ half ionization potential, IP/2) of PCP is located in the well structured region of the S_1-transition. By comparison of the one and two-color REMPI-spectrum the IP can be determined. Figure 2D shows that the one-color limit lies between 31500 and 31545 cm^{-1}. Under consideration of the ionization threshold reduction (ΔE_{IP}) due to the extraction field ($\Delta E_{IP}=\alpha\sqrt{F} \sim 60$ [cm^{-1}]; with F is the field strength (240 [$^V/_{cm}$] in our setup and the constant $\alpha \sim 4$ for aromatics) the IP is 63110 ± 60 cm^{-1} (7.824 ± 0.007 eV, literature value: 7.8 ± 0.1 eV [10]).

ACKNOWLEDGEMENT

The author is deeply indebted to Dr. U.Boesl, Prof. E.W.Schlag, Prof. D.Lenoir, Prof. E.R.Rohwer and Prof. A.Kettrup for the ongoing support. Further on the author thanks Prof. F.Vögtle and S.Laufenberg for disposal of a PCP-sample and motivating discussions.

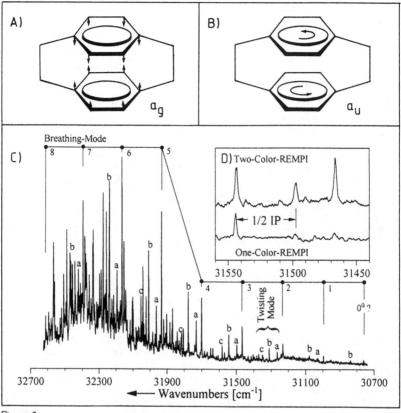

Figure 2

a)Relative atomic displacements due to the $S_1 \leftarrow S_0$ geometry shift and the a_g-breathing mode.
b)Relative atomic displacements of the twisting a_u-vibrational mode. c)Two-color REMPI-spectrum of the S_1-transition of jet cooled [2.2]-paracyclophane. The indicated progression is due to the a_g-breathing mode, induced by the $S_1 \leftarrow S_0$ geometrical shift. As second active mode the a_u-twist mode is observed (a,b,c in figure). d)Two-color(top) and one-color (bottom) REMPI-spectrum of PCP in the region of the one-color limit. The ionization potential is $7.824 \pm 0.007\ eV$.

REFERENCES

[1] a) B.Nickel, J.Hertzberg, *Chem. Phys.* 132 (1989) 219; b) B.Nickel, J.Hertzberg; *ibid.* 235
c) J.Wirz in *Excited States in Organic Chemistry and Biochemistry*, Ed.: B.Pullman, N.Goldblum (1977) D.Reidel Publishing Company, Dordrecht (Holland) 283
[2] a) A.Ron, M.Noble, E.K.C.Lee, *Chem. Phys.* 83 (1984) 215; b) T.-L.Shen, J.E.Jackson,J.H.Yeh, D.G.Nocera, G.E.Leroi, *Chem. Phys. Lett.* 191 (1992) 149
[3] R.Zimmermann, D.Lenoir, A.Kettrup, H.Nagel, U.Boesl, *Proceedings of the 26th Symp. (Internat.) on Combustion*, The Combustion Instiute (1996) in press
[4] a) U.Boesl, *J. Chem. Phys.* 95 (1991) 29; b) C.Weickhardt, R.Zimmermann K.-W.Schramm, U.Boesl E.W.Schlag, *Rapid Commun. Mass Spectrom.* 8 (1994) 381
[5] J.W.Barton in *Organic Chemistry, A Series of Monographs: Nonbenzenoid Aromatics*, Ed.: A.T.Blomquist, J.P.Snyder (1969) Academic Press, New York 33 ff.
[6] R.Zimmermann, *J.Molec.Structure* 337 (1996) 35
[7] G.Marconi, *Chem. Phys. Lett.* 169 (1990) 617
[8] R.Zimmermann, U.Boesl to be published
[9] a) F. Vögtle in *Cyclophan-Chemie*, Teubner Studienbücher Chemie (1990) B.G. Teubner Stuttgart b) S.Canuto, M.C.Zerner *Chem. Phys. Lett.* 571 (1989) 353
[10] S.Pignataro, V.Mancini, J.N.A.Ridyard, H.J.Lempka; *Chem. Commun.* (1971) 142

Resonance Photoelectron Spectroscopy (ZEKE) of Negatively Charged Molecules and Molecular Cluster: Spectral Resolution

Carsten Bäßmann, Gerhard Drechsler, Rainer Käsmeier, Ulrich Boesl

Institut für Physikalische und Theoretische Chemie, Technische Universität München
Lichtenbergstraße 4, 85747 Garching , Germany

Abstract. Spectroscopy on negatively charged molecular systems offers two major advantages: Anions may be mass separated prior to spectroscopy and experiments involving photodetachment enable access to neutral states. The corresponding methods are anion mass spectrometry and anion photoelectron spectroscopy. A highlight of mass selective anion photoelectron spectroscopy is high resolution anion-ZEKE spectroscopy. Some examples will be presented to illustrate the spectral resolution achieved in our group.

Anion mass spectrometry allows a large variety of experiments of otherwise hardly or not accessible molecular systems by supplying efficient anion sources and anion mass separation tools. High anion yields may be achieved by laser induced electron emission from metal surfaces with subsequent electron attachment to neutral molecules seeded or formed and cooled within a supersonic beam. A well resolved mass separation is available in the space focus of the pulsed ion source of a time-of-flight mass analyzer. Focussing a second pulsed laser into this space focus and choosing a well defined time delay between pulsed ion source and laser pulse results in mass selective photoelectron detachment.

Analyzing the energy of these photodetached electrons allows to perform anion photoelectron spectroscopy. In opposite to conventional photoelectron spectroscopy, the molecular final state is the neutral ground state: thus, mass selective anion PES gives information about the ground state of neutral molecules. These molecules may be short lived species, intermediates of chemical reactions or substances which are not availbale as pure samples (e.g. molecular clusters) and whose neutral ground state usually is hardly or not at all accessible to conventional spectroscopy. An enhancement of time-of-flight photoelectron spectroscopy is anion-ZEKE spectroscopy (ZEKE for Zero Kinetic Energy Electrons). Resolution has been increased from some 20 cm^{-1} for anion-photoelectron spectroscopy to 1,5 cm^{-1} for anion-ZEKE spectroscopy in our group.

Mass selected photoelectron spectroscopy of anions has been performed by several authors, already /1/. While ZEKE spectroscopy (neutral - cation transition) has been developed in our institute /2/, anion-ZEKE spectroscopy was introduced by Neumark and coworkers /3/. Our group succeeded in improvement of the resolution and applying it to difficult systems such as the iodine water van der Waals complex. This is possible due to the good cooling efficiency of our anion source and the special procedure to detect zero-kinetic electrons /4-6/. Sytems we have investigated or are still investigating by these methods (i) are metal carbon hydrides (i.e. intermediate complexes of the catalytic reaction of acetylene and iron/4,5/), (ii) molecular clusters (i.e. halogen water clusters, in particular $I(H_2O)/6/$), (iii) halogen oxyde compounds (i.e. ClO_x-isomers with relevance in stratospheric chemistry). In this paper, we want to demosntrate the resolution achieved for anion-ZEKE spectroscopy of different molecular systems, i.e. OH, FeO and $I(H_2O)$.

Two special means resulted in the improvement of resolution: (i) decelerating of the electrons by an electric stopping pulse after photodetaching the fast anions and (ii) delayed extraction of the zero-kinetic energy electrons after detchment. The effect of the time delay on the ZEKE resolution is shown in figure1. With increasing time delay (from 100 ns to 250 ns) the resolution increased from 2.6 cm^{-1} to 1.5 cm^{-1} demonstrated at the Q3(1)-line (see figure 2) of the OH$^-$-anion-ZEKE spectrum. The decrease of the integrated intensity is larger than linear. This can be explained by Wigner's threshold law which predicts a $E^{1/2}$ energy dependence of the detachment yield of emitted s-wave electrons. In figure 2, the anion-ZEKE spectrum of OH$^-$ around the vertical anion-neutral transition (the R3(0)-line) is shown. The cold and well resolved spectrum allows an unambiguous assignment. The rotational constants determined from this spectrum coincide with values in the literature. However, the line intensities do not justify the theoretical models applied to the photodetachment applied to OH$^-$ / 7/. More elaborate theories are necessary.

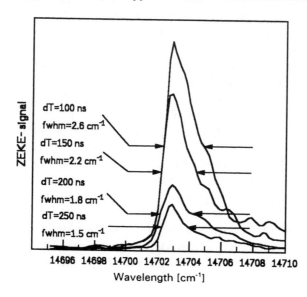

FIGURE 1. Single rotational line of the OH$^-$-ZEKE spectrum: the effect of time delayed electron extraction on the resolution.

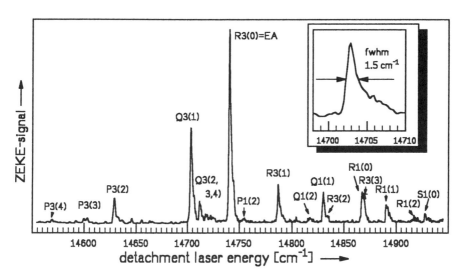

FIGURE 2. Anion-ZEKE spectrum of OH⁻; single rotational lines of the transition OH⁻ (X $^1\Sigma^+$) - OH(X $^2\Pi_{3/2}$). The energy of the line R3(0) (transition J = 0 (OH⁻) - J = 3/2 (OH)) corresponds to the vertical detachment energy (electron affinity EA).

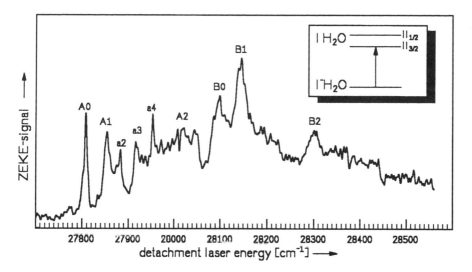

FIGURE 3. Anion-ZEKE spectrum of I(H₂O)⁻. The final state of the transition displayed in the figure corresponds to the $\Pi_{3/2}$-spin orbit component of atomic iodine. In the neutral I(H₂O) cluster this state splits into two components resulting in band systems A and B.

In figure 3, the anion-ZEKE spectrum of the iodine water cluster is presented. The final neutral state corresponds to the 3/2-spin orbit component of atomic iodine. In the cluster, the degeneracy of this component is lifted giving rise to the A and B band system in figure 3. Several vibrational modes (e.g. I-H$_2$O van der Waals stretching mode with 45 cm^{-1} (A1-A0; B1-B0)) appear. But also the rotational constant A (rotation around the I-O-axis) is large enough (>15 cm^{-1}) to induce resolvable rotational structure. The instrumental resolution was 2 cm^{-1} in figure 3. The narrowest line width was 3 cm^{-1} which is probably due to unresolved rotational substructure (rotation around the axis vertical to the I-O-axis). In figure 4, the FeO anion-ZEKE spectrum of the vibrationless FeO$^-$($^4\Lambda_{7/2}$) - FeO($^5\Lambda_4$) transition is displayed. The spectral resolution is sufficient to reveal rotational fine structure. By comparing experiment and simulation the anionic rotational constant has been deduced: B''(FeO$^-$) = 0.5075 cm^{-1} ±0.0005 cm^{-1}. This is in disagreement with values from autodetachment spectra /8/ (see second simulation in figure 4). For our simulations, the accurate rotational constant of neutral FeO /9/ has been used. In addition, a rotational temperature of 35 K has been deduced. This is a very low value considering how FeO is formed in our anion source. Water seeded in the supersonic beam reacted with Fe-atoms desorbed by laser from an iron needle in front of the nozzle.

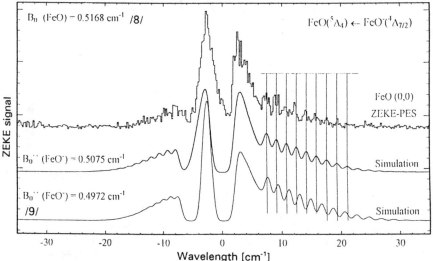

FIGURE 4. Anion-ZEKE spectrum of the transition between the lowest vibrationless states of anionic and neutral FeO. The resolution (<1.5 cm^{-1}) is sufficient to resolve rotational fine strucutre and determine the rotational constant of the FeO-anion.

1. Johnson M.A., Alexander M.L., Lineberger W.C., Chem. Phys. Lett. **112**, 285 (1984)
2. Müller-Dethlefs K., Schlag E.W., Ann. Rev. Phys. Chem. **42**, 109 (1991)
3. Kitsopoulos T.N., Waller I.M., Loeser J.G., Neumark D.M., Chem. Phys. Lett. **159**, 300 (1989)
4. Drechsler G., Bäßmann C., Boesl U., Schlag E.W., Z. Naturforsch. **49a**, 1256 (1994)
5. Drechsler G., Bäßmann C., Boesl U., Schlag E.W., J. Mol. Spectrosc. **348**, 337 (1995)
6. Bäßmann G., Boesl U., Yang.D., Drechsler G., Schlag E.W., Int. J. Mass Spectrom. Ion Proc. accept.
7. Schulz P.A., Mead R.D., Jones P.L., Lineberger W.C., J. Chem. Phys. **77**, 1153 (1982)
8. Andersen T., Lykke K.R., Neumark D.M., Lineberger W.C., **86**, 1858 (1987)
9. Merer A.J., Annu. Rev. Phys. Chem. **40**, 407 (1989)

Dissociation of the Rydberg States of CaCl
Investigated by Ion-dip Spectroscopy

Jian Li

Department of Chemistry, Tsinghua University, Beijing 100084

Yaoming Liu, Hui Ma

Department of Physics, Tsinghua University, Beijing 100084

Jason Clevenger, David Moss, Robert W. Field

Department of Chemistry, MIT, MA02139

Introduction

The alkaline earth monohalides (MX) have uniquely simple electronic structures. They are well described as simple three-particle systems consisting of one nonbonding electron plus two closed shell atomic ions, M^{2+} and X^-. MX molecules serve as a prototype for the electronic structure and electron-nuclei energy exchange processes of more complex dipolar core molecules. CaCl has $D_0^0(32990 \text{ cm}^{-1})<<\text{IP}(48489 \text{ cm}^{-1})$. The early onset of predissociation adds difficulties in probing the whole Rydberg series. Previous results on CaCl are mainly the study of the low electronic states [1-7]. Christopher Gittins and Nicole Harris, two former members of Prof. Field's group in MIT, observed several Rydberg states in $n^*<4$ region by direct fluorescence detection and $v=1$ $n^*=16-40$ autoionizing Rydberg states by REMPI-TOF spectroscopy[8,9]. Several series have not been detected either at low n^* or in autoionizing region. Both fluorescence and REMPI-TOF detections, however, failed to detect the Rydberg states with intermediate n^*. Those states are believed to be dominated by a fast predissociation process. It's important to observe those states to make a smooth connection from the lowest states to the states observed at and above $n^*>16$ and get a complete picture. They can also provide an understanding of the mechanism for predissociation of the highly polar core.

We report here initial results of an ion-dip spectroscopy study on CaCl predissociated Rydberg states.

Experiment

Jet-cooled CaCl is produced by the reaction of Ca plasma with chloroform. The PUMP laser selectively populates CaCl molecules to an individual rotational level ($v''=0$ and 1, N'') of the $D^2\Sigma^+$ state. Part of the CaCl molecules in the D state are further ionized by an IONIZATION laser, which is the second harmonic of a Nd:YAG laser. The amplitude of the ion signal indicates the CaCl population in the D state. A third laser (PROBE) excites CaCl in the D state to higher Rydberg states. Since most of the Rydberg states dissociate immediately, the probe transition leads to a depletion of D state population, causing a dip in the ion signal. By scanning the PROBE laser, an ion-dip spectrum is obtained which represents the transitions to the high Rydberg states from the D state.

The PUMP and PROBE lasers are simultaneously pumped by a Nd:YAG laser. The PROBE and IONIZATION laser pulses are delayed by about 10ns and 25ns respectively relative to the PUMP shot to avoid temporal overlapping of the three pulses. These delays are important to ensure the observed ion-dips are truly due to transitions from D state to Rydberg states.

The $CaCl^+$ ions are collected by a pulsed extraction field and detected by the time-of-flight spectrometer using dual micro-channel plates. Two isotopes of chlorine, ^{35}Cl and ^{37}Cl, were resolved by our time-of-flight spectrometer.

Results and Discussion

We recorded spectra via both v=0 and v=1 $D^2\Sigma^+$ intermediate state. Spectra of both isotopes were resolved and recorded. The R branch of the D-X transition is well resolved and fully assigned.

So far we have observed several Σ states in the n*=6-8 region. Figure 1 is a term value plot of the v=0 7.84 $^2\Sigma^+$ for $Ca^{37}Cl$. The label at the left corner of each spectrum indicates the intermediate state transition pumped. The term value is determined by the term value of the intermediate level plus the calibrated probe frequency. There is one line coincide for two intermediate level with N differing by 2. It's the typical Σ-Σ transition pattern. Therefore the electronic assignment was obtained. Since both isotopes were resolved, the absolute vibrational numbering was determined unambiguously in terms of the isotope shift. All the spectra recorded were

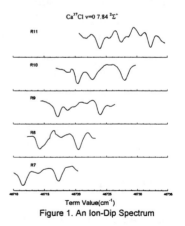

Figure 1. An Ion-Dip Spectrum

rotationally well resolved as well. Effective principal quantum numbers and rotational constants were evaluated for each state as well as the linewidth of the probe-induced dip structures. In order to make sure the observed linewidth was not affected by power broadening, a power effect study was carried out. The observed dip structures didn't show an obvious change in linewidth except the dip became ambiguous at low power. We have arranged all the assigned states into three series. The series are indicated by the effective principal quantum number module one. States within the series have almost the identical quantum defects. These results are listed in Table 1.

Table 1. Evaluated molecular constants and measured linewidth.

Series	n*	v	T_v(cm^{-1})	B_v(cm^{-1})	Linewidth (cm^{-1})	Lifetime (ps)
0.18	6.194	1	46068.65(7)	0.16177(13)	0.65±0.15	8
	8.178	1	47288.23(12)	0.15973(25)	1.1±0.3	5
0.45	6.416	1	46263.15(4)	0.16269(10)	0.6±0.1	9
	7.456	0	46512.46(3)	0.16146(18)	0.95±0.1	6
	7.451	1	46592.36(13)	0.16068(24)	1.15±0.2	5
0.84	7.843	0	46702.37(3)	0.16368(17)	0.95±0.1	6
	7.846	1	47147.16(18)	0.16184(34)	1.05±0.15	5

1σ uncertainties are in parentheses.

The dissociation process of CaCl to neutral atoms requires the transfer of two electrons onto the Ca^{2+} ion. First, the Rydberg electron must enter the spatial region of the 4sσ orbital on the Ca^{2+}. Second, while the Rydberg electron is near the Ca^{2+}, a Cl$^-$ 3pπ or 3pσ electron must transfer to the Ca 4sσ orbital. The predissociation rate, Γ, of such two-electron transfer processes are determined by the square of the product of an interelectronic $1/r_{12}$ matrix element and a vibrational overlap factor between the continuum of the repulsive curve with the Rydberg state,

$$\Gamma \alpha \mid <Ca^04s,Ca^04s\mid 1/r_{12}\mid CaCl\ n*l\lambda,\ Cl^-\ 3p\sigma\ or\ 3p\pi> \times <V_{n*l\pi}\mid \varepsilon_{Ca,Cl}>\mid^2$$

The n* and v dependence of the predissociation rate within a given Σ or Π series is determined by the shape of the repulsive Ca(^{1}S)+Cl(^{2}P) $^2\Sigma^+$ or $^2\Pi$ diabatic potential curves. Once the shapes of the two repulsive diabatic potential curves are known, all electronic, vibrational and rotational factors are calculable and the observed predissociation rates will provide information about the intra-core partial-l characters of each n*lλ Rydberg series.

The predissociation rate can be measured directly from the width of a rovibronic transition in the ion-dip spectrum. From the linewidth information we can estimate the lifetime. The evaluated lifetime were also listed in Table 1. So far the obtained linewidth information is not enough to determine the exact repulsive potential curves. However, the observation of the linewidth varied from 0.6cm^{-1} to 1.15cm^{-1} demonstrates a change in predissociation rate. And, the fact that several Σ states of the same n* but different v are missing in our ion-dip spectra may suggest how the repulsive curves cross the Rydberg potential curves.

Since CaCl is very similar to CaF we can compare the results of CaCl to the core-penetrating Rydberg series of CaF which are well studied[10]. The three observed Σ series of CaCl are 0.18, 0.45 and 0.84 $^2\Sigma^+$. They are close to those of CaF, 0.18, 0.55 and 0.88 $^2\Sigma^+$ series and we think we observed all the core-penetrating Σ series for CaCl. There are two Π core-penetrating series in CaF but no Π state was observed in this ion-dip experiment. A possible reason is that Cl 3pσ orbitals are localized along the internuclear axis hence have larger overlap with the Ca 4s orbital. Therefore, the Π states may dissociate slower than the Σ states and can't be detected by ion-dip detection. The $^2\Delta$ Rydberg states are also expected to have a weak predissociation trend, nor are they accessible via the current D$^2\Sigma^+$ state pumping. New experiment are being designed to detect the missing states.

Acknowledgment

We would like to thank Kevin Cunningham for helpful discussions. We also thank National Science Foundation for financial support.

Reference

1. P.J. Domaille, T.C. Steimle, and D.O. Harris, J. Chem. Phys. , **68**, 4977(1978).
2. L.E. Berg, L. Klynning , and H. Martin, Chem. Phys. Lett. , **54**, 357(1978).
3. L.E. Berg, L. Klynning , and H. Martin, Physica Scripta **21**, 173(1980).
4. L.E. Berg, L. Klynning , and H. Martin, Physica Scripta **22**, 216(1980).
5. L.E. Berg, L. Klynning , and H. Martin, Physica Scripta **22**, 221(1980).
6. W.E. Ernst, J.O. Schroder, U. Buck, J. Kesper, T. Seelemann, L.E. Berg, and H. Martin, J. Mol. Spectrosc., **117**, 342(1986).
7. A. Pereira, Physica Scripta, **34**, 788(1986).
8. Christopher M. Gittins, Ph.D. thesis, Masachusetts Institute of Technology, 1995.
9. Nicole Harris, Ph.D. thesis, Masachusetts Institute of Technology, 1995.
10. J.M. Berg, J.E. Murphy, N.A. Harris, and R.W. Field, Phys. Rev. A , **48**, 3012(1993).

The pump-probe femtosecond VUV REMPI technique applied to the study of intramolecular relaxation of high-lying molecular states.

Alexey V. Baklanov[*], Lars Karlsson[x], Ulf Sassenberg[x]
Anders Persson[#] and Claes-Göran Wahlström[#].
[*]Institute of Chemical Kinetics & Combustion, Institutskaya 3, Novosibirsk, 630090, Russia.
[x]Stockholm University, Department of Physics, Box 6730, S-113 85, Stockholm, Sweden.
[#]Lund Institute of Technology/Atomic Physics, Box 118, S-221 00 Lund, Sweden.

Abstract. The pump-probe VUV (1+1')REMPI technique has been used for a study of the femtosecond dynamics of the allyl iodide (AI), benzyl iodide (BI) and toluene (Tol) molecules excited in the VUV region. The VUV radiation was produced by non-resonant quintupling of Ti:S laser radiation in xenon gas. The lifetimes of the excited ($h\nu \approx 7.8$ eV and 8.1 eV) states of these molecules were found to lie between 100 and 500 fs. The nature of the processes governing these decay times is discussed.

In a recent work[1] the pump-probe femtosecond VUV REMPI technique was shown to be applicable to the study of the dynamics of high-lying molecular states. In this work this technique is used for the measurements of the dynamics of VUV excited allyl iodide, benzyl iodide and toluene molecules..

The experimental set-up is shown in fig. 1. A femtosecond Ti:S laser at Lund Institute of Technology described by Svanberg et al.[2] has been used in this work. The characteristics of the laser pulse were: time duration - 150 fs; wavelengths in different experiments -767 and 792 nm; spectral fwhm - 8 nm; energy - 5-15 mJ. The laser beam was split into two approximately equal parts, one of which (pump beam) was used for generation of the VUV radiation. This was produced by non-resonant frequency quintupling of the fundamental output from a Ti:S laser in xenon. The fundamental radiation of Ti:S laser was focused (15 cm focal length) into a cell containing Xe at a pressure of 30 Torr. The cell (a simple cylindrical tube) had a MgF_2 collimating lens arranged as the output window. The collimated output VUV radiation was directed into the ionization cell (IC_1) containing the vapor of the substance under study by means of complete internal reflection in a MgF_2 prism. The other part of the fundamental (probe beam) was time delayed with a movable translation stage (MTS) and directed into the same IC_1 cell through a set of MgF_2 prisms. Pump and probe beams were arranged to spatially overlap in the IC_1 cell. It must be pointed out that part of the fundamental radiation necessarily also passes through the Xe cell and leaks into the IC_1 cell together with the VUV component. But this did not affect the experiment. This is so because the pulse enters the IC_1 cell before the VUV pulse (approximately 7 ps) due to different refraction indices of the various MgF_2 optical elements (input window and output lens of IC_1 and prism) at the two wavelengths (VUV and fundamental) involved. So this fraction of the fundamental is not capable to ionize any VUV excited molecules by itself.

Details of the ion current measurements and the VUV intensity control are described elsewhere[3].

In fig. 2 the ionization signal of benzyl iodide (P_{BI}=30 mTorr) is shown as a function of the delay time between the pump (λ_{pump} =158 nm; $h\nu \approx 8.1$ eV) and probe (λ_{probe} =792 nm; $h\nu \approx 1.57$ eV) pulses. Due to nonlinear conversion of the fundamental wavelength in Xe odd harmonics are generated. However the optics is only transparent for wavelengths longer than what corresponds to the seventh harmonic. For the fifth harmonic it is necessary to remove absorbing air from the optical path for example by purging with a transparent gas (Ar). The third harmonic is always penetrating. The signal presented in fig. 2 disappeared when the VUV radition was removed which indicates that this was caused by ionization of VUV (5th harmonic) excited molecules. The experimental time profile was fitted (solid line in the figure) by a convolution of the exponential decay with the gaussian cross-correlation function. The decay time of the excited state of benzyl iodide was found to be τ=480±60 fs. In fig. 3 the ionization signal of toluene (P_{Tol}=250 mTorr) is shown as a function of the delay time between the pump (λ_{pump}=154 nm; $h\nu \approx 8.3$ eV) and probe (λ_{probe} =767 nm; $h\nu \approx 1.61$ eV) pulses. The result of the fit in this case was τ=200±40 fs. Analogous measurements for allyl iodide (λ_{pump}=154 nm, λ_{probe} =767 nm) give τ=120±60 fs. The power dependence of the signal on the probe pulse intensity was also measured. For this purpose the measurements were made by attenuating the probe pulse energy with filters. For all three cases the degree of the power dependence was found to be near 1. Therefore one quantum of the fundamental radiation was sufficient to ionize the excited states of the molecules under study. This is also in agreement with known ionization potential data (for toluene[4] IP=8.82±0.01 eV, for benzyl iodide IP is probably lower than for toluene in accordance with the behaviour of other halogen containing derivatives of toluene[4] and for allyl iodide Carrapt et al.[5] calculate IP_{ad}=9.24 eV) .

The nature of the observed signal and the decay processes are questionable. In none of the investigated three cases was it possible to characterize the excited states. Nevertheless the nature of the processes governing the observed decay time can be considered from the general point of view. Formally the observable kinetics of the signal decay corresponds to a molecular transition from a state with higher ionization cross-section σ_{ion} at the particular wavelength of the probing pulse to a state with lower σ_{ion}. It is possible to imagine several processes which are relevant for the excitation mechanism and subsequent evolution:

1) - "vertical" photoabsorption, leaving the molecule in an excited state with the same geometry as the ground state (coherent wave packet[6]),

2) - electronic and/or vibrational relaxation resulting in an incoherent molecular state,

3) - photodissociation and resulting products.

The energy (1.6 eV) of the probing quantum of the fundamental is substantially less than IP_{ad} for all these molecules. So the efficiency of the ionization will be strongly dependent on the onset value (or term value) of the transient electronic state. The intramolecular relaxation processes transfer the molecules into electronic states with generally lower onsets. So it is natural to anticipate a substantial drop in the σ_{ion} value in a relaxed state (2) in comparison with the initial state (1), in particular if this relaxed state is the ground electronic state. Therefore the observed decay time can be referred to the relaxation process.

The toluene molecule is of particular interest for further discussions.

Shimada et al.[7] have established that toluene when excited with the nearby wavelength λ=158 nm (E^*=7.85 eV) dissociates with a characteristic time of about 10 ns - probably from the ground electronic state. So we conclude that the very much shorter time (200 fs) we observe is due to intramolecular relaxation processes. At this point it is difficult to tell whether this corresponds to relaxation to the ground electronic state or to some transient electronically excited state. For this purpose it will be necessary to probe the ground state of toluene possibly with a shorter wavelength. At the present probing wavelength used (767 nm) the absorption from the electronic ground state of toluene is very small.

In an earlier study Wiesenfeld and Greene[8] excited a 4d Rydberg state at 7.94 eV (which is to be compared with our 8.1 eV) in toluene via a resonant two-photon process. They concluded that the decay time of their Rydberg state was equal to 170±20 fs.

As to the iodine containing molecules we can not exclude the possibility that the decay times are affected by fast dissociation processes. Cheng et al.[9] observed both the decay of the transient ionization signal of UV excited iodobenzene and the simulataneous accompanying atomic iodine appearing during the decay. Zewail and co-workers[10] have studied methyl iodide by applying the pump-probe technique and drawn the conclusion that the decay time of the ionization signal is governed by predissociation of the excited Rydberg states.

Further experiments are therefore necessary to determine which decay channel is the dominating one in AI as well as in BI.

In conclusion we believe that the femtosecond VUV REMPI technique will turn out to be a useful method in the study of the dynamics of the high-lying valence and Rydberg molecular states.

This work was financially supported by the Royal Swedish Academy of Sciences and in part by the Russian Foundation of Fundamental Research.

References

1 A.V. Baklanov, V.P. Maltsev, L. Karlsson, U. Sassenberg and A. Persson, *Faraday Transactions*, 1996, **92**, 1681.

2 S. Svanberg, J. Larsson, A. Persson, and C.-G. Wahlström, *Physica Scripta*, 1994, **49**, 187.

3 A.V. Baklanov, V.P. Maltsev, L. Karlsson, B. Lindgren and U. Sassenberg, *Chem. Phys.*, 1994, **184**, 357.

4 Handbook of Chemistry and Physics, 73rd Edition, 1992-93, Ed. by D.R. Lide (CRC Press), p.10-229.

5 P.-A. Carrapt, J. M. Riveros and D. Stahl, *Int. J. Mass Spectrom. Ion. Proc.*, 1991, **110**, R1.

6 A.H. Zewail, *J. Phys. Chem.*, 1993, **97**, 12427.

7 T. Shimada, Y. Ojima, N. Nakashima, Y. Izava, and C. Yamanaka, *J. Phys. Chem.*, 1992, **96**, 6298.

8 J.M. Wiesenfeld and B.I. Greene, *Phys. Rev. Lett*, 1983, **51**, 1745.

9 P.Y. Cheng, D. Zhong and A.H. Zewail, *Chem. Phys. Lett.*, 1995, **237**, 399.

10 M.H.M. Janssen, M. Dantus, H.Guo and A.H. Zewail, *Chem. Phys. Lett.*, 1993, **214,** 281.

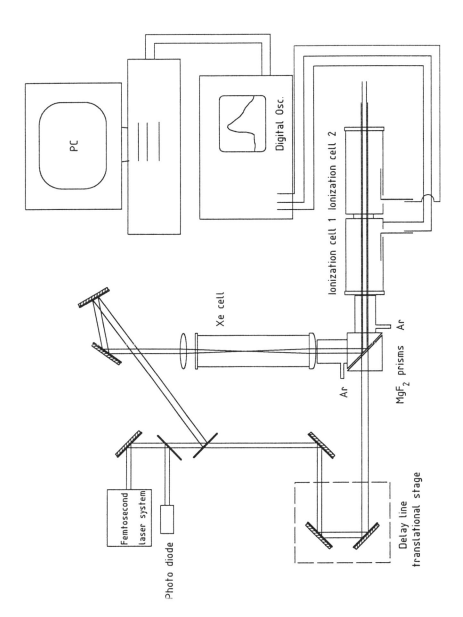

Fig 1. Experimental set-up.

415

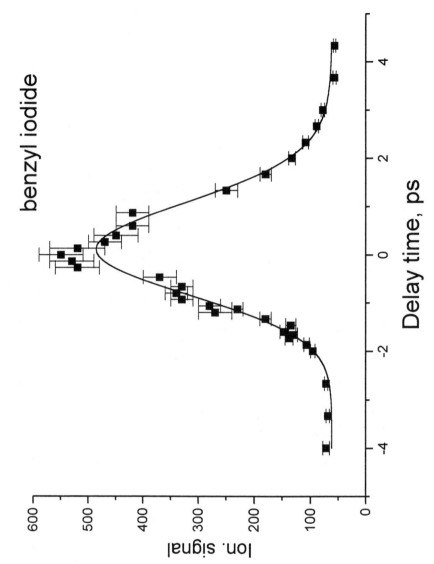

Fig.2. Ionization signal of the VUV excited BI molecules plotted against the delay time between pump and probe pulses. The solid line is the fitted curve, which is the decaying exponent convoluted with the gaussian cross-correlation function.

POSTER SESSION III: APPLICATIONS

Photon Burst Mass Spectrometry for the Measurement of ^{85}Kr at Ambient Levels

C. S. Hansen[a], W. M. Fairbank, Jr.[a], E. P. Chamberlin[b],
N. S. Nogar[b], X.-J. Pan[a], H. Oona[b] and B. L. Fearey[c]

[a]*Physics Department, Colorado State University, Fort Collins CO 80523-1875 USA*
[b]*CST Division, MS J565 Los Alamos National Laboratory, Los Alamos NM 87545 USA*
[c]*NIS Division, MS J565 Los Alamos National Laboratory, Los Alamos NM 87545 USA*

Abstract. Photon Burst Mass Spectrometry is being developed for direct measurement of noble gas isotope ratios in the 10^{-9} to 10^{-13} range. First measurements of a ^{85}Kr sample with a 6×10^{-9} abundance are reported. Incremental improvements in detection efficiency and stray light should make possible measurements in the near term of ^{85}Kr at the ambient atmospheric abundance of 10^{-11}, with a detection limit of 10^{-13}.

INTRODUCTION

Noble gas radioisotopes, with inert properties and half-lives ranging from several days to hundreds of thousands of years, are ideally suited for use in a number of geological dating, tracer and monitoring applications in both the basic and applied sciences (1). For example, ^{85}Kr, the most abundant of these isotopes in the atmosphere, is primarily a fission product with a half life of 10.7 years. It can be used for nuclear monitoring, atmospheric transport and dating young ground water up to 40 years. Unfortunately low elemental concentrations and isotopic abundance, compounded by low energy decay modes, for most of these isotopes has made routine measurements a daunting task.

Photon Burst Mass Spectrometry (PBMS) is a laser based detection technique which combines the high isotopic selectivity (10^{10}) of photon burst detection with the moderate mass selectivity (10^{5}) of a mass spectrometer (2-4). This technique relies on the detection of a "burst" of spontaneously emitted photons produced by the interaction of a single two-level atom with a resonant laser beam. Fairly isolated recycling transitions, accessible to existing lasers, can be found for atoms and ions of half the elements of the periodic table, including the noble gases (2). The photon burst transition used for Kr isotopes is between the $1s_5$ and $2p_9$ states. In ^{85}Kr and ^{83}Kr, because of a non-zero nuclear spin, the F"=13/2 to F'=15/2 hyperfine transition is used.

Preliminary tests of our PBMS instrument, demonstrating an extrapolated detection limit of 10^{-8} isotopic abundance, have been reported previously (5). Here we present actual measurements of ^{85}Kr in a commercial sample with an isotopic abundance of 6 x 10^{-9}. Projected enhancements should make the instrument capable of measuring ^{85}Kr and other radioactive noble gas isotopes at ambient atmospheric abundances ($\leq$10^{-11}).

APPARATUS

The major components of the photon burst mass spectrometer are shown in Fig. 1. Kr ions are produced in a microwave ion source and imaged through a 126° magnetic sector. A deceleration lens slows the ions to 2000 eV before they are prepared in the correct neutral Kr quantum state by charge exchange in a Rb vapor. A narrow linewidth, external-cavity, tunable diode laser (MicraLase ML-02B) resonantly excites the photon burst transition in the atoms. The anti-collinear geometry of the atom and laser beams ensures a long region of overlap and also reduced Doppler width due do velocity compression in fast beams. Spontaneous emission from the Kr atoms is collected and counted in ten photon burst detectors with as high efficiency as possible (75% and 3%, respectively), and analyzed for bursts by a series of digital electronic circuits.

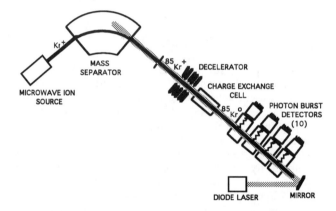

FIGURE 1. Schematic diagram of the Photon Burst Mass Spectrometry apparatus.

RESULTS AND DISCUSSION

The procedure for the measurement of the 6x10^{-9} isotopic abundance sample is as follows. A gas manifold at the source allows switching and mixing of multiple gas samples, including one with ^{85}Kr at 10ppm in nitrogen (10^{-2} ^{85}Kr abundance), one with pure Kr at an abundance of 6x10^{-9}, and a third with nitrogen buffer gas. The PBMS system was first tuned up with approximately 50% N$_2$, 50% Kr from the low level sample, and a small amount from the high level ^{85}Kr sample. The ^{85}Kr abundance in this mixture was 1.6x10^{-6}, calibrated by fluorescence measurements on the high level sample at a measurable ^{85}Kr current. The leak valve to the high level sample was then closed off, so that the ^{85}Kr abundance in the mixture decayed rapidly to 6×10^{-9}.

Fluorescence and photon burst signals were measured for 6-10 100 second periods consisting of 50 seconds with the laser in resonance with the ^{85}Kr atoms followed by 50 seconds with the final energy of the atomic beam shifted by 30 V (equivalent to 650 MHz laser detuning). The raw count rates for total fluorescence on one photomultiplier tube (Tube 9) and for a burst of ≥ 1 photon in 3 of the 9 detectors (Coincidence 3) are shown on the left in Fig. 2 for the first of three such runs (Run K). In resonant periods 1-4 final tuning of the system was done. The data from periods 4-7 were approximately constant and were used to get a calibration at 1.6×10^{-6} abundance. When the leak valve was closed at the end of period 7, the ^{85}Kr abundance decayed exponentially to the low level within one 100 second cycle. The differences between the on-resonance coincidence 3 rate in one 50 second period and the average the off-resonance rates in the two adjacent periods are plotted in the graph on the right. Fluctuations in individual 100 second measurements are large, but the average of these six 100 sec measurements, 37(21) bursts/sec (solid line and gray area representing the uncertainty) is in reasonable agreement with the expected value, 53 bursts/sec (dotted line).

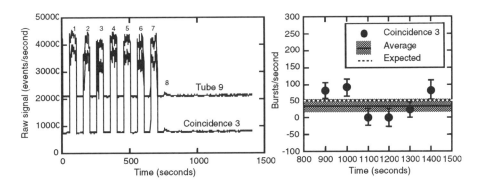

FIGURE 2. Measurement of ^{85}Kr at 6×10^{-9} abundance: raw data before and after closing the leak valve (left) and difference between on and off resonance (right).

The final results for all three measurement runs are summarized in Table 1. All but one of the measured values is within one standard deviation of the expected values for 6×10^{-9} abundance. One pleasant surprise in these results is the fact that measurements of ^{85}Kr at an abundance of 6×10^{-9} were possible to make using the fluorescence recorded in a single detector. A signal of about 50 counts/sec was found in a background of 21,000 counts/second. The high stability of the diode laser and the ion beam is responsible for this success. The background rate in the burst mode (coincidence 3) was somewhat lower (8500 bursts/sec), but the results were not appreciably better. It was not possible to use the burst method to great

TABLE 1. Summary of measurements of ^{85}Kr at 6×10^{-9} abundance

Data run	Integration time(sec)	Coincidence 3 (bursts/sec) Measured	Expected	Tube 9 (counts/sec) Measured	Expected
K	650	37(21)	53	45(24)	40
L	1000	60(16)	57	88(20)	42
M	950	64(31)	54	60(17)	46

advantage in this case because the average burst size per resonant atom was small (~2.2 counts total, or 0.25 per detector).

In order to reach the desired goal of ^{85}Kr measurements at the ambient atmospheric abundance (1×10^{-11}), it is necessary to increase the quantum efficiency of the detectors and reduce the stray light background. Several new low gain (10^3-10^4) detectors have been tested and shown to be capable of more efficient (at least 6-15%) detection of single photons at 811nm using a charge-sensitive preamplifier. These include a hybrid photomultiplier tube (DEP PPO275B), an intensified photodiode (Intevac 280AD) and an avalanche photodiode (Advanced Photonics). With this change and a 4-8 times reduction in stray light, measurements of atmospheric ^{85}Kr samples with 10-30% statistical accuracy in samples of modest size is expected (6). The projected minimum detectable abundance is ~10^{-13}.

ACKNOWLEDGMENTS

This work was under the auspices of the U. S. Department of Energy under Contract No. W-7405-ENG-36.

REFERENCES

1. B. E. Lehmann, B. E. and Loosli, H. H., "Use of noble gas radioisotopes for environmental research" in *Resonance Ionization Spectroscopy 1984*, G. S. Hurst and M. G. Payne, editors, Bristol, Institute of Physics, 1984, pp. 219-226.
2. Fairbank, W. M., Jr., *Nucl. Instrum. Methods B* **29**, 407-414 (1987).
3. LaBelle, R. D., Fairbank, W. M., Jr. and Keller, R. A., *Phys. Rev. A* **40**, 5430-5433 (1989).
4. LaBelle, R. D., Hansen, C. S., Mankowski, M. M. and Fairbank, W. M., Jr., "Isotopically Selective Atom Counting Using Photon Burst Detection," *Phys. Rev. A*, in press.
5. Hansen, C. S., Pan, X.-J., Fairbank, W. M., Jr., Oona, H., Chamberlin, E. P., Nogar, N. S. and Fearey, B. L., "Photon burst mass spectrometry--ultrasensitive detection of rare isotopes," *Proc. Soc. Photo-Opt. Instr. Eng.* **2385**, 136-139 (1995).
6. Hansen, C. S., Ph. D. Thesis, Colorado State University, 1995.

Development of a New Ti:Sapphire Laser System for Femtosecond Laser Ionization at kHz Repetition Rates

M.A. Dugan

Clark-MXR, Inc.
7300 West Huron River Drive
Dexter, MI 48130

M. L. Pacholski, K. F. Willey, R. M. Braun and N. Winograd

Department of Chemistry
The Pennsylvania State University
184 Materials Research Institute, Research Park
University Park, PA 16802

Abstract. We have developed a new laser system which operates at a repetition rate of one kHz and generates 3.5 mJ/pulse with 85 fs pulse widths at 800 nm. It is composed of a self-mode locked oscillator followed by a pulse stretcher, a regenerative amplifier and post-amplifier which are pumped by frequency doubled Nd:YAG lasers, and a compressor. The new post-amplification stage has a four-pass, nearly collinear geometry. This allows for improved power and shorter pulses while still providing high repetition rates. Fast repetition rates are essential for imaging experiments using a time of flight (TOF) mass spectrometer since a fast rate decreases the probability of sample drift during image acquisition and decreases the amount of time needed per image.

LASER SYSTEM

The lasing process is initiated by a cw Argon Ion laser pumping a Ti:Sapphire crystal at 3 W at all lines. The oscillator self-mode locks due to Kerr lens modelocking (1) and produces an output of 3 nJ with a 60 fs pulse width. The pulse width is then stretched to 300 ps by multiple passes on a 1400 lines per millimeter grating.(figure 1) The stretched pulse is amplified to approximately 1.5 mJ/pulse in a Ti:Sapphire crystal which is pumped with a frequency doubled,

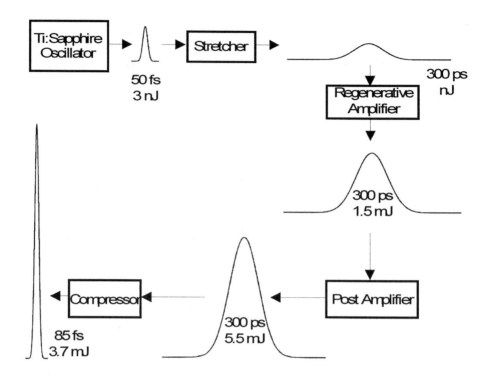

FIGURE 1. Overview of laser system pulses

Q-switched Nd:YAG laser running at an average power of 8.5 W. A Pockels cell injects pulses into the regenerative amplifier cavity and dumps the selected amplified pulse. This output is then directed into the post-amplifier by the isolation stage of the stretcher. The light is amplified to 5.5 mJ/pulse by passing four times, nearly collinearly through a third Ti:Sapphire crystal which is pumped with 14 W average power from a doubled Nd:YAG. The beam from the post-amplifier is expanded with a telescope and directed into the compressor. After compression the fundamental can be used directly or can be frequency doubled or tripled.

This newly designed laser system has a pulse energy of 3.7 mJ with a pulse width of 85 fs at a 1 kHz repetition rate. Without a post-amplifier, the pulse energy is 1.5 mJ and the pulse width is 150 fs at the fundamental. The original plans for this laser system called for a bow-tie configuration (which is typically used in Clark-MXR's 10 Hz systems) in the post-amplification stage. However, at these high powers, the very first pass of the laser is close to the saturation intensity for the Ti:Sapphire crystal. These powers also generate thermal gradients causing thermal lensing within the crystal which can result in poor mode quality and diffraction of the beam. The nearly collinear design helps minimize these problems, since each beam, passing through about the same point, is experiencing the same aberrations. In spite of this, there were still deleterious effects from thermal lensing in the crystal, so gain reshaping was employed by using a 532 nm pump beam which was smaller than the 800 nm beam. This amplified the most intense parts of the 800 nm beam and therefore eliminated many of the negative qualities of the beam which existed at the edges of the gaussian profile. At energy densities near the saturation point, the pump fluence dominates the amount of amplification that can occur. Ideally, the pump beam would be as small as possible, however, a compromise must be reached so the thermal gradients are not too large. The increased power of this system will lead to an increase in doubling and in tripling power.

POSTIONIZATION AND MASS SPECTROMETRY

The time of flight instrument used for our studies has been specially designed to incorporate laser ionization experiments. It is equipped with a leak valve to facilitate gas phase experiments. Moreover, we have incorporated a liquid nitrogen cooled sample stage which removes the background gases which are present and slows the sublimation rate of volatile components of the sample. This is critical to postionization studies.(2) The system is also equipped with a time to digital converter (TDC) and an analog detector which receive signal from a dual microchannel plate assembly. Previous experiments in our laboratory, on a similar machine, indicated that the detection system (TDC) was being saturated due to the large number of ions produced in postionization studies. The analog detector eliminates this problem and counts all of the ions hitting the detector giving a more quantitative representation of the mass spectrum.

The shorter pulses of this new laser increase the probability of outrunning neutral dissociation events and should aid in the detection of intact molecular ions.(3) Previous power dependence studies indicated that the ionization volume was not saturated. However, preliminary experiments with the new laser system indicate that the ionization volume is beginning to approach saturation.

A very fast repetition rate is essential to chemical imaging using mass spectrometry. It not only saves time by requiring less time per image, but also decreases the likelihood of sample drift during image acquisition which further improves the spatial resolution of imaging experiments. If we consider an area of 20 μm by 20 μm with 20000 pixels and an ion dose of 1% (static limit) with a Ga^+ beam at 60 pA, then a typical secondary ion mass spectrometry (SIMS) experiment with a primary ion pulse of 10 ns and a repetition rate of 5 kHz takes 33 minutes to reach the static limit. Our goal is to reach the static limit as quickly as possible so as to gain the most information in the shortest amount of time. In comparison, a laser running at 10 Hz with a primary ion pulse of 500 ns would require 5.5 hours to reach the static limit. However, a femtosecond laser with a repetition rate of 1 kHz and a primary ion pulse of 500 ns only requires 3.2 minutes for the same area and conditions. A short primary ion beam pulse is necessary in SIMS because the pulse length is what determines the mass resolution. In laser postionization studies, the mass resolution is determined by the ability of the mass spectrometer to compensate for the energy dispersion introduced by the finite width of the laser beam so a much larger ion pulse can be used.

Images and spectra have already been obtained using this laser system and instrument. The image in figure 2 shows a map of pyrene butyric acid on silver beads. The image is 200 μm by 200 μm and was obtained by postionizing the neutral particles with 266 nm light that were produced as a result of Ga$^+$ ion bombardment. This map represents the m/z 215 ion (M-CH$_2$CH$_2$COOH) where the brightest areas are most intense. The postionized spectra show significant quantities of the molecular ion and fragments at m/z 215. In comparison, SIMS spectra show much fragmentation and very little intact molecular ion.

FIGURE 2. Map of fragment ion of pyrene butyric acid on ~70 μm silver beads.

CONCLUSION

The new laser system and time of flight mass spectrometer have resulted in improved chemical imaging capabilities. In addition, the increased peak power of the laser is leading to a further understanding of the mechanisms which govern the ionization process in the gas phase.

The authors gratefully acknowledge the financial support of the National Science Foundation and the National Institute of Health.

REFERENCES

1. Captain, H. C., and Mourning, M. M., *Optics & Photonics News* **5**, 20–28 (1994).
2. Wood, M. C., Zhou, Y., Brummel, C. L., and Winograd, N., *Anal Chem* **66**, 2425–2432 (1994).
3. Weinkauf, R., Aicher, P., Wesley, G., Grotemeyer, J., and Schlag, E.W., *J Phys Chem* **98** 8381 (1994).

RIS-TOFMS System to Search for Double Beta Decay of ^{136}Xe

S.Sasaki, H.Tawara, E.Shibamura*and M.Miyajima

National Laboratory for High Energy Physics (KEK), Oho 1-1, Tsukuba, Ibaraki 305, Japan
*Saitama College of Health, Kamiokubo 519, Urawa, Saitama 338, Japan

Abstract In order to search for the double beta decay of ^{136}Xe through quantitative measurements of numbers of its daughter nuclei, a RIS-TOFMS system was developed. Methods to absolutely measure efficiency of the system are described.

INTRODUCTION

In order to search for nuclear double beta decay (DBD) of ^{136}Xe, an experimental study based on collection of ^{136}Ba in (liquid and/or gaseous) Xe (1) is in progress, where ^{136}Ba is a daughter nucleus of ^{136}Xe in the DBD. The ^{136}Ba can be collected on an electrode under an electric field since it remains as a stable positive ion in Xe after the decay (2). For determination of a half-life of the DBD from the number of ^{136}Ba collected in a given period, positive ion collectors (PIC) and a time of flight mass spectrometer (TOFMS) were developed, where the resonance ionization spectroscopy (RIS) is used as a method to ionize Ba. To obtain absolute number of collected Ba, it is necessary to determine the efficiency of 1) collection of Ba ions on the electrode in PIC, 2) liberation of Ba atoms from the surface, 3) sampling and ionization of the liberated atoms with a laser beam of RIS, 4) transmission of the ions in TOFMS and 5) detection of the ions with its detector. While the efficiency of 1) and 5) are determined independently, those of 2) and 3) are measured using TOFMS with known efficiency for the transmission and the detection. TOFMS must be also designed to attain high efficiency. In this article, we introduce the details of the TOFMS system and describe results for the transmission efficiency of ions obtained using Xe which is ionized by RIS.

TIME OF FLIGHT MASS SPECTROMETER

Figure 1 shows a system of TOFMS with an ion gun. The gun produces Ba ions of 0.1 to 10nA in current and deposits desired amounts of Ba on a surface of an octahedron electrode by electrical pulse shutters. The electrode is attached to a repeller RP by rotating its surface, from which Ba atoms are ablated by a Nd-YAG laser beam with an angle of $45°$ against the surface. The electrode system to accelerate ions consists of RP and two grid plates, G1 and G2. Two guard-rings, R1 and R2, are also introduced at the middle in RP-G1 and in G1-G2, respectively, to make electric fields uniform. The distance is 20mm in RP-G1 and 10mm in G1-G2. Distortion of the field is calculated to be within 0.1% at 6mm from the central axis (3). The grid plate has an outer diameter of 150mmϕ and an inner 20mmϕ, on which gold-coated W wires (50µm in dia.) are strung with a 0.5mm spacing. A free flight region of 748mm is formed between G2 and the third grid plate G3, the outer diameter of which is 140mmϕ and the inner 100 mmϕ. On G3 gold-coated W wires (0.1mmϕ in dia.) are strung with a 1mm spacing. The open-area ratio in those grid plates are 91%. A two-stage microchannel plate (MCP) with an effective area of 44.2cm^2 (75mmϕ in dia.) is placed after G3. The apparatus is evacuated to below $1x10^{-8}$ Torr with two turbo-molecular pumps.

MEASUREMENT

In order to measure transmission efficiency in TOFMS, Xe was filled into the system. The resonant ionization of Xe was made through a level of 6p[3/2]$_2$ with a beam (~0.7mJ/pulse, 252.484 nm) from a dye laser pumped by the 3rd harmonics of a Nd-YAG laser, where a frequency doubler (BBO) was used. The pressure of Xe was measured with a spinning-rotor gauge. The electrode configuration was slightly modified in the measurement. A stainless-steel plate with the same dimension as that of the MCP was installed as a collector CL in place of the MCP. A repeller having a sampling electrode SM of 12mmϕ on the center and a guard-ring with a central hole of 12mmϕ were placed instead of RP and R1, respectively. Signals from SM and CL are fed to charge-sensitive preamplifiers, further amplified with main amplifiers with a

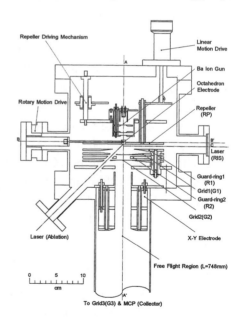

Figure1.   A system of TOFMS.

time constant of 10μs, and analyzed by a two-parameter multichannel analyzer (MCA). Before measurements the amplifier systems had been calibrated in terms of numbers of unit charge within 0.7% by measuring signals of alpha-particles with known energies in a gridded ionization chamber. Transmission efficiency F_{tra} is determined as a ratio of the charges measured at CL to those at SM.

A typical result obtained with MCA is shown in Fig.2. While numbers of charges at SM and CL fluctuate widely, a linear correlation between them can be clearly seen. F_{tra} was obtained from the slope in least-square fittings to the data. $F_{tra}s$ measured under different conditions are shown in Fig.3 as a function of ratio of the field in G1-G2 (E_{G1-G2}) to that in RP-G1 (E_{RP-G1}), Z. F_{tra} tends to decrease at Z<2 and saturate at Z>2. The theoretical calculation for electron transmission in a gridded ionization chamber (4) gives a value of 1.92 as a minimum Z to prevent electron trapping due to grid wires, and would explain well this tendency. F_{tra} depends also on pressure and position of a laser beam LL, where the distance from RP is (67.0-LL) mm. F_{tra} averaged over saturated values for different conditions (shown as □,△ and ● in Fig.3) is 0.72±0.01, 0.77±0.02 and 0.74±0.01, respectively. The pressure dependence of F_{tra} was measured at LL=61.0mm at pressures from 1x10^{-6} to 1.5x10^{-5} Torr. F_{tra} was found to be almost constant at pressures below 7x10^{-6} Torr and decrease slightly at the higher pressures. In the case of the dependence on LL, F_{tra} varies from 0.65 to 0.78 as indicated in Fig.4.

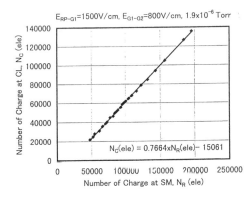

Figure 2. Numbers of charges measured at collector and at sampling electrode.

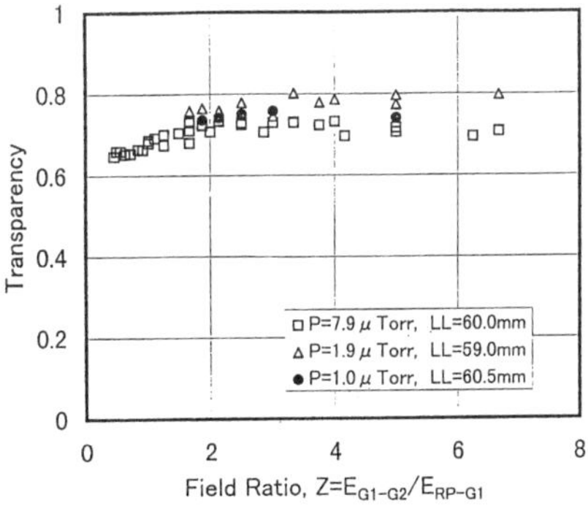

Figure 3. Transmission efficiency measured as a function of filed ratio.

DISCUSSION

The dependence of F_{tra} on Z does not basically vary for Z>2 at any field conditions. F_{tra} does not strongly depend on pressure in the range of pressures used in the present study. On the other hand, F_{tra} varies steeply at a smaller value of LL, but becomes almost constant for the larger value, namely at a position close to RP, as shown in Fig.4. The reason is not presently known, but may be related to variation of charges induced on SM by ions in RP-G1 and/or to losses of ions caused by the trapping with R1 and G1. In the plot of charges measured at CL to those at SM, a non-linearity is always observed as if there existed additional charges on SM. The non-linearity is almost independent of fields conditions, but

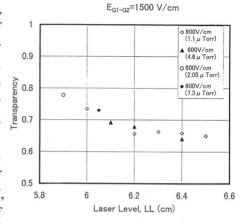

Figure 4. Transmission efficiency measured as a function of laser beam position (LL).

slightly depends on both pressure and LL. The induced charges on SM by ion-pairs created along a laser beam outside the effective region of SM in RP-G1 should be taken into account. The non-linearity seems not to cause in the case of the ionization localized near SM (or RP) like that due to ablation on surface by laser shots. The transmission efficiency of this TOFMS system is 0.73 ± 0.07 by only considering the variation of F_{tra} against LL. The relatively larger error in F_{tra} may be improved by understanding the induction of charges existing in RP-G1 on SM in detail.

Independent measurements for detection efficiency of MCP were performed by injecting single Ar ions produced by alpha-particles into MCPs. The detection efficiency of MCP was measured to be 59 %, which is closed to its intrinsic open-area ratio (65%). Assuming no difference of the detection efficiency of MCP between lighter ions and the heavier, the total efficiency of the TOFMS system is evaluated to be 43 %.

REFFERENCE

1.Miyajima,M. Sasaki,S. and Tawara,H., *IEEE Trans.* **NS41**, 835-839 (1994).

2.Moe,M. and Vogel,P., *"Double beta decay"*, *UCI-Neutrino* **94-5**, 1994.

3.Albriton,D.L., et.al., *Technical Report (Georgia Institute of Technology)*, 1967.

4.Bunemann,O., Cranshaw,T.E. and Harvey,J.A., *Can. J. Res.*, 27 191-206 (1949).

A Novel Two-Step Excitation Scheme for the Analysis of Lead by Laser Enhanced Ionization in a Flame

K.L. Riter, O.I. Matveev, B.C. Castle, B.W. Smith, and J.D. Winefordner*

*Department of Chemistry, University of Florida, P.O. Box 117200, Gainesville, FL 32611-7200

Abstract. Various two-step excitation schemes were examined for the analysis of lead by laser enhanced ionization in a flame. It was found that the scheme employing the transitions $6p^2\,^3P_0 \rightarrow 7s\,^3P_1^0$ ($\lambda_1 = 283.3$ nm) and $7s\,^3P_1^0 \rightarrow 9p\,^3P_1$ ($\lambda_2 = 509.0$ nm) produced the largest enhancement of the ionization signal. This scheme was subsequently applied to the analysis of aqueous lead solutions and blood lead standards. Detection limits of 5.2 pg/mL (52 fg absolute) for aqueous lead solutions and 4.2 pg/mL (42 fg absolute) for lead in diluted whole blood were obtained.

INTRODUCTION

The determination of lead concentrations has long been a subject of interest because of the detrimental effects of lead on the human body. Of particular interest is the determination of trace lead concentrations in human blood. Previously, researchers have reported low limits of detection for lead using two-step laser enhanced ionization (LEI) in flames (1-3). All employed a two-step optical excitation scheme where the first dye laser was tuned to the transition $6p^2\,^3P_0 \rightarrow 7s\,^3P_1^0$ ($\lambda_1 = 283.3$ nm) and the second dye laser was tuned to the transition $7s\,^3P_1^0 \rightarrow 8p\,^3D_2$ ($\lambda_2 = 600.2$ nm).

In this work, four different two-step excitation schemes for lead were examined. The scheme producing the greatest ionization enhancement was subsequently used for the determination of lead concentrations in aqueous solutions and blood.

EXPERIMENTAL

The four different two-step excitation schemes for laser enhanced ionization of lead are shown in Figure 1. All four excitation schemes employed the transition $6p^2\,^3P_0 \rightarrow 7s\,^3P_1^0$ (283.3 nm) as the first step. The four different second steps involved the

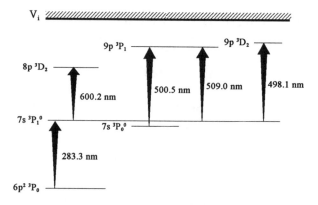

FIGURE 1. Different two-step excitation schemes used for LEI of lead.

transitions $7s\,^3P_1^0 \rightarrow 8p\,^3D_2$ (600.2 nm), $7s\,^3P_1^0 \rightarrow 9p\,^3P_1$ (509.0 nm), $7s\,^3P_1^0 \rightarrow 9p\,^3D_2$ (498.1 nm), and $7s\,^3P_0^0 \rightarrow 9p\,^3P_1$ (500.5 nm).

The experimental setup used is shown in Figure 2 and described in detail elsewhere (4). A XeCl excimer laser (Lambda Physik, LPX-240i) was used to pump two dye lasers. One dye laser (Lambda Physik, Scanmate 1) contained Coumarin 153 (Lambda Physik) and its output was frequency doubled with a BBO III crystal (Lambda Physik) to provide optical excitation at 283.3 nm. A linewidth of 0.15 cm^{-1}, typical pulse energy of 20 µJ, and approximate beam diameter of 2 mm were measured for the first dye laser. A second dye laser (Laser Photonics, DL-14) contained Rhodamine B (Lambda Physik) to provide optical excitation at 600.2 nm and Coumarin 307 (Lambda Physik) to provide optical excitation at 498.1 nm, 500.5 nm, and 509.0 nm. A linewidth of 0.4 cm^{-1}, typical pulse energy of 1.2 mJ, and approximate beam diameter of 5 mm were measured for the second dye laser. Both lasers were pumped at a repetition rate of 30 Hz.

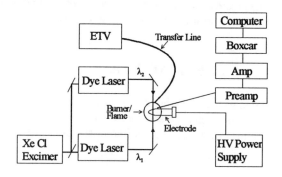

FIGURE 2. Block diagram of experimental setup for LEI.

Aqueous lead standards were prepared by diluting a stock solution (1000 mg/L) with ultra pure 2% nitric acid. Blood samples were prepared by diluting 50 μL of whole blood into 1 mL of ultra pure water. The blood samples were obtained from the Centers for Disease Control (CDC) and are part of their Blood Lead Laboratory Reference System (BLLRS) program. The samples analyzed included pool ID # 694 (0.7 μg/dL), 192 (3.9 μg/dL), 1291 (10.6 μg/dL), 0191 (19.3 μg/dL), and 1092 (61.6 μg/dL).

RESULTS AND DISCUSSION

Results from the comparison of the different excitation schemes for lead are summarized in Table 1. The enhancement factor reported is the enhancement of the ionization signal observed when the second excitation step is added to the first step or enhancement factor = (signal for two-step / signal for one-step). The largest enhancement factor of 1522 was observed when the transition $7s\ ^3P_1^0 \to 9p\ ^3P_1$ (509.0 nm) was used as the second step. This is almost an eight fold improvement over the transition $7s\ ^3P_1^0 \to 8p\ ^3D_2$ (600.2 nm) which resulted in an enhancement factor of 194. An enhancement factor of only 111 was observed for the $7s\ ^3P_1^0 \to 9p\ ^3D_2$ (498.1 nm) transition and an enhancement factor of 37 was observed for the $7s\ ^3P_0^0 \to 9p\ ^3P_1$ (500.5 nm) transition. Therefore, the $7s\ ^3P_1^0 \to 9p\ ^3P_1$ (509.0 nm) transition was chosen as the second step for LEI of lead.

Analytical curves for both aqueous lead and blood lead exhibited good linearity and are shown in Figure 3. For aqueous lead standards, it was found that the addition of 10 μL of 100 μg/L Na (as NaCl) solution was necessary. The NaCl acted as a physical carrier and restored linearity at low concentrations of lead. The addition of NaCl was not necessary for the blood samples. For aqueous lead, the slope of the analytical curve was found to be 0.0522 ± 0.0002 C/(μg/L). A 3σ detection limit of 5.2 ng/L (52 fg absolute) was calculated for aqueous lead.

For lead in diluted blood, the slope of the analytical curve was found to be 0.0641 ± 0.0004 C/(μg/L). A 3 σ detection limit of 4.2 ng/L (42 fg absolute) was calculated

TABLE 1. Enhancement factors for different second steps for LEI of lead

λ_2(nm)	Transition	Enhancement Factor
498.1	$7s\ P_1 \to 9p\ D$	111
500.5	$7s\ P_0^0 \to 9p\ P_1$	37
509.0	$7s\ ^3P_1^0 \to 9p\ ^3P_1$	1522
600.2	$7s\ ^3P_1^0 \to 8p\ ^3D_2$	194

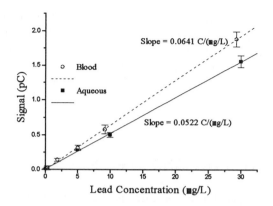

FIGURE 3. Analytical curves for aqueous lead and blood lead samples.

for the diluted blood samples. This corresponds to a detection limit of 89 ng/L (890 fg absolute) for lead in whole blood.

From this work, the novel two-step excitation scheme involving the transitions $6p^2$ $^3P_0 \rightarrow 7s\ ^3P_1^0$ ($\lambda_1 = 283.3$ nm) and $7s\ ^3P_1^0 \rightarrow 9p\ ^3P_1$ ($\lambda_2 = 509.0$ nm) was found to result in a greater enhancement of the LEI signal than the scheme previously used. This new scheme was also found to be useful when applied to the analysis of lead in aqueous solutions and blood samples.

ACKNOWLEDGEMENTS

This work was supported by a grant from the National Institutes of Health (5-R01-GM49638-03).

REFERENCES

1. Turk, G.C., DeVoe, J.R., and Travis, J.C., *Anal. Chem.* **54**, 643-645 (1982).
2. Omenetto, N., Smith, B.W., and Hart, L.P., *Fresenius' J. Anal. Chem.* **324**, 683-697 (1986).
3. Marunkov, A.G., and Chekalin, N.V., *J. Anal. Chem.* USSR (Eng. Transl.) **42**, 506-508 (1987).
4. Riter, K.L., Clevenger, W.L., Mordoh, L.S., Matveev, O.I., Smith, B.W., and Winefordner, J.D., *J. Anal. At. Spectrom.*, accepted April 1996.

Electrospray Ionization Source for Highly Sensitive Resonance and Laser Enhanced Ionization Analysis

O.I. Matveev, I.B. Gornushkin, W.L. Clevenger, B.W. Smith, and J.D. Winefordner

Department of Chemistry, University of Florida, Gainesville, FL 32611-7200 USA

Abstract. A method is described involving electrospray nebulization of liquid samples on the surface of an atomizer or vaporizer, which can enable quantitative and even deposition and also matrix modification of the solid residue of the sample. Ultralow limits of detection of atoms can be achieved in combination with laser excited fluorescence, RIS and LEI methods of atomic and molecular analysis with ES deposition of the sample on a wire atomizer or vaporizer.

Resonance ionization spectroscopy (RIS) has been successfully used for the detection of single atoms within the volume illuminated by laser radiation (1). However, the detection of single atoms in the bulk of a real sample remains unsolved. One of the possible ways to solve this problem is to combine RIS or laser enhanced ionization (LEI) with a preliminary electrolytic deposition of the analyte from a solution to the surface of the electrode-collector. The electrode then serves as an atomizer of the deposited sample. As was shown in (2), very low limits of detection, at the level of 10^5-10^6 atoms in the bulk of the sample, can be obtained by using such a combination. Unfortunately, electrolytic deposition from a solution has several disadvantages. First, only few elements can be deposited on the electrode. For example, alkali and alkaline-earth metals cannot be directly deposited from water solutions due to their high reactivity. Second, it is very difficult to provide selective electrolytic deposition of an analyte when a complex multielement solution such as seawater is used. Also, the majority of negative ions cannot be deposited.

In this work, a new, simple and more universal method of preliminary sample deposition and matrix modification for subsequent RIS and LEI in a flame or laser excited atomic fluorescence spectrometry with electrothermal atomization (LEAFS-ETA) analysis is described. The method is based on the electrospray (ES) of ions from solutions (3).

EXPERIMENTAL

A stainless steel capillary with 0.1 mm i.d. and 0.2 mm o.d. was attached to a 5 mL glass syringe filled with the working solution. The syringe was operated by a syringe pump at a flow rate of 10 μL/min.

ES-LEI in flame. As seen in Figure 1, the electrode-collector for LEI was made from a 2 mm × 50 mm loop of 0.3 mm d. iridium wire. The loop was placed near the capillary tip and supplied with a high negative potential (-6-15 kV). Positively charged droplets of the solution were sprayed (in open air) from the capillary tip

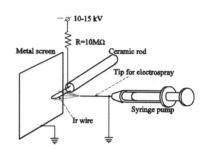

Figure 1. Sample Introduction.

and collected on the loop as either charged microdroplets, charged hydrated clusters, or ions. A grounded metallic screen behind the loop electrically repelled some energetic particles back to the collector, increasing the collection efficiency. Two elements, Sr and Cs, were chosen to study the analytical characteristics of the method. Working solutions of Sr and Cs of 10^{-8}-10^{-4} M were used. The final solutions contained 90% ethanol and 10% water to provide efficient formation of free ions with the fast evaporation of volatile solvent.

After sample deposition, the loop was inserted into an air-acetylene flame (Figure 2). A new LEI scheme for Sr with 515.7 nm as the second excitation step instead of 496.6 nm provided a 1.5-fold increase in the analytical signal. A single excitation step was used for Cs at 455.5 nm. The signal was detected by a water-cooled HV electrode, a charge sensitive preamplifier (Avangard, Inc., Russia), an amplifier and boxcar integrator (SR 560 and

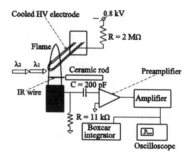

Figure 2. Detection system.

SR250, Stanford Research Inc., USA), and an oscilloscope (2430A, Tektronix, USA).

ES-LEAFS-ETA. The electrode-collector for LEAFS-ETA was a small graphite cup with inner and outer diameters of 4 mm and 6 mm, respectively, and a height of 6 mm. The cup was held between two spring-loaded graphite rods and grounded. The capillary was inserted into the cup so that the gap between the capillary tip and the inner surface of the cup was 2 mm. A positive potential (~3 kV) was then applied to the capillary, and the solution was sprayed to the inner surface of the graphite cup. Solutions of Ag in methanol (0.1-1000 ng/mL) were used.

After the sample was deposited, the graphite cup was heated by electric current to an atomization temperature of $\sim 1800^\circ$ C. The fluorescence of silver was excited ($\lambda_{ex} = 228$ nm) by the dye laser above the cup and collected ($\lambda_{em} = 338$ nm) at a right angle relative to the laser beam. The collection optics included two biconvex lenses, a set of colored glass filters and a monochromator (Digikrom, CVI Laser Corp., USA). The detection system included a PMT, an amplifier and boxcar integrator (SR 440 and SR265, Stanford Research Inc., USA), and a PC.

RESULTS AND DISCUSSION

ES-LEAFS-ETA. Calibration plots for Ag (Figure 3b, curves 1b and 2b) were constructed by using both ES deposition and pipetting of 10 µL samples. The two calibration plots nearly coincided at concentrations above 10 ppb. Below this point, the plot obtained with the ES deposition deviated from linearity seemingly because of memory effects in the syringe needle. The RSD for the ES deposition was 5-10 %, slightly better than for the pipetting, probably because of more uniform distribution of the deposit over the graphite collector.

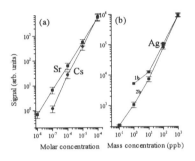

Figure 3. Calibration curves for (a) ES- RIS and (b) ES-LEAFS.

Enrichment of cations. The possibility of using ES deposition on the electrode for matrix modification was examined using ICP-MS analysis of the deposited residue. After 2 min of ES deposition (at a working voltage of 10 and 15 kV) of 10^{-3} M solutions of $SrI_2 \cdot 6H_2O$ and CsI, the residue was dissolved in 4 mL water (Cs) or 5% HNO_3 (Sr). Comparison of the original reference solutions to the solutions containing ES residue showed the average ratios Cs/I and Sr/I to be correspondingly 2 and 3.5 times higher due to positive ion enrichment. Within the error of measurement, this result did not depend on the working voltage. The efficiency of electrosprayed Cs and Sr ion collection by the Ir wire was found to be 35-50%. This indicates that ES could provide enrichment of cations and matrix modification, and implies the feasibility of metal ES deposition from different types of solutions, such as seawater.
LEI flame analysis of Sr and Cs. An acceptable linearity of the calibration curve (see Figure 3a) was observed for Sr. For Cs, nonlinear behavior was found, as is typical in flame spectroscopy (4). Signals for all concentrations were measured three times after 1 min of ES deposition of sample. The behavior of the calibration curves was also checked when 5 µL of the standard solutions were pipetted onto the Ir wire. It was found that there was no distinguishable difference in the behavior of the

calibration curve. Considering the recent literature (5) on ES-MS, it should be mentioned that the shape of the calibration curve obtained in this experiment is different. The authors of (5) did not observe linear behavior of the calibration curve, and between 10^{-4} -10^{-5} M concentration there was a three order of magnitude drop in the signal. Comparing the results in Figure 3a to those in (5), one can conclude that for small concentrations of positive ions in solution, effective linearity of the ES process still exists. However, when these solutions with small concentration were electrosprayed, the relative amount of free (bare) ionic component in the gas phase decreased much faster than the concentration of ions in the original solution. It is interesting to note that with pipetting it was impossible to deposit on the surface of the Ir wire more than 5 μL of solution, but with electrospray residue a much larger volume (200-300 μL) of solution could be easily deposited.

It is clear that due to the physical nature of ES, an even distribution of deposited atoms along the collecting electrode should be expected, which was clearly observed from a red SrOH molecular emission in the flame after insertion of the Ir wire. After pipetting, this emission was seen only from the end of the wire loop where the drop of solution was deposited, but after ES, a uniform emission from the whole wire was observed.

After 10 minutes of deposition of 10^{-8} M (876 ppt for Sr and 1.33 ppb for Cs) solutions and LEI measurements of the signal and noise, the 3σ limits of detection of Sr and Cs in water solutions were found to be 5 and 28 ppt correspondingly.

ACKNOWLEDGEMENTS

This work was supported by a grant from the US Department of Energy (DOE-DE-FGO5-88-ER13881).

REFERENCES

(1) Hurst, G.S., Nayfeh, M.H., Young, J.P., *Appl. Phys. Lett.*, **30**, 229-231(1977).
(2) Trautman, N., "Ultratrace Analysis of Long-Lived Radioisotopes in the Environment," in *AIP Conference Proceedings No. 329 (RIS 94)*, 1994, pp. 243-250.
(3) Kebarle, P., Tang, L., *Anal. Chem.*, **65**, 972A-986A(1993).
(4) Poluektov, N.S., *Techniques in Flame Photometric Analysis*, Princeton, NJ: D. Van Nostrand Company, Inc, 1966.
(5) Agnes, G.R., Horlick, G., *Appl. Spectrosc.*, **48**, 649-654(1994).

Investigation of optimal photoionization schemes for Sm by multi-step resonance ionization

Hyungki Cha, Kyuseok Song, and Jongmin Lee

*Laboratory for Quantum Optics, Korea Atomic Energy Research Institute,
P.O. Box 105, Yusong, Taejon, Korea 305-600*

Abstract. Excited states of Sm atoms are investigated by using multi-color resonance enhanced multiphoton ionization spectroscopy. Among the ionization signals one observed at 577.86 nm is regarded as the most efficient excited state if an 1-color 3-photon scheme is applied. Meanwhile an observed level located at 587.42 nm is regarded as the most efficient state if one uses a 2-color scheme. For 2-color scheme a level located at 573.50 nm from this first excited state is one of the best second excited state for the optimal photoionization scheme. Based on this ionization scheme various concentrations of standard solutions for samarium are determined. The minimum amount of sample which can be detected by a 2-color scheme is determined as 200 fg. The detection sensitivity is limited mainly due to the pollution of the graphite atomizer.

Introduction

Resonance enhanced multiphoton ionization spectroscopy (REMPI) has been utilized for two decades for the identification of excited state energy levels and optimal photoionization schemes (1-2). Saloman has published a series of papers on optimal ionization schemes for various elements. But optimal photoionization schemes for lanthanide series elements are not well established yet (3). The investigation of an optimal photoionization scheme for Sm is important for the purpose of trace analysis. There are only a few data available for the excited states of Sm (4-5). Recently two papers on the investigation of high-lying even-parity levels of Sm have been published (6-7). But these data were not enough to pinpoint the best photoionization scheme in the trace analysis, because they only identified new states without verifying the ionization schemes, e.g. 1-photon or 2-photon resonance. Here we present results of a more detailed spectroscopic study on excited states of Sm. In addition results of trace determination for Sm standard solution are discussed.

Experiment

The atomic beam generation of Sm is performed by a resistive heating method with a tungsten filament atomizer. About 10 mg of samarium metal were put into the tungsten atomizer which is composed of two filaments. The atomizer is

© 1997 American Institute of Physics
439

heated by 60 A electric current to the temperature about 600 °C which corresponds to a samarium vapor pressure of ~10^{-4} Torr. For detection of samarium ions a time-of-flight mass spectrometer (TOFMS) with a microchannel plate (MCP) were used. The MCP output signal was then amplified and registered by a computer via a Boxcar integrator and stored in the computer. A Nd:YAG laser (Spectra Physics model GCR-230) was used to pump two dye lasers (Lambda Physik model Scanmate2). The dye lasers are filled with Rhodamine 590 and Rhodamine 610 dye solutions for 568-600 nm. Wavelength calibration was performed with the use of Burleigh WA 4500 Pulsed Wavemeter. The delay between the two dye laser beams was set to almost zero.

The measurement was performed with graphite crucible atomizer for the convenience of sample exchange. The Sm standard solutions were prepared from 1000 ppm samarium AAS standard (Johnson Matthey SpecPure) by stepwise dilution with pure water. Each measurement started with putting the sample of 20 μl into the crucible by a micro-pipette. The sample solution was dried by IR-lamp. At the crucible temperature 1000 °C the lasers and ion detection were switched on. The ion signal was accumulated up to the crucible temperature 2000 °C, then ion accumulation was stopped and crucible is cleaned by heating up to 2400 °C.

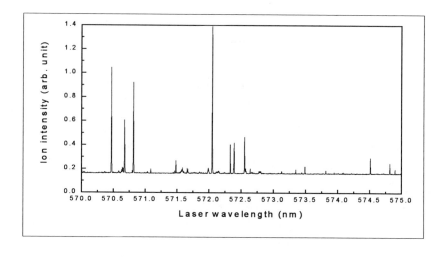

Fig. 1. 1-color 3-photon ionization spectrum of Sm (570-575 nm)

Results and discussions

Identification of the optimal photoionization scheme

The ground state of the Sm atom has $^{7}F_{1}$ term. But there are five low energy excited states with the J value of 2 to 6. Since Sm has a very complicated electron configuration, a definite assignment of excited states is difficult, and therefore, only J values are considered for the level identification. Fig. 1 is an example of an 1-color 3-photon ionization spectrum for one of Sm isotopes (m/e=152). Several intense lines were observed, but identification of the ionization scheme for these lines were necessary. When the same spectral region was studied with 3 times higher laser energy, many more transition lines were additionally observed compared to that in the Fig. 1. New lines observed with high laser energy were tentatively assigned as multiphoton resonance transitions (2-photon or 3-photon). With these investigations five intense lines (572.02 nm, 577.86nm, 577.91nm, 580.85nm and 587.42nm) were chosen as first excited states in 2-color schemes. The assignment for observed lines are listed in Table 1.

Table 1. Result of 1-color 3-photon ionization of Sm in the 568-603 nm region.

Laser Wavelength, nm	Ion Signal, arb. units	Energy Levels, cm^{-1}	Scheme
569.95	0.17		
570.45	0.17		
570.62	0.08		
570.675*	0.10	293-17810	$\omega_1+2\omega_1$
570.80	0.16		
571.97	0.08		
572.02*	0.35	293-17770	$\omega_1+2\omega_1$
572.32	0.08		
572.54	0.12	0-17462**	$\omega_1+2\omega_1$
573.47	0.08		
574.50	0.11		
574.81	0.08		
577.86	1.04		
577.91	1.10		
578.99	0.10		
580.19	0.99		
580.55	0.09		
580.85	0.10	293-17505**	$\omega_1+2\omega_1$
586.48	0.08		
586.58	0.10		
587.11	0.72		
587.42*	0.15	812-17831	$\omega_1+2\omega_1$
587.62	0.42		
589.57	0.12	812-17770**	$\omega_1+2\omega_1$
591.59	0.10		
592.88	0.08		
598.38	0.08		
599.74	0.08		

*Lines known from literature.,   **Energy levels known from literature.

Determination of Sm in standard solution using 2-color ionization schemes.

For the determination of Sm following a 2-color 3-photon ionization scheme of type $\omega_1+\omega_2+\omega_{1,2}$ was selected.

$$
\begin{array}{ccccccc}
& 587.42\text{nm} & & 573.50\text{nm} & & 587.42\ \text{nm or } 573.50\text{nm} & \\
4f^6 6s^2\ {}^7F_2 & \rightarrow & 4f^6 6s6p\ {}^3F_3^{\circ} & \rightarrow & ???? & \rightarrow & \text{continuum}
\end{array}
$$

$$
812 \quad \rightarrow \quad 17831\ \text{cm}^{-1} \quad \rightarrow \quad 35263\ \text{cm}^{-1} \quad \rightarrow \quad \text{continuum}
$$

The analytical signal measured with standard solution by using above photoionization scheme was listed in Table 2.

Table 2. Result of determination of Sm by using 2-color 3-photon resonance ionization.

Contents, ppb	Amount of Sm, pg	Ion signal, arb unit	Standard deviation
10	200	436000	0.05
1	20	48000	0.13
0.1	2	7630	0.07
0.01	0.2	2620	0.17
0.001	0.02	1720	0.11
pure water	0	2260	0.36
"clean blank"	0	20	0.43

As it can be seen from the table above the signal from samples with samarium content lower than 0.01 ppb or 200 fg is approximately the same. Taking into account that a similar signal intensity as 200 fg sample is obtained for pure water, while much smaller signal is obtained for thermally cleaned blank, one can attribute a major limitation for detection sensitivity is the contamination of graphite crucible.

Conclusions

The optimal multi-step photoionization schemes are investigated. For two-color experiment several efficient ionization schemes are identified. Among those $\omega_1 + \omega_2 + \omega_{1,2}$ schemes are mainly considered and $\omega_1 = 587.42$ nm, $\omega_2 = 573.50$ nm were determined. The efficiency of each ionization scheme is tested by measuring ion signals of standard solutions. The minimum amount which can be detected in 2-color 3-photon ionization scheme was 200 fg. More detailed results on Sm including identification of autoionization states will be published soon.

References

1. V.S.Letokhov, "Laser Photoionization Spectroscopy", Academic Press Inc. (1987)
2. R.C.Estler and N.S.Nogar, Spectrochim. Acta., 48B, 663 (1993)
3. E.B.Saloman, Spectrochim. Acta., 48B, 1130 (1993) and references therein
4. A.C.Parr and M.GIngram, J.Opt. Soc. Am., 65, 613 (1975)
5. P.Brix and H.Kopfermann, Zeitschrift fur Physik., 126, 364 (1949)
6. L.Jia, C.Jing, Z.Zhou, and F.Lin, J.Opt.Soc.Am. B., 10, 1317 (1993)
7. T.Jayasekharan, M.A.N.Razvi, and G.L.Bahle, J.Opt.Soc.Am. B., 13, 641 (1996)

Laser-induced Fluorescence as an Approach to Single Molecules Detection in Fluid Solution

W. Y. Ma, J. Xiong, K. L. Wen, J. Wei, L. Zhang, D. Y. Chen

Department of Physics, Tsinghua University,
Beijing 100084, P. R. China

Abstract. We report here the ultrasensitive detection of Rhodamine-6G (R6G) molecules in a flowing aqueous sample based on laser-induced fluorescence (LIF) combined with mode-locked pulsed excitation and time-correlated single-photon counting. At the detection limit of 3.8×10^{-14} M, the probability of a R6G molecule being present in the detector's probed volume of 12.6 pL was about 0.28.

INTRODUCTION

The ability to detect single molecules in flowing samples as they transit a focused laser beam with fluorescence detection has a variety of application in the areas of analytical chemistry, environment science, bioanalytical and medicine.

Work in the area of single molecules detection (SMD) in solution using laser-induced fluorescence was initiated by Dovichi et al (1) in 1983 utilizing the 514.5 nm continuous wave output of an Ar^+ laser for aqueous R6G solution. Until 1990 Shera et al (2) first realized the single R6G molecules detection by a high repetitively pulsd Nd:YAG laser with output wavelength of 532 nm and time-gated discrimination of fluorescence photons. Castro et al (3) began to apply the SMD technique to the analysis of molecules of biological interest.

This paper describes preliminary experiments of detection R6G molecules with our own apparatus designed for detecting single molecules in fluid solution by laser-induced fluorescence combined with mode-locked pulsed excitation and time-correlated single-photon counting.

EXPERIMENTAL SECTION

The experimental diagram is shown in Fig. 1. The excitation source was a

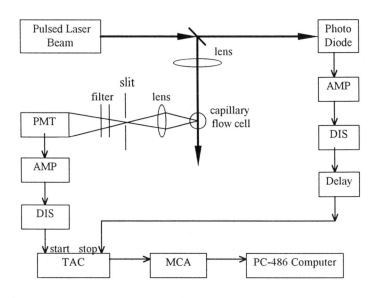

FIGURE 1. Experimental diagram of the detection apparatus.

shortpulse (~200 ps), high repetition rate (82 MHz) mode-locked argon ion laser operated at 514.5 nm. The laser was split into two beams. One was focused through a 17 mm focal-length lens into a capillary flow cell with an average power of 20 mW. The laser beam waist at 1/e intensity was 40 μm in diameter. Fluorescence was collected at 90° using a ×40, 0.65 NA microscope objective. A slit (0.4 mm) located in the image plane of the collection objective limited the probe volume to 12.6 pL. Two bandpass filters (560 nm center wavelength, 40 nm FWHM) were located behind the slit. Fluorescence was detected by a fast response PMT detector (R928 photomultiplier tube). The PMT signal was amplified by a fast amplifier (AMP), and then shaped by a discriminator (DIS) that provided the start pulse for a time-to-amplitude converter (TAC). The second laser beam was detected by a photodiode whose output was amplified and shaped and finally fed to the gate to stop the input of the TAC with appropriate time delay. The time-gated TAC output was counted by a multichannel analyzer (MCA). All data were acquired and processed automatically by a PC-486 computer.

RESULTS AND DISCUSSION

In our experiment the aqueous R6G solutions were analyzed. The difficulty to identify the passage of individual R6G molecules is the background more strong than the fluorescent signal of R6G. The background may come from several sources, such as Rayleigh scattered light, light from fluorescence of the flow cell and the collection optics, and Raman scattering light from the solvent. A carefully designed optical system can largely eliminate first two background, and the time-correlated single-photon counting electronics was used to discriminate the prompt Raman scattering light background.

Fresh solutions of aqueous R6G ranging in concentration from 1×10^{-8} M to 5×10^{-13} M were prepared by serial dilution of a 1×10^{-6} M stock solution in deionized water. No apparent reduction in fluorescent signal due to absorption of R6G by the glass wall was observed within 48 h of the experiment.

Our flow cell was a thin wall capillary with an internal diameter of 500 μm. The aqueous R6G solutions were analyzed and the resulting calibration curve was linear in concentration from 1×10^{-10} M to 1×10^{-12} M. Blank measurements were made with the deionized water. Time spectra obtained with our apparatus from a concentration 5×10^{-13} M of R6G solution and the blank of deionized water are shown in Fig. 2. A strong peak at the time of the laser pulse (t = 0) was due primarily to Raman scattering photons. The prompt peak was followed by an exponential decline in the intensity of the fluorescent signal with the lifetime characteristic of R6G. The Raman scattering background can be rejected by setting a time-window that allows only the delayed fluorescent signal of R6G molecules to pass. The position and the width of the time-window are important for retaining the majority of the desired fluorescent signal and eliminating the background as large as possible. In the experiment, a width 3 ns of time-window was placed at the position 1.5 ns behind the prompt peak.

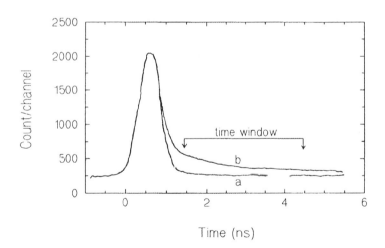

FIGURE 2. Time spectra for (a) the blank sample of deionized water, and (b) the concentration 5×10^{-13} M of R6G solution. The time-window that was used to reject prompt scattering is shown. The measurement time of every time-delay curve was 50 second.

The detection limit is defined as the concentration yielding a signal equal to 3 standard deviations (3σ) of the blank. Every time-delay curve was accumulated for 50 seconds, and the average signal counts within the time window for 6 times measurements were 4425 in the concentration 5×10^{-13} M of R6G solution, and σ of the blank was 111. So the detection limit (D. L.) of R6G molecules was

$$D.L.(3\sigma) = (5 \times 10^{-13} M)\frac{3 \times 111}{4425}$$

$$= 3.8 \times 10^{-14} M$$

The probability of a R6G molecule being present in the detector's probe volume was about 0.28 at the detection limit. According to Poisson distribution law, the the probability of two or more R6G molecules being present in the probe volume simultaneously was about 3%.

We are interested in the detection of single molecules in fluid solution and its applications in bioanalytical, medicine and environment science. A projection of making further improvements has been made for the realization of this goal.

REFERENCES

1. Dovichi, N. J., Martin, J. C., Jett, J. H., and Keller, R. A., *Science* **219**, 845-847 (1983).
2. Shera, E. B., Seitzinger, N. K., Davis, L. M., Keller, R. A., and Soper, S. A., *Chem. Phys. Lett.* **174**, 553-557 (1990).
3. Castro, A., and Shera E. B., *Applied Optics*, **34**, 3218-3222 (1995).

The Development of Laser Method of Rare Isotope Al-26 Detection as Applied for Environmental Problems.

S.A. Aseyev, A.V. Kunetz, V.S. Letokhov

Institute of Spectroscopy, Russian Academy of Sciences,
142092 Troitzk, Moscow Region, Russia

Abstract. The problems concerning Al-detection are briefly discussed. A simple one-step scheme of laser photoionization of metastable aluminum atoms is suggested for detecting this element. The metastable Al atoms can be prepared by charge-exchange of fast Al^+ with alkali metal vapor.

Aluminum is a widespread chemical element: the content by mass in the Earth's crust is approximately 7.5 %. There is only one stable isotope, namely, ^{27}Al. The long-lived radioisotope is ^{26}Al with half-life of $7.16 * 10^5$ yrs. Other aluminum radioisotopes have half-lives shorter than 7 min. ^{26}Al is generated from the spallation by cosmic rays of argon in the atmosphere. At present, it's relative concentration reaches the value of $[^{26}Al]/[^{27}Al] \sim 10^{-14}$ (1).

Chemical compounds of aluminum are widely used in medicine as antiseptic, as alum, as antacid, etc. (2). But Al is dangerous element. For example, it is considered to cause severe diseases such as osteopathy, or anemia, etc. (3). Therefore the development of reliable methods for it's measurements is vitally important in evaluation of aluminum effects in health and disease.

There are two main approaches to investigations of Al-distribution within complex biological systems. The first way is based on direct measurement of the stable isotope in biological materials. In this approach one is faced with the problems of detecting aluminum in the ppb range. One

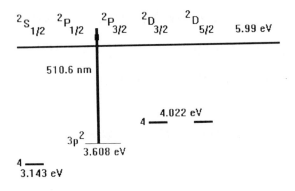

FIGURE 1. The diagram of energy levels of neutral aluminum.

must be able to accomplish collection, storage, processing and final measurement without contamination from this ubiquitous element. Different analytical techniques (in the main, without laser use) for such analysis and their main problems are described in (3). In the first chapter of that review much attention is given to the avoidence of contamination during analysis. Several groups have used laser methods based on RIS to the detecting aluminum [these are compiled in (4)].

The second approach uses isotopic tracer - ^{26}Al. Its increasing availability now opens the way for such study of biochemistry. In experiments presented in (2), orange juice with $[^{26}Al]/[^{27}Al] \cong 10^{-3}$ was ingested by adults; then during definite time interval after injection of aluminum citrate, samples were taken and analyzed with accelerator mass spectrometry (AMS). One of the important problems of ^{26}Al-detecting is connected with isobaric background ($[^{26}Mg]/[^{27}Al] \sim 4 * 10^{-2}$). In order to do effective isobaric separation, AMS uses negative ions: unlike aluminum, magnesium don't form negative ion.

We propose a simple RIS scheme which, in principal, can be used for Al-analysis. The diagram of levels of aluminum is depicted in fig. 1. Al has

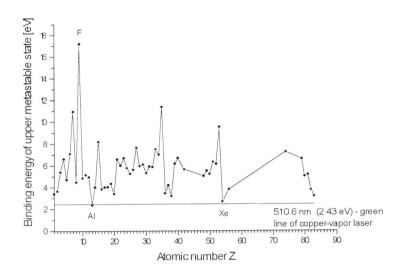

FIGURE 2. Binding energy of upper metastable state vs atomic number Z.

two metastable states: $3s^2 3p$ ($3^2P_{3/2}$) with binding energy of 5.97 eV and $3s3p^2$ with binding energy of 2.38 eV.

The upper metastable level - $3s3p^2$ is rather interesting. The point is that, according to data contained in (5), it has minimal binding energy among other neutral atoms (see fig. 2). In this chart bindings energies of upper metastable levels of different atoms are plotted as function of atomic numbers Z. For alkali metals ionizing potentials are laid off. For atoms with atomic numbers $42 < Z < 48$, $54 < Z < 56$, $56 < Z < 74$ and $74 < Z < 79$ data on metastable states are unknown to us. From this graph it's clear, that, neglecting molecular components in the beam, it's possible to effect Z-selective photoionization of aluminum by use of one laser. For this purpose the green line ($\lambda \cong 510.6$ nm) of copper vapor laser may be applied. Such investigations are presently in progress. The metastable Al atoms can be produced by charge-exchange between fast Al^+ ions and alkali metal vapor.

At first, we measured the cross section σ of total charge-exchange between fast Al^+ (4 keV) and potassium vapor. The sputtered ion source was used for production of fast ions. Krypton support gas was fed into the source for plasma formation. In order to separate Al^+ from Kr^+ and Kr^{2+} in the beam, simple velocity filter was mounted between ion source and charge-exchange cell. We measured the signal against the charge-exchange cell thickness, which was the function of temperature of vapor. The charge-exchange cross-section is $\sigma \cong 2 * 10^{-15}$ cm^2. For further investigations we plan to use the ion source based on surface ionization.

In conjunction with mass-spectrometry technique (time-of-flight) proposed simple photoionization scheme can be used for Al-analysis in samples with moderate relative concentrations of rare isotope. For aluminum, the stable isotope is heavier than rare isotope. Therefore the device used for pulsed fast Al-beam generation in time-of-flight method must ensure a sharp leading edge. Also this will make it possible to suppress the noise connected with inelastic collisions to a certain degree (more frequently, the rare isotope is heavier than stable isotope and tail connected with abundant isotope would contribute to measurements in time-of-flight technique).

References

1. Kilius, L.R., et al, *Nature* **282**, 488 (1979).
2. Hohl, Ch., et al, *Nucl. Instr. and Meth. B* **94**, 478 (1994).
3. Gitelman, H.J., *Aluminum and Health - a Crytical Review*, New York: Dekker, 1987.
4. Saloman, E.B., *Spectrochimica Acta B* **46**, 319 (1991).
5. Radtzig, A.A., and Smirnov, B.M., *Parameters of Atoms and Atomic Ions - Reference Book*, Moscow, 1986.

Measurement of the Efficiency of a Laser Ion Source with Helium Jet Transport, Laser Ablation and RIS

A. Barrera[a], W. M. Fairbank, Jr.[a] and H. K. Carter[b]

[a]Physics Department, Colorado State University, Fort Collins CO 80523-1875 USA
[b]ORISE, P. O. Box 117, Oak Ridge TN 37831 USA

Abstract. The efficiencies for the various processes in a new laser ion source, which has been proposed for study of radioactive nuclei in previously inaccessible regimes, have been measured off-line with radioactive atoms of ^{67}Ga and ^{111}In. High efficiencies for evaporation from a filament into helium gas (100%), helium jet transport (40-50%), and laser ablation (often 80-90%) were obtained. The efficiency for resonance ionization was lower than expected, ~0.1% for ^{67}Ga and ~0.03% for ^{111}In in most runs, although much higher RIS efficiency (up to 2%) was obtained in ^{111}In with resonant laser ablation. A possible explanation for the disappointing RIS efficiency is discussed.

INTRODUCTION

The development of a new experimental method such as a laser ion source is required in order to extend on-line laser spectroscopy of radioactive atoms to previously inaccessible regimes, e.g., nuclei farther from stability, which may be created at a future radioactive ion beam facility, or nuclei of refractory elements, which cannot be rapidly and efficiently evaporated and ionized in conventional ion sources. One laser ion source proposed for these applications combines helium jet transport of reaction products, laser ablation of the deposit, resonance ionization in vacuum of the desired element, and acceleration into a mass spectrometer (1,2). In this paper off-line measurements of the efficiency for these individual processes using the radioactive atoms of ^{67}Ga and ^{111}In are presented and discussed.

APPARATUS

The laser ion source test apparatus used for these measurements is shown in Fig. 1. To simulate the on-line production of nuclei by fusion, fission or spallation reactions and the stopping of these nuclei in helium gas, radioactive atoms are released into helium gas by heating a rhenium filament. These atoms are originally

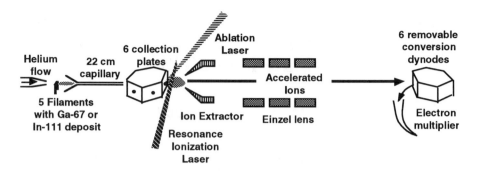

FIGURE 1. Schematic diagram of the laser ion source test apparatus (not to scale).

deposited on the filament by evaporation of a drop of a commercial HCl solution containing ^{67}Ga or ^{111}In at ~1% abundance.

Helium gas at 180 torr in the evaporation chamber flows past the filament and is directed by a 35° cone into a stainless steel capillary of 1 mm inside diameter and 22 cm length (except 35 cm length in the first ten ^{67}Ga runs). The gas containing the seeded Ga or In atoms impacts a Re, W, or Ta plate in the target chamber. The Ga or In atoms stick to the target plate and the helium gas is pumped away by a diffusion pump. The target chamber pressure is 10^{-1} torr when the helium is flowing. All parts of the vacuum system, except the small o-rings for the capillary and one rotatable fitting in the evaporation chamber, are made with high vacuum copper gasket seals to ensure high purity in the helium gas. This solves a problem of irreproducible results in preliminary tests of helium jet transport, which were attributed to a dirty and leaky vacuum system (2,3).

In the proposed on-line laser ion source, the target plate is immediately (to <1 sec) rotated into a differentially pumped high vacuum region for laser ablation (LA) and resonance ionization (RIS). Since the isotopes in the off-line tests were long lived (several days), a simpler procedure requiring fewer pumps could be used. Samples on 5 filaments in a carousel were heated sequentially and deposits made on 5 targets on a second carousel. In most runs the 5 target plates and 5 filaments were removed and their radioactivity counted, to determine separate evaporation and transport efficiencies, prior to the LA and RIS steps. After reinsertion of the target plates, the chamber was pumped down to 10^{-7} torr.

The helium jet deposits of diameter of ~0.7 mm were ablated by harmonic beams from a Nd:YAG laser with 0.1-0.25 mm FWHM diameter and wavelength 532 nm or 355 nm. The peak intensity used was $< 2 \times 10^8$ W/cm^2 for both wavelengths. In the last ten ^{111}In runs, resonant ablation with 304 nm and 608 nm dye laser beams was also tried. Each deposit was ablated in several spots (generally five with the larger beams), and hundreds to thousands of laser shots per spot were used.

Gallium and indium atoms in the ablated plume were resonantly ionized with a two-photon (2ω, ω) scheme, where the resonant $5p^2P_{1/2}$-$5d^2D_{3/2}$ transitions were at 287.4 nm and 303.94 nm, respectively. The RIS beam, of diameter ~0.6 mm FWHM passed through the plume parallel to the target surface and centered ~0.6 mm from the surface. The optimum time delay was ~0.5 μsec for In. According to Monte Carlo simulations (4), ~10% of the ablated atoms should be within FWHM of the RIS beam in this configuration. The typical average fluence, 14 mJ/mm^2, was about 10x the measured saturation fluence for Ga and In. Thus the ionization probability should have been ~100% for ground state atoms within the laser beam FWHM.

Ions were extracted and focused electrostatically over a distance of ~1 m to a conversion dynode plate on a third carousel. Electrons released from the conversion dynode were focused into a channeltron. Time-of-flight spectra for the ions were obtained by recording the amplified output of the channeltron with a digital oscilloscope or transient recorder, triggered by light from the RIS laser. The five conversion dynodes and the five target plates were removed after the LARIS process and their radioactivity counted. From this data the ablation, RIS and total efficiencies were determined.

RESULTS AND DISCUSSION

The evaporation efficiency was found to be very high (~100%) in every run, except one in which the filament burned. The helium jet transport efficiency for ^{67}Ga was quite consistent (20-40%) after the first few ^{67}Ga runs, with higher values occurring toward the end, as improvements were made. For the ^{111}In runs, where changes were minor, a relatively constant 50% efficiency was measured. These are quite impressive results considering that no aerosols were used to enhance transport.

The measured efficiencies for laser ablation varied from low values to near 100% for ^{67}Ga, with a median value of ~50%. For ^{111}In, the results were somewhat more consistent, with the majority of runs above 60%. Variation in these values is expected due to the small ablation laser spot size compared to the deposit diameter and the crude method of moving the ablation laser around. Nevertheless, it is clear from this data that high efficiency for ablation (>90%) is achievable. For ablation spots in which a clear RIS signal was seen, the In or Ga peak decayed in a few seconds to 20 seconds. This indicates that ablation of the submonolayer of In and Ga atoms occurs in tens to hundreds of laser shots. This is an important observation for future application with very short-lived isotopes.

All these results were consistent with expectations (1). On the other hand, the RIS efficiencies, shown in Fig. 2, were disappointingly low. Based on ground state population (>50%) and beam overlap (~10%) considerations, >5% RIS efficiency was expected. Typical values for ^{67}Ga and ^{111}In were ~0.1% and 0.03%, respectively. On the positive side, much higher efficiency (up to 2%) was obtained in the last ten runs in ^{111}In when resonant ablation was used.

Based on the black color observed for larger helium jet deposits of stable Ga HCl solutions and GaO peaks observed in SIMS spectra of the deposits, the deposits are probably Ga_2O_3 and In_2O_3, rather than elemental metals or chlorides. It is possible that the oxide molecules are not completely dissociated in the ablation process, so that the fraction of atoms available to be resonantly ionized is low. The dissociation energies of GaO and InO are ~3.6 eV and ~3.3 eV, respectively. Single 532 nm (2.3 eV) photons cannot dissociate these molecules, and 355 nm photons (3.5 eV) may not be sufficient to reach dissociating states in InO even though it exceeds the dissociation limit. For example it has been found in studies of ablation of oxidized Cu with various excimer lasers, that 193 nm photons (6.4 eV)

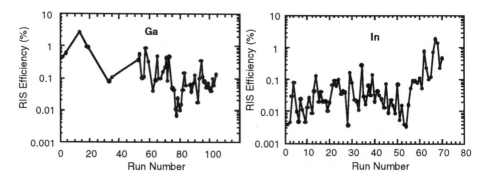

FIGURE 2. Measured RIS efficiency for ^{67}Ga (left) and ^{111}In (right).

were required before large atomic fractions of Cu were detected even though the dissociation energy of CuO is only ~ 2 eV (5). It is possible that the much higher RIS efficiency in ^{111}In with resonant ablation was due to greater photodissociation of InO molecules by the higher intensity 286 nm (4.3 eV) photons.

In conclusion, high efficiencies for evaporation (100%), helium jet transport (40-50%), and laser ablation (often 80-90%) have been demonstrated in this work, confirming expectations for a proposed laser ion source. The efficiency for resonance ionization was generally much lower than the expected 1-10%. However, the possibility of using an ablation wavelength that dissociates molecular species leaves room for optimism that this remaining problem can be solved in the future. The high RIS efficiency (up to 2%) observed with the simpler resonant laser ablation process is also intriguing.

ACKNOWLEDGMENTS

This work was supported in part by funds from the UNISOR consortium.

REFERENCES

1. Fairbank, W. M., Jr. and Carter, H. K., *Nucl. Instr. and Meth.* **B26**, 357-361 (1987).
2. Fairbank, W. M., Jr., Barrera, A., Carter, H. K., Newton, K. R. and Trivedi, A. C., "Resonant Laser Ion Sources for High Elemental Selectivity, " in *Proceedings of the Workshop on the Production and Use of Intense Radioactive Beams at the Isospin Laboratory*, J. D. Garrett, editor, Oak Ridge: ORISE, 1992, pp. 131-136.
3. Trivedi, A. C., M. S. thesis, Mississippi State University, 1991.
4. Barrera, A., Ph. D. Thesis, Colorado State University, 1996.
5. Dreyfus, R. W., *J. Appl. Phys.* **69**, 1721-1729 (1991).

APPENDIX

A Seat-of-the Pants Approach to RIS Theory

W. M. Fairbank, Jr.

Physics Department, Colorado State University, Fort Collins CO 80523-1875 USA

Abstract. This paper is intended as an experimentalist's guide to Resonance Ionization Spectroscopy with broadband pulsed lasers. The basic equations and rules of thumb for one-photon resonances and ionization are presented. Sources of information on energy levels and cross sections are given. The role of magnetic sublevels in populations, rates and isotopic biases is also discussed.

INTRODUCTION

The theory of Resonance Ionization Spectroscopy (RIS) has been presented in various books and papers (e.g., 1-4). In some cases it can be quite complicated. In the majority of RIS applications, however, the RIS process is quite simple: broadband pulsed lasers are used, which fully saturate one to three resonant one-photon transitions, and the final ionizing step is generally into the continuum. The purpose of this paper is to summarize the basic equations, rules of thumb and sources of information which an experimentalist might need to design a RIS experiment of this type. The one important complication discussed is the effect of magnetic sublevels, which leads to isotopic biases in broadband RIS.

RESONANT TRANSITIONS

The key feature of Resonance Ionization Spectroscopy is the use of resonant transitions between bound atomic or molecular states. Only atomic systems will be discussed here. Generally a RIS user planning a new experiment begins with a table of energy levels. Sources of this information are listed in Table 1. Although somewhat dated, the energy level tables by Moore (5), W. Martin et al.(6), and Blaise and Wyart (7) contain most of the needed information. Experimental and theoretical data for specific RIS schemes in 33 elements are available in the published RIS data sheets by Saloman (8). In addition, up-to-date energy levels for 37 elements (more in the future) can be found on the World Wide Web (Table 1).

An understanding of selection rules in atomic transitions is essential for practical use of the energy level tables. The angular momentum quantum numbers in atoms and electric dipole selection rules involving these quantum numbers are summarized in Table 2. It is important to note that the parity selection rule (odd↔even only) and the total angular momentum selection rules ($\Delta J=0,\pm 1$ and $\Delta F=0,\pm 1$ in isotopes with

TABLE 1. Sources of data on atomic energy levels.

Charlotte E. Moore	Atomic Energy Levels, Vols. 1,2,3, NSRDS-NBS-35 (Ref. 5)
Wiese and W. Martin	Atomic Energy Levels –the Rare Earths, NSRDS-NBS- (Ref. 6)
Blaise and Wyart	Atomic Energy Levels–Actinides (Ref. 7)
E. B. Saloman	RIS Data Sheets (Ref. 8)
W. Martin	Recent review of atomic data sources (Ref. 9)
J. Phys. Chem. Ref. Data:	Additional tables of energy levels for selected elements
NIST tables on the WEB:	http://aeldata.phy.nist.gov/archive/data.html
List of other WEB sites:	http://plasma-gate.weizmann.ac.il/API.html

nuclear spin) are rigorous. On the other hand, the spin rule, $\Delta S=0$, although quite rigorous for low Z elements, is easily broken for high Z elements (e.g., the major $\Delta S=\pm 1$ transitions in Hg are typically an order of magnitude weaker than the corresponding $\Delta S=0$ transitions). Similarly, in complex atoms with multiple valence electrons, the selection rule that only one electron can change state is often broken due to widespread configuration mixing in these atoms. In some of the energy level tables referenced above the calculated configuration mixture for each energy level is given. It is then possible to search for additional candidate transitions allowed by configuration mixing.

A common misunderstanding concerns a supposed ΔL selection rule. In atoms with one valence electron the parity is $(-1)^L$, so one may write the parity selection rule as $\Delta L=\pm 1$. In atoms with multiple valence electrons, there is no ΔL selection rule. The 286 nm transition in tin discussed below is an example of a strong $\Delta L=0$ transition, from the $^3P_0$ ground state to the $^3P_1$ excited state.

The transition rate or probability, A_{21}, for spontaneous decay from excited state

TABLE 2. Atomic angular momentum quantum numbers and selection rules.

Quantum numbers (vector addition, L-S coupling):

Single electron	l_i=0,1,2, ...	s_i=1/2		
All electrons	$L=\Sigma l_i$	$S=\Sigma s_i$	$J=L+S$	m_J=–J to J step 1
Whole atom (l≠0)	J	l=nuclear spin	$F=J+l$	m_F=–F to F step 1

Electric dipole selection rules:

Parity : $p=(-1)^{\Sigma l_i}$ (algebraic sum)	$\Delta p=\pm 1$ (odd $\leftrightarrow$ even only)
J, m_J:	$\Delta J=0,\pm 1$ (0 $\leftrightarrow$ 0) and $\Delta m_J=0,\pm 1$ (0 $\leftrightarrow$ 0 for ΔJ=0)
F, m_F:	$\Delta F=0,\pm 1$ (0 $\leftrightarrow$ 0) and $\Delta m_F=0,\pm 1$ (0 $\leftrightarrow$ 0 for ΔF=0)
S:	$\Delta S=0$ (rigid low Z; broken high Z)
only one electron changes its state	(broken in complex atoms)
L:	no ΔL rule!

TABLE 3. Sources of data on atomic transition probabilities.

Fuhr and Wiese	principal lines most elements	Hand. Chem. Phys. (Ref. 10)
Wiese et al.	H through Ca	(Refs. 11,12)
G. Martin, Fuhr et al.	Sc through Ni	(Refs. 13, 14)
Morton	H through Ge	(Ref. 15)
Corliss and Bozman	70 elements	(Ref. 16)
Kurucz and Peytremann	gf values He through Ni	(Ref. 17)
E. B. Saloman	RIS Data Sheets	(Ref. 8)
W. Martin	Review of atomic data sources	(Ref. 9)
NIST tables on the WEB:	http://aeldata.phy.nist.gov/archive/data.html (8 elements)	
List of other WEB sites:	http://plasma-gate.weizmann.ac.il/API.html	

2 to lower state 1, is typically in the range of 10^6-10^8 s^{-1} for allowed transitions in the visible or ultraviolet region. A summary of useful sources of transition probabilities is given in Table 3. For the principal lines of all elements (8300 lines), the most convenient source is the Handbook of Chemistry and Physics (10). Additional tables for Z=1-20 (11,12) Z=21-28 (13,14) and Z=1-32 (15) are more complete. Less accurate values (but useful for rough estimates) covering lines in 70 elements are given in the tables of Corliss and Bozman (16). Semiempirical calculations of gf values, which can be converted to A_{21} values, are also available for Z=2-28 in the tables of Kurucz and Peytremann (17). Tables on the World Wide Web (Table 3) are not extensive yet but are gradually being expanded. Values for specific RIS schemes are given in the RIS data sheets of Saloman (8). Additional references can be found in the review of W. Martin (9).

The excitation and stimulated emission rates (s^{-1}) for a transition resonance, W_{12} and W_{21}, respectively, can be calculated from the tabulated A_{21} values by (18)

$$W_{12} = g(\nu)\frac{I}{h\nu}\frac{g_2}{g_1}\frac{\lambda^2 A_{21}}{8\pi} = \frac{g_2}{g_1}W_{21} \tag{1}$$

where λ is the wavelength (cm), $\nu=c/\lambda$, I is the laser intensity (W/cm^2), $g(\nu)$ is the normalized lineshape function (s) and, for *unpolarized* light, $g_i=2J_i+1$ are the degeneracy factors for the lower and upper state. For a Gaussian or Lorentzian lineshape at line center, $\nu=\nu_0$,

$$g(\nu_0) = \sqrt{\frac{4\ln 2}{\pi}}\frac{1}{\Delta\nu} \quad \text{or} \quad g(\nu_0) = \frac{2}{\pi\Delta\nu} \tag{2}$$

where $\Delta\nu$ is the FWHM linewidth of the laser or transition (s^{-1}). The corresponding cross sections are

$$\sigma_{12} = \frac{g_2}{g_1}\sigma_{21} = W_{12}\frac{h\nu}{I} = g(\nu)\frac{g_2}{g_1}\frac{\lambda^2 A_{21}}{8\pi} = \frac{g_2}{g_1}\frac{\lambda^2 A_{21}}{4\pi^2 \Delta\nu} \qquad (3)$$

where the last expression assumes a Lorentzian lineshape, and $\Delta\nu$ is the full experimental linewidth (laser plus atomic).

A simple expression for the laser pulse energy, E_{sat}, which is required to saturate the resonance transition can be derived under the following conditions: (1) the laser linewidth is large compared to W_{12} (so Rabi oscillations are averaged out), and (2) the rate of higher excitation or ionization is small compared to W_{12}. From a simple solution of the rate equations, E_{sat} can be expressed as:

$$E_{sat} = \frac{h\nu}{\sigma_{12} + \sigma_{21}} \qquad (4)$$

The sum in the denominator accounts for the competition of stimulated emission with excitation. Typical values for σ_{12} and σ_{21} are $\sim 10^{-13}$ cm^2 if $\Delta\nu \sim 10^{10}$ s^{-1} for a multimode laser. Thus $E_{sat} \sim 1$ μJ/cm^2. The pulse energy density E in a typical RIS experiment is on the order of (1mJ)/(0.1cm^2) or $\sim 10^4$ E_{sat}. Therefore resonant transitions are often easily saturated.

When the resonant transition(s) are fully saturated ($E \gg E_{sat}$), the populations of interacting ground and excited m_J states become equal early in the laser pulse, and the fraction of the population, ß, in the final excited state, f, becomes for *unpolarized* light,

$$\beta = \frac{N_f}{N_1 + N_2 + ... + N_f} = \frac{g_f}{g_1 + g_2 + ... + g_f} \qquad (5)$$

This is the fraction of the atoms which are available for ionization.

With *polarized* laser light the above equations are only approximate. The density matrix equations or rate equations need to be solved for the complete manifold of lower and upper m-levels. Complications such as optical pumping and different transition rates, including some forbidden m–level transitions, can occur. For *linearly polarized* light, the m_J selection rule for π polarization, $\Delta m_J = 0$, requires that the same upper and lower m_J states be involved in the transition. Thus for isotopes with zero nuclear spin, the above equations can be used with $g_1 = g_2 = g_3....$ This leads to ß=1/2 for single-resonance RIS and ß=1/3 for double-resonance RIS, etc., unless multiple fine structure states lie within the laser bandwidth. The analysis is more complicated in odd-mass isotopes due to unresolved hyperfine structure, as discussed below.

IONIZATION STEP

The ionization step of the RIS process is conceptually simpler than the resonance step because a tunable laser is usually not required. However, experimental or theoretical data on ionization cross sections or the energies of suitable autoionizing

TABLE 4. Suggested rules of thumb for ionization cross sections.

Dependence on photon energy:

Generally high near threshold and then fall off to higher energy above threshold

Electric dipole selection rules:

s-electrons: low cross sections
p- and d-electrons: higher cross sections

---

states in the continuum, which might have enhanced cross sections, are scarce. The RIS data sheets of Saloman contain theoretical ionization cross sections for specific RIS schemes in 33 elements (8).

In the absence of reliable ionization cross section data, I have found the following rules of thumb, summarized in Table 4, to be valuable as a guide for planning RIS experiments. First, ionization cross sections are generally high near threshold and then fall off to higher energy. Thus, if there is a choice of a scheme in which the ionization photon goes just above threshold or far above threshold, the former is favored, barring other considerations. Second, ionization cross sections are generally larger when a p or d electron is removed by ionization rather than a s electron. These rules are not expected to be reliable for complex atoms, such as uranium, where a large number of autoionizing states are likely to exist in the continuum.

The first rule is illustrated by calculations for ionization of ns electrons in hydrogen and the 6p electron in Cs, which are presented in refs. (2,3). A sharp rise at threshold and power law falloff is seen. Confirming experimental data are shown in the case of Cs. The calculated cross sections for H are higher at threshold and fall off faster above threshold for higher Rydberg states with larger n.

As a general check on these rules of thumb, the theoretical ionization cross sections for most of the RIS schemes in the RIS data sheets (8) are presented in Fig. 1 as a function of ΔE, the excess energy in the final photon above the ionization threshold, and the type of electron removed, s, p or d. While there is significant scatter in the data (the hydrogen data suggest that some scaling with n and threshold energy might help), the rules of thumb are confirmed. A general fall-off is seen at higher ΔE values, roughly an order of magnitude per 1.7 eV. Typical p- and d-electron ionization cross sections just above threshold are about 10^{-17} cm^2, whereas s-electron values (dark circles) are about an order of magnitude lower.

When the resonance step is saturated, the fraction of atoms ionized for a pulse of energy density E is

$$N_i = 1 - e^{-\beta\sigma(E/h\nu)} = 1 - e^{-E/E_{sat}} \qquad (6)$$

where

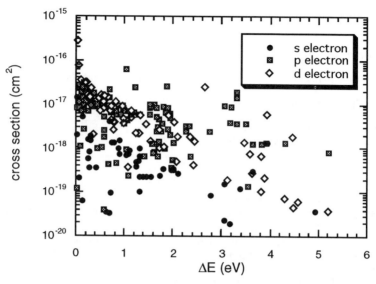

FIGURE 1. Ionization cross section vs. above threshold energy for atoms listed in the RIS data tables (8).

$$E_{sat} = \frac{h\nu}{\beta\sigma_i} \tag{7}$$

is the laser pulse energy required to ionize 63% of the atoms, σ_i is the ionization cross section and ß is the fraction of atoms in the final excited state. Note that the factor of β is often omitted in theoretical discussions (2,8). For example, the values of E_{sat} given in the RIS tables (8) should be multiplied by 1/ß, which is ≈ 2 in single resonance schemes. This may also lead to an error in experimental determinations of ionization cross sections.

m-LEVEL EFFECTS, ANOMALOUS ISOTOPE RATIOS

In the above discussion, the effects of magnetic sublevels have been lumped into the degeneracy factors g_i. This is appropriate for *unpolarized* light, but it is not a complete description for excitation and ionization with *polarized* laser light.

Consider, for example, the single-resonance RIS scheme shown in Fig. 2, which is an example of a scheme with a J=0 to J=1 transition. Specifically this is the $^3P_0$–$^3P_1$ transition in tin at 286.3 nm, which has been investigated in detail both experimentally (19) and theoretically (4). The calculated ionization rates shown in italics on the diagram are in s^{-1} for I in W/cm^2.

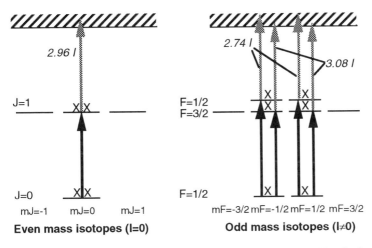

FIGURE 2. Origin of the odd-even effect in the 286.3 nm transition in Sn.

With linear polarized laser light, the population in the ground $J=0$, $m_J=0$ state of even isotopes is transferred only to the $m_J=0$ sublevel of the $J=1$ state. Under saturated conditions half the population is in each $m_J=0$ state, and the excited state population is $\beta=1/2$. In contrast, for unpolarized excitation the $m_J=\pm1$ excited states would also be equally populated, and the total excited state fraction would be $\beta=g_2/(g_1+g_2)=3/4$. So polarization makes a big difference in the fraction of the population available to be ionized.

Due to the hyperfine structure associated with nuclear spin $I=1/2$ in the stable odd isotopes of tin, there are two excited F states, with a small separation, which may be less than the laser bandwidth Linear polarized excitation populates only $m_F=\pm1/2$ excited sublevels, but in this case 2/3 of the population is in the excited states under saturated resonance conditions. Thus there is 50% more population available for ionization in the odd mass isotopes than in the even mass isotopes when linear polarization is used. Note that the calculated ionization rates (4) for different magnetic sublevels are not the same (e.g., 2.74 I and 3.08 I for the $F=1/2$ and $F=3/2$ hyperfine states), but that the average happens to be close to that, 2.96 I, for the even isotopes ($J=1$, $m_J=0$ state). Thus when the ionization step is not saturated but the resonance step is saturated, there is a substantial difference, ~50%, in the RIS rates for even and odd isotopes. This theoretical prediction has been verified by experimental measurements (20).

Experimentally this odd-even effect in broadband RIS has now been observed in lesser and greater degrees in a number of atomic systems, although there are also many systems in which the effect is small. There are several ways to reduce the odd-even effect in practical RIS measurements. These include reducing the resonance laser intensity well below saturation, increasing the ionization intensity to full saturation so that all isotopes are ionized, depolarizing the laser as discussed above, tuning the laser off resonance, and using a short laser pulse with $\Delta t \ll 1/$(hyperfine structure). These methods do not all work in all cases.

In some extreme cases, (21) excitation can be totally forbidden by selection rules in the even isotopes, but allowed in the odd isotopes. An example is the $J=0 \rightarrow J=1 \rightarrow J=1$ double resonance RIS scheme, where the second step is forbidden in even isotopes because $m_J=0 \leftrightarrow m_J=0$ when $\Delta J=0$ but is allowed in odd isotopes because $m_F \neq 0$. Such schemes have potential for use in isotope separation.

REFERENCES

1. Hurst, G. S. and Payne, M. G., Principles and Applications of Resonance Ionization Spectroscopy, Bristol: Adam Hilger, 1988.
2. Letokhov, V. S., *Laser Photoionization Spectroscopy*, Orlando: Academic Press, 1987.
3. Paisner, J. A. and Solarz, R. W., "Laser Photoionization Spectroscopy," Ch. 3 in *Laser Spectroscopy and Its Applications*, Radziemski, L. J., Solarz, R. W. and Paisner, J. A., eds., New York: M. Dekker, 1987.
4. Lambropoulos, P. and Lyras, Y., Phys. Rev. A**40**, 2199-2203 (1989).
5. Moore, C. E., *Atomic Energy Levels*, Vols. 1,2,3, Washington: U. S. Government Printing Office, 1971, Nat. Stand. Ref. Data Ser. NSRDS-NBS-35.
6. Martin, W. C., Zalubas, R. and Hagan, L., *Atomic Energy Levels--The Rare-Earth Elements*, Washington: U. S. Govt. Printing Office, 1978, Nat. Stand. Ref. Data Ser. NSRDS-NBS-60.
7. Blaise, J. and Wyart, J.-F., *Energy Levels and Atomic Spectra of Actinides*, Paris: Tables de Constantes, 1992, International Tables of Selected Constants, Vol. 20.
8. Saloman, E. B., *Spectrochim. Acta* **45B**, 37-83 (1990); **46B**, 319-378 (1991), **47B**, 517-543 (1992) and **48B**, 1139-1203 (1993).
9. Martin, W. L., "Sources of Atomic Spectroscopic Data for Astrophysics," Ch. 8 in *Atomic and Molecular Data for Space Astronomy*, Smith, P. L. and Wiese, W. L. eds., No. 407 of "Lecture Notes in Physics", Springer-Verlag, 1992; updated in *Reports on Astronomy* **22A**, 105-134, (1994).
10. Fuhr, J. R. and Wiese, W. L., "Atomic Transition Probabilities," in *Handbook of Chemistry and Physics*, Lide, D. R. ed., 71st ed., Boca Raton, FL: CRC Press, 1990.
11. Wiese, W. L., Smith, M. W. and Glennon, B. M., *Atomic Transition Probabilities H through Ne–A Critical Data Compilation*, Vol. I , Washington: U. S. Govt. Printing Office, 1966, Nat. Stand. Ref. Data Ser. NSRDS-NBS-4.
12. Wiese, W. L., Smith, M. W. and Miles, B. M., *Atomic Transition Probabilities Na through A–A Critical Data Compilation*, Vol. II, Washington: U. S. Govt. Printing Office, 1969, Nat. Stand. Ref. Data Ser. NSRDS-NBS-22.
13. Martin, G. A., Fuhr, J. R. and Wiese, W. L., Atomic Transition Probabilities–Scandium through Nickel, *J. Phys. Chem. Ref. Data* **16**, Suppl. 3 (1988).
14. Fuhr, J. R., Martin, G. A. and Wiese, W. L., Atomic Transition Probabilities–Iron through Manganese, *J. Phys. Chem. Ref. Data* **17**, Suppl. 4 (1988).
15. Morton, D. C., Astrophys. J. Suppl. Ser. **77**, 119-202 (1991).
16. Corliss, C. H., and Bozman, W. R., Experimental Transition Probabilities for Spectral Lines of Seventy Elements, Washington: U. S. Govt. Printing Office, 1962, Natl. Bur. Stand. Monograph 53.
17. Kurucz, R. L. and Peytremann, E. "A Table of Semiempirical gf Values," Parts 1-3, Smithsonian Astrophysical Observatory Special Report 362, Cambridge MA, 1975.
18. Verdeyen, J. T., *Laser Electronics*, 3rd ed., Englewood Cliffs, NJ: Prentice Hall, 1995, p. 186.
19. Fairbank, W. M., Jr., Spaar, M. T., Parks, J. E. and Hutchinson, J. M. R., Phys. Rev. A**40**, 2195-2198 (1989).
20. Xiong, X. Hutchinson, J. M. R., Fassett, J. and Fairbank, W. M., Jr., "Measurement of the odd-even effect in the resonance ionization of tin as a function of laser intensity,", in *Resonance Ionization Spectroscopy 1992*, C. M. Miller and J. E. Parks, editors, Bristol: Institute of Physics, 1992, pp. 123-126.
21. Balling, L. C. and Wright, J. J., Appl. Phys. Lett. **29**, 411 (1976).

467

AIP Conference Proceedings

Title	L.C. Number	ISBN
No. 356 The Future of Accelerator Physics (Austin, TX 1994)	96-83292	1-56396-541-0
No. 357 10th Topical Workshop on Proton-Antiproton Collider Physics (Batavia, IL 1995)	95-83078	1-56396-543-7
No. 358 The Second NREL Conference on Thermophotovoltaic Generation of Electricity	95-83335	1-56396-509-7
No. 359 Workshops and Particles and Fields and Phenomenology of Fundamental Interactions (Puebla, Mexico 1995)	96-85996	1-56396-548-8
No. 360 The Physics of Electronic and Atomic Collisions XIX International Conference (Whistler, Canada, 1995)	95-83671	1-56396-440-6
No. 361 Space Technology and Applications International Forum (Albuquerque, NM 1996)	95-83440	1-56396-568-2
No. 362 Two-Center Effects in Ion-Atom Collisions (Lincoln, NE 1994)	96-83379	1-56396-342-6
No. 363 Phenomena in Ionized Gases XXII ICPIG (Hoboken, NJ, 1995)	96-83294	1-56396-550-X
No. 364 Fast Elementary Processes in Chemical and Biological Systems (Villeneuve d'Ascq, France, 1995)	96-83624	1-56396-564-X
No. 365 Latin-American School of Physics XXX ELAF Group Theory and Its Applications (México City, México, 1995)	96-83489	1-56396-567-4
No. 366 High Velocity Neutron Stars and Gamma-Ray Bursts (La Jolla, CA 1995)	96-84067	1-56396-593-3
No. 367 Micro Bunches Workshop (Upton, NY, 1995)	96-83482	1-56396-555-0
No. 368 Acoustic Particle Velocity Sensors: Design, Performance and Applications (Mystic, CT, 1995)	96-83548	1-56396-549-6
No. 369 Laser Interaction and Related Plasma Phenomena (Osaka, Japan 1995)	96-85009	1-56396-445-7
No. 370 Shock Compression of Condensed Matter-1995 (Seattle, WA 1995)	96-84595	1-56396-566-6
No. 371 Sixth Quantum 1/f Noise and Other Low Frequency Fluctuations in Electronic Devices Symposium (St. Louis, MO, 1994)	96-84200	1-56396-410-4

	Title	L.C. Number	ISBN
No. 372	Beam Dynamics and Technology Issues for + - Colliders 9th Advanced ICFA Beam Dynamics Workshop (Montauk, NY, 1995)	96-84189	1-56396-554-2
No. 373	Stress-Induced Phenomena in Metallization (Palo Alto, CA 1995)	96-84949	1-56396-439-2
No. 374	High Energy Solar Physics (Greenbelt, MD 1995)	96-84513	1-56396-542-9
No. 375	Chaotic, Fractal, and Nonlinear Signal Processing (Mystic, CT 1995)	96-85356	1-56396-443-0
No. 376	Chaos and the Changing Nature of Science and Medicine: An Introduction (Mobile, AL 1995)	96-85220	1-56396-442-2
No. 377	Space Charge Dominated Beams and Applications of High Brightness Beams (Bloomington, IN 1995)	96-85165	1-56396-625-7
No. 379	Physical Origin of Homochirality in Life (Santa Monica, CA 1995)	96-86631	1-56396-507-0
No. 378	Surfaces, Vacuum, and Their Applications (Cancun, Mexico 1994)	96-85594	1-56396-418-X
No. 380	Production and Neutralization of Negative Ions and Beams / Production and Application of Light Negative Ions (Upton, NY 1995)	96-86435	1-56396-565-8
No. 381	Atomic Processes in Plasmas (San Francisco, CA 1996)	96-86304	1-56396-552-6
No. 382	Solar Wind Eight (Dana Point, CA 1995)	96-86447	1-56396-551-8
No. 383	Workshop on the Earth's Trapped Particle Environment (Taos, NM 1994)	96-86619	1-56396-540-2
No. 384	Gamma-Ray Bursts (Huntsville, AL 1995)	96-79458	1-56396-685-9
No. 385	Robotic Exploration Close to the Sun: Scientific Basis (Marlboro, MA 1996)	96-79560	1-56396-618-2
No. 386	Spectral Line Shapes, Volume 9 13th ICSLS (Firenze, Italy 1996)		1-56396-656-5
No. 387	Space Technology and Applications International Forum (Albuquerque, NM 1997)	96-80254	1-56396-679-4 (Case set) 1-56396-691-3
No. 388	Resonance Ionization Spectroscopy 1996 Eighth International Symposium (State College, PA 1996)	96-80324	1-56396-611-5